INSTITUT FRANÇAIS DU PÉTROLE
ÉCOLE NATIONALE SUPÉRIEURE DU PÉTROLE ET DES MOTEURS

DRILLING

DATA

HANDBOOK

GULF PUBLISHING COMPANY
BOOK DIVISION
HOUSTON, LONDON, PARIS, TOKYO

ÉDITIONS TECHNIP
27, RUE GINOUX, 75737 PARIS, Cedex 15

D
622.3382'028
ECO

This Edition, 1980
Gulf Publishing Company
Book Division
Houston, Texas

© 1978. Éditions Technip, Paris
and Institut Français du Pétrole, Rueil-Malmaison

ISBN 0-87201-204-2
Library of Congress Catalog Card No. 79-56343

Printed in France
by Imprimerie Chirat, 42540 Saint-Just-la-Pendue

INTRODUCTION
TO THE FIRST ENGLISH-LANGUAGE EDITION
OF THE DRILLING DATA HANDBOOK

The great success elicited by the French edition entitled *Formulaire du Foreur* has resulted in the publication of a fifth edition. Because of this and of the many requests we have received from non-French-speaking drillers, this English version is now available under the title **Drilling Data Handbook.**

The primary purpose of this book is to provide useful information in a compact and easily accessible form for use in drilling and production work.

This first edition includes, in section A, several conversion tables which may be helpful to those who want to use the International System of Units (SI).

Sections B through M give drilling equipment characteristics according to the latest API standards, or according to manufacturer specifications. Also included are technical drilling data and procedures for drilling muds, pressure losses, cementing, directional drilling, kick-off control, etc.

This handbook is as accurate as possible but it can be improved with the help of the users. Any corrections or suggestions are cordially invited to make it work better for you.

Refer suggestions to : ENSPM - CESFEG, BP 311, 92506 Rueil-Malmaison Cedex (France).

IMPORTANT NOTICE

• Measurements and formulas in this handbook are given in legal units: MKSA or International System (Système international : SI).

On oil fields the MKpS system is still commonly used. Its units have only a 2 % difference from the equivalent units of the International System (SI).

For example :

THE LEGAL UNIT FOR PRESSURE IS THE PASCAL. The bar, a multiple of the pascal (1 bar = 10^5 pascals), which is a convenient unit, is authorised.

$$1 \text{ bar} = 1,02 \text{ kgf/cm}^2$$
$$1 \text{ kgf/cm}^2 = 0,98 \text{ bar}$$

THE LEGAL UNIT FOR FORCE IS THE NEWTON (N)

$$1 \text{ decanewton (daN)} = 1,02 \text{ kilogramme force (kgf)}$$
$$1 \text{ kgf} = 0,98 \text{ daN}$$
$$10^3 \text{ daN} = 1,02 \text{ tf (tonne-force)}$$
$$1 \text{ tf} = 0,98 \ 10^3 \text{ daN}$$

THE LEGAL UNIT FOR TORQUE IS THE METRE-NEWTON (m.N). A multiple of this unit is usualy used, i.e. the metre-decanewton (m.daN)

$$1 \text{ m.daN} = 1,02 \text{ m.kgf}$$
$$1 \text{ m.kgf} = 0,98 \text{ m.daN}$$

KEEP THESE EQUIVALENCES IN MIND

IN THIS HANDBOOK THE COMMA IS USED TO INDICATE THE DECIMAL POINT IN BOTH ENGLISH AND METRIC VALUES

For example :

3,125 in instead of 3.125 in
1 325,4 kg instead of 1325.4 kg

A CONVERSION TABLES AND GENERAL DATA

Summary

THE INTERNATIONAL SYSTEM OF UNITS (SI)

	Quantity	INTERNATIONAL SYSTEM SI or MKSA			SUBMULTIPLE SYSTEM CGS			OTHER MULTIPLES AND SUBMULTIPLES OF SI
		Dimension	Unit	Symbol	Unit	Symbol	SI equivalent	
BASE UNITS	Length	L	metre	m	centimetre	cm	10^{-2} m	10^3 m = 1 km ; 10^{-3} m = 1 mm ; 10^{-6} m = 1 micron (μ) ; 10^{-10} m = 1 angström (Å) ;
	Mass	M	kilogramme	kg	gramme	g	10^{-3} kg	10^3 kg = 1 tonne (t) ; 10^2 kg = 1 quintal (q)
	Time	T	second	s	second	s	s	3 600 s = 1 heure (h) ; 60 s = 1 minute (min) ; 10^{-3} s = 1 milliseconde (ms)
	Electric current		ampere	A				10^{-3} A = 1 milliampère (mA) ; 10^{-6} A = 1 microampère (μA) ;
	Thermodynamic Temperature		kelvin	K				Temperature Celsius (°C) is the difference between two thermodynamic temperatures $T-T_0$ where T_0 = 273.15 kelvins
	Amount of substance		mole	mol				
	Luminous intensity		candela	cd				
SUPPLE-MENTARY	Angle, plane		radian	rad				
	Angle, solid		stéradian	sr				
DERIVED UNITS	Area	L^2	square metre	m^2	square centimetre	cm^2	$10^{-4}\,m^2$	$10^4 m^2$ = 1 hectare (1 ha) ; $10^2 m^2$ = 1 are (a) ;
	Volume	L^3	cubic metre	m^3	cubic centimetre	cm^3	$10^{-6}\,m^3$	$10^{-3} m^3$ = 1 dm^3 = 1 litre (l)
	Force	MLT^{-2}	newton	N	dyne	dyn	10^{-5} N	10^3 N = 10^2 daN : 10^3 N = 1 kilonewton (kN) ; 10 N = 1 décanewton (daN)
	Energy - Work Heat	ML^2T^{-2}	joule	J	erg	erg	10^{-7} J	10^3 J = 1 kilonewton-km (kN·km) ; 10 J = 1 daN · m ; 1 J = 1 Nm ; 4,18 J = 1 calorie (cal) ; 4,18 10^9 J = 1 thermie (th)
	Power	ML^2T^{-3}	watt	W	erg par seconde	erg/s	10^{-7} W	10^6 W = 1 mégawatt (1 MW)
	Velocity linear	LT^{-1}	metre per second	m/s	centimetre per second	cm/s	10^{-2} m/s	$\dfrac{1}{3,6}$ m/s = 1 km/h ; $\dfrac{1\,852}{3\,600}$ m/s = 1 knot
	Acceleration	LT^{-2}	metre per second squared	m/s^2	gal	cm/s^2	$10^{-2}\,m/s^2$	
	Pressure	$ML^{-1}T^{-2}$	pascal	Pa	barye	dyn/cm^2	10^{-1} Pa	10^7 Pa = 1 hectobar (hbar) ; 10^5 Pa = 1 bar ; 10^2 Pa = 1 millibar (mbar)
	Viscosity, dynamic	$ML^{-1}T^{-1}$	pascal-second	Pa.s	poise	P	10^{-1} Pa.s	10^{-3} Pa. s = 1 centipoise (cP)
	Viscosity kinematic	L^2T^{-1}	metre squared per second	m^2/s	stokes	St	$10^{-4}\,m^2/s$	$10^{-4} m^2/s$ = 1 centistoke (cSt)

OBSOLETE SYSTEMS OF UNITS

QUANTITY	Dimension	Symbol	MTS					MkpS				
			UNIT	S	MkpS	CGS	MKSA (SI)	UNIT	S	MTS	CGS	MKSA (SI)
Length	L	l	metre	m	1	10^2	1	metre	m	1	10^2	1
Mass	M	m	tonne	t	$\frac{1}{9.81} \times 10^3$	10^6	10^3			9.81×10^{-3}	9.81×10^3	9.81
Area	L^2	S	square metre	m²	1	10^4	1	square metre	m²	1	10^4	1
Volume	L^3	V	cubic metre	m³	1	10^6	1	cubic metre	m³	1	10^6	1
Time	T	t	second	s	1	1	1	second	s	1	1	1
Force	MLT^{-2}	F	sthène	sn	$\frac{1}{9.81} \times 10^3$	10^8	10^3	kilogramme force	kgf	9.81×10^{-3}	9.81×10^5	9.81
Energy-work	ML^2T^{-2}	E, σ ou W	kilojoule	kJ	$\frac{1}{9.81} \times 10^3$	10^{10}	10^3	kilogrammetre	kgm	9.81×10^{-3}	9.81×10^7	9.81
Power	ML^2T^{-3}	P	kilowatt	kW	$\frac{1}{9.81} \times 10^3$	10^{10}	10^3	kilogrammetre per second	kgm s	9.81×10^{-3}	9.81×10^7	9.81
Velocity	LT^{-1}	v	metre per second	m/s	1	10^2	1	metre per second	m s	1	10^2	1
Acceleration	LT^{-2}	γ	metre per second per second	m/s²		10^2	1	metre per second per second	m s²		10^2	1
Pressure	$ML^{-1}T^{-2}$	p	pièze	pz	$\frac{1}{9.81} \times 10^3$	10^4	10^3	kilogramme force per square metre	kgf.m²	9.81×10^{-3}	98.1	9.81
Dynamic viscosity	$ML^{-1}T^{-1}$	η ou μ	myriapoise	maP	$\frac{1}{9.81} \times 10^3$	10^4	10^3			9.81×10^{-3}	98.1	9.81
Kinematic viscosity	L^2T^{-1}	ν	myriastokes	maSt	1	10^4	1	myriastokes	maSt	1	10^4	1
Heat quantity	L^2T^{-1}	Q	thermie	th	10^3	10^6	4.18×10^6	kilocalorie or millithermie	kcal / mth	10^{-3}	10^3	4.18×10^3

IMPORTANT NOTICE

• Measurements and formulas in this handbook are given in legal units: MKSA or International System (Système international : SI).

On oil fields the MKpS system is still commonly used. Its units have only a 2 % difference from the equivalent units of the International System (SI).

For example :

THE LEGAL UNIT FOR PRESSURE IS THE PASCAL. The bar, a multiple of the pascal (1 bar = 10^5 pascals), which is a convenient unit, is authorised.

$$1 \text{ bar} = 1,02 \text{ kgf/cm}^2$$
$$1 \text{ kgf/cm}^2 = 0,98 \text{ bar}$$

THE LEGAL UNIT FOR FORCE IS THE NEWTON (N)

$$1 \text{ decanewton (daN)} = 1,02 \text{ kilogramme force (kgf)}$$
$$1 \text{ kgf} = 0,98 \text{ daN}$$
$$10^3 \text{ daN} = 1,02 \text{ tf (tonne-force)}$$
$$1 \text{ tf} = 0,98 \text{ } 10^3 \text{ daN}$$

THE LEGAL UNIT FOR TORQUE IS THE METRE-NEWTON (m.N). A multiple of this unit is usualy used, i.e. the metre-decanewton (m.daN)

$$1 \text{ m.daN} = 1,02 \text{ m.kgf}$$
$$1 \text{ m.kgf} = 0,98 \text{ m.daN}$$

KEEP THESE EQUIVALENCES IN MIND

IN THIS HANDBOOK THE COMMA IS USED TO INDICATE THE DECIMAL POINT IN BOTH ENGLISH AND METRIC VALUES

For example :

3,125 in instead of 3.125 in
1 325,4 kg instead of 1325.4 kg

CONVERSION TABLE
(MKpSA System into SI System)
Tonne-force (tf) into 10^3 decanewtons (10^3 daN)
Kilogrammes-force per square centimetre (kgf/cm^2) into bars
Metre-kilogrammes-force (m.kgf) into metre-decanewtons (m.daN)

	0	1	2	3	4	5	6	7	8	9
0	0	0,981	1,962	2,943	3,924	4,905	5,886	6,867	7,848	8,829
10	9,810	10,79	11,77	12,75	13,73	14,72	15,70	16,68	17,66	18,64
20	19,62	20,60	21,58	22,56	23,54	24,52	25,50	26,48	27,46	28,44
30	29,43	30,41	31,39	32,37	33,35	34,33	35,32	36,30	37,28	38,26
40	39,24	40,22	41,20	42,18	43,16	44,14	45,13	46,11	47,09	48,07
50	49,05	50,03	51,01	51,99	52,97	53,95	54,93	55,91	56,89	57,88
60	58,86	59,84	60,82	61,80	62,78	63,76	64,74	65,72	66,70	67,68
70	68,67	69,65	70,63	71,61	72,59	73,57	74,55	75,53	76,51	77,49
80	78,48	79,46	80,44	81,42	82,40	83,38	84,36	85,34	86,32	87,30
90	88,29	89,27	90,25	91,23	92,21	93,19	94,17	95,15	96,13	97,11
100	98,10	99,08	100,06	101,04	102,02	103,00	103,98	104,96	105,94	106,92
110	107,91	108,89	109,87	110,85	111,83	112,81	113,79	114,77	115,75	116,73
120	117,72	118,70	119,68	120,66	121,64	122,62	123,60	124,58	125,56	126,54
130	127,53	128,51	129,49	130,47	131,45	132,43	133,41	134,39	135,37	136,35
140	137,34	138,32	139,30	140,28	141,26	142,24	143,22	144,20	145,18	146,16
150	147,15	148,13	149,11	150,09	151,07	152,05	153,03	154,01	154,99	155,97
160	156,96	157,94	158,92	159,90	160,88	161,86	162,84	163,82	164,80	165,78
170	166,77	167,75	168,73	169,71	170,69	171,67	172,65	173,63	174,61	175,59
180	176,58	177,56	178,54	179,52	180,50	181,48	182,46	183,44	184,42	185,40
190	186,39	187,37	188,35	189,33	190,31	191,29	192,27	193,25	194,23	195,21
200	196,20	197,18	198,16	199,14	200,12	201,10	202,08	203,06	204,04	205,02
210	206,01	206,99	207,97	208,95	209,93	210,91	211,89	212,87	213,85	214,83
220	215,82	216,80	217,78	218,76	219,74	220,72	221,70	222,68	223,66	224,64
230	225,63	226,61	227,59	228,57	229,55	230,53	231,51	232,49	233,47	234,45
240	235,44	236,42	237,40	238,38	239,36	240,34	241,32	242,30	243,28	244,26
250	245,25	246,23	247,21	248,19	249,17	250,15	251,13	252,11	253,09	254,07
260	255,06	256,04	257,02	258,00	258,98	259,96	260,94	261,92	262,90	263,88
270	264,87	265,85	266,83	267,81	268,79	269,77	270,75	271,73	272,71	273,69
280	274,68	275,66	276,64	277,62	278,60	279,58	280,56	281,54	282,52	283,50
290	284,49	285,47	286,45	287,43	288,41	289,39	290,37	291,35	292,33	293,31
300	294,30	295,28	296,26	297,24	298,22	299,20	300,18	301,16	302,14	303,12
310	304,11	305,09	306,07	307,05	308,03	309,01	309,99	310,97	311,95	312,93
320	313,92	314,90	315,88	316,86	317,84	318,82	319,80	320,78	321,76	322,74
330	323,73	324,71	325,69	326,67	327,65	328,63	329,61	330,59	331,57	332,55
340	333,54	334,52	335,50	336,48	337,46	338,44	339,42	340,40	341,38	342,36
350	343,35	344,33	345,31	346,29	347,27	348,25	349,23	350,21	351,19	352,17
360	353,16	354,14	355,12	356,10	357,08	358,06	359,04	360,02	361,00	361,98
370	362,97	363,95	364,93	365,91	366,89	367,87	368,85	369,83	370,81	371,79
380	372,78	373,76	374,74	375,72	376,70	377,68	378,66	379,64	380,62	381,60
390	382,59	383,57	384,55	385,53	386,51	387,49	388,47	389,45	390,43	391,41
400	392,40	393,38	394,36	395,34	396,32	397,30	398,28	399,26	400,24	401,22
410	402,21	403,19	404,17	405,15	406,13	407,11	408,09	409,07	410,05	411,03
420	412,02	413,00	413,98	414,96	415,94	416,92	417,90	418,88	419,86	420,84
430	421,83	422,81	423,79	424,77	425,75	426,73	427,71	428,69	429,67	430,65
440	431,64	432,62	433,60	434,58	435,56	436,54	437,52	438,50	439,48	440,46
450	441,45	442,43	443,41	444,39	445,37	446,35	447,33	448,31	449,29	450,27
460	451,26	452,24	453,22	454,20	455,18	456,16	457,14	458,12	459,10	460,08
470	461,07	462,05	463,03	464,01	464,99	465,97	466,95	467,93	468,91	469,89
480	470,88	471,86	472,84	473,82	474,80	475,78	476,76	477,74	478,72	479,70
490	480,69	481,67	482,65	483,63	484,61	485,59	486,57	487,55	488,53	489,51
—	0	1	2	3	4	5	6	7	8	9

DECIMAL MULTIPLES
AND SUBMULTIPLES OF A UNIT

MULTIPLES

UNIT MULTIPLIER	Prefix to put before the name of the unit	Symbol to put before the unit symbol
$10^{12} = 1\ 000\ 000\ 000\ 000$	téra	T
$10^{9} = 1\ 000\ 000\ 000$	giga	G
$10^{6} = 1\ 000\ 000$	méga	M
$10^{3} = 1\ 000$	kilo	k
$10^{2} = 100$	hecto	h
$10^{1} = 10$	déca	da

SUBMULTIPLES

UNIT MULTIPLIER	Prefix to put before the name of the unit	Symbol to put before the unit symbol
$10^{-1} = 0,1$	déci	d
$10^{-2} = 0,01$	centi	c
$10^{-3} = 0,001$	milli	m
$10^{-6} = 0,000\ 001$	micro	μ
$10^{-9} = 0,000\ 000\ 001$	nano	n
$10^{-12} = 0,000\ 000\ 000\ 001$	pico	p
$10^{-15} = 0,000\ 000\ 000\ 000\ 001$	femto	f
$10^{-18} = 0,000\ 000\ 000\ 000\ 000\ 001$	atto	a

Examples : 1 megametre (Mm) = 10^6 metres (m),
1 micrometre (μm) (micron or μ) = 10^{-6} metres (m),
1 kilonewton (kN) = 10^3 newton (N).

CONVERSION FACTORS
ENGLISH INTO METRIC UNITS

to obtain ✓		multiply the number of ✓		by ✓
0.0393701	Millimètres	Inches (pouces)	in	25.4
3.28084	Mètres	Feet (pieds)	ft	0.3048
1.09361	Mètres	Yards	yd	0.9144
0.621373	Kilomètres	Statute miles (milles terrestres)		1.60934
0.539613	Kilomètres	Nautical miles (UK) (milles marins anglais)		1.85318
0.539957	Kilomètres	Nautical miles (milles marins-autres pays)		1.852
0.155	Centimètres carrés	Square inches (pouces carrés)	in², sq. in	6.4516
10.7639	Mètres carrés	Square feet (pieds carrés)	ft², sq. ft	0.0929
2.47105	Hectares	Acres		0.404686
0.386102	Kilomètres carrés	Square miles (milles carrés)	sq. mile	2.58999
0.0610236	Centimètres cubes	Cubic inches (pouces cubes)	in², cu. in	16.3871
0.0353147	Décimètres cubes	Cubic feet (pieds cubes)	ft², cu. ft	28.3168
0.264178	Décimètres cubes	Gallons (US)	gal (US)	3.78533
0.219976	Décimètres cubes	Gallons (UK)	gal (UK)	4.54596
35.3147	Mètres cubes	Cubic feet	ft², cu. ft	0.0283168
6.28994	Mètres cubes	Barrels (US) (barils)	bbl	0.158984
150.959	Mètres cubes par heure	Barrels per day (barils par jour)	bbl/day	0.00662433
15.4324	Grammes-force	Grains-force	grf	0.0647989
0.035274	Grammes-force	Ounces-force (onces-force)	ozf	28.3495
2.20462	Kilogrammes-force	Pounds-force (livres-force)	lbf	0.453592
0.224809	Newtons	Pounds-force	lbf	4.44822
0.0234534	Kilogrammes-force	Sacks (cement)		42.6377
1.10231	Tonnes-force	Short tons-force (t.-force USA)	sh tonf	0.907185
0.984204	Tonnes-force	Long tons-force		1.01605
0.671971	Kilogrammes-force par mètre	Pounds-force per foot	lbf/ft	1.48816
8.34523	Kilogrammes-force par décimètre cube	Pounds-force per gallon (US)	lbf/gal	0.119829
62.4278	Kilogrammes-force par décimètre cube	Pounds-force per cubic foot	lbf/ft³	0.0160185
0.3505	Kilogrammes-force par mètre cube	Pounds-force per barrel	lbf/bbl	2.85307
0.0805214	Litres par mètre	Gallons (US) per foot	gal/ft	12.4191
14.5038	Bars	Pounds-force per square inch	lbf/in², psi	0.0689476
14.2233	Kilogrammes-force par centimètre carré	Pounds-force per square inch	lbf/in², psi	0.070307
0.711167	Kilogrammes-force par millimètre carré	Short tons-force per square inch	tonf/in²	1.40614
102.408	Kilogrammes-force par millimètre carré	Short tons-force per square foot	tonf ft²	0.00976486
0.737561	Joules	Feet-pounds-force	ft. lbf	1.35582
7.23301	Kilogrammètres	Feet-pounds-force	ft. lbf	0.138255
0.737562	Mètres-newtons	Feet-pounds-force	ft. lbf	1.35582
0.684944	Tonnes-force-kilomètres	Short tons-force-miles		1.45997
0.00134102	Watts	Horse powers	hp	745.7
0.98632	Chevaux-vapeur	Horse powers	hp	1.01387
0.000947813	Joules	British thermal units	Btu	1055.06
3.96707	Kilocalories	British thermal units	Btu	0.252075
0.368553	Kilocalories par mètre carré	British thermal units per square foot	Btu/ft²	2.71331
1.79943	Kilocalories par kilogramme	British thermal units per pound	Btu/lb	0.55573
0.112335	Kilocalories par mètre cube	British thermal units per cubic foot	Btu/ft³	8.90196
$°C \cdot \frac{9}{5} - 32$	Degrés Celcius	Degrees Fahrenheit	°F	$(°F - 32)\,\frac{5}{9}$
5.61448	Mètre cube par mètre cube	Cubic feet per barrel (US)	ft³/bbl	0.178111
0.042	Litres par mètre cube	Gallons (US) per barrel (US)	gal/bbl	23.8095
by ↖	↖ multiply the number of	↖ to obtain		

CONVERSION FACTORS
METRIC INTO ENGLISH UNITS

ENGLISH AND AMERICAN UNITS

NAME	SYMBOL	VALUE	
		relative	in metric units
Length			(metre)
Inch (pouce)	in		0,02540
Foot (pied)	ft	12 in	0,30479
Yard	yd	3 ft	0,91438
Fathom (brasse)	—	2 yd	1,82876
Mile (statute)	—	1760 yd	1609,3149
Mile (nautical)	—	2029 yd	1853,1232
Area			(square metre)
Square inch	sq. in.		0,00064513
Square foot	sq. ft.	144 sq. in.	0,0928997
Acre		4840 sq. yd.	4046,81
Square mile (stat.)	sq. mile	640 acres	259,0 ha
Volume			(cubic metre)
Cubic inch	cu. in.		0,000016386
Cubic foot	cu. ft.	1728 cu. in.	0,02831531
Capacity			(litre)
Gallon (US)	gal. (US)		3,785
Barrel of oil	bbl	42 gal. (US)	158,98
Mass			(kilogramme)
Ounce	oz		0,02835
Pound (livre)	lb	16 oz	0,453593
Ton (short ton)	sh tn	2000 lb	907,1853

MISCELLANEOUS CONSTANTS

0,0764 = air density in lb/ft³ at 60° F and 14,6 psia,
14,691 = normal atmospheric pressure (76 cm Hg) in psi,
32,174 = gravity acceleration in ft/s² (980,665 cm/s²),
550 = number of lb. ft/s in one horse power (hp),
778,2 = number of lb. in one Btu,
62,43 = water density in lbf/cu. ft at 4° C,
8,345 = water density in lbf/gal at 4° C,
°C + 273,16 = °K (Kelvin),
°F + 459,69 = °R (Rankine).

DECIMAL AND METRIC EQUIVALENTS OF INCHES AND FRACTIONS OF AN INCH

Fraction of an inch	Decimal equivalent	mm	Fraction of an inch	Decimal equivalent	mm
1/64	0,015625	0,40	33/64	0,515625	13,10
1/32	0,03125	0,79	17/32	0,53125	13,49
3/64	0,046875	1,19	35/64	0,546875	13,89
1/16	0,0625	1,59	9/16	0,5625	14,29
5/64	0,078125	1,98	37/64	0,578125	14,68
3/32	0,09375	2,38	19/32	0,59375	15,08
7/64	0,109375	2,78	39/64	0,609375	15,48
1/8	0,125	3,175	5/8	0,625	15,875
9/64	0,140625	3,57	41/64	0,640625	16,27
5/32	0,15625	3,97	21/32	0,65625	16,67
11/64	0,171875	4,37	43/64	0,671875	17,07
3/16	0,1875	4,76	11/16	0,6875	17,46
13/64	0,203125	5,16	45/64	0,703125	17,86
7/32	0,21875	5,56	23/32	0,71875	18,26
15/64	0,234375	5,95	47/64	0,734375	18,65
1/4	0,25	6,35	3/4	0,75	19,05
17/64	0,265625	6,75	49/64	0,765625	19,45
9/32	0,28125	7,14	25/32	0,78125	19,84
19/64	0,296875	7,54	51/64	0,796875	20,24
5/16	0,3125	7,94	13/16	0,8125	20,64
21/64	0,328125	8,33	53/64	0,828125	21,03
11/32	0,34375	8,73	27/32	0,84375	21,43
23/64	0,359375	9,13	55/64	0,859375	21,83
3/8	0,375	9,525	7/8	0,875	22,225
25/64	0,390625	9,92	57/64	0,890625	22,62
13/32	0,40625	10,32	29/32	0,90625	23,02
27/64	0,421875	10,72	59/64	0,921875	23,42
7/16	0,4375	11,11	15/16	0,9375	23,81
29/64	0,453125	11,51	61/64	0,953125	24,21
15/32	0,46875	11,91	31/32	0,96875	24,61
31/64	0,484375	12,30	63/64	0,984375	25,00
1/2	0,5	12,70	1	1	25,40

CONVERSION TABLE INCHES INTO MILLIMETRES
from 0 to 26 15/16 in

Pouces	0	1/16	1/8	3/16	1/4	5/16	3/8	7/16	1/2	9/16	5/8	11/16	3/4	13/16	7/8	15/16
0	0	1.6	3.2	4.8	6.3	7.9	9.5	11.1	12.7	14.3	15.9	17.5	19.0	20.6	22.2	23.8
1	25.4	27.0	28.6	30.2	31.7	33.3	34.9	36.5	38.1	39.7	41.3	42.9	44.4	46.0	47.6	49.2
2	50.8	52.4	54.0	55.6	57.1	58.7	60.3	61.9	63.5	65.1	66.7	68.3	69.8	71.4	73.0	74.6
3	76.2	77.8	79.4	81.0	82.5	84.1	85.7	87.3	88.9	90.5	92.1	93.7	95.2	96.8	98.4	100.0
4	101.6	103.2	104.8	106.4	107.9	109.5	111.1	112.7	114.3	115.9	117.5	119.1	120.6	122.2	123.8	125.4
5	127.0	128.6	130.2	131.8	133.3	134.9	136.5	138.1	139.7	141.3	142.9	144.5	146.0	147.6	149.2	150.8
6	152.4	154.0	155.6	157.2	158.7	160.3	161.9	163.5	165.1	166.7	168.3	169.9	171.4	173.0	174.6	176.2
7	177.8	179.4	181.0	182.6	184.1	185.7	187.3	188.9	190.5	192.1	193.7	195.3	196.8	198.4	200.0	201.6
8	203.2	204.8	206.4	208.0	209.5	211.1	212.7	214.3	215.9	217.5	219.1	220.7	222.2	223.8	225.4	227.0
9	228.6	230.2	231.8	233.4	234.9	236.5	238.1	239.7	241.3	242.9	244.5	246.1	247.6	249.2	250.8	252.4
10	254.0	255.6	257.2	258.8	260.3	261.9	263.5	265.1	266.7	268.3	269.9	271.5	273.0	274.6	276.2	277.8
11	279.4	281.0	282.6	284.2	285.7	287.3	288.9	290.5	292.1	293.7	295.3	296.9	298.4	300.0	301.6	303.2
12	304.8	306.4	308.0	309.6	311.1	312.7	314.3	315.9	317.5	319.1	320.7	322.3	323.8	325.4	327.0	328.6
13	330.2	331.8	333.4	335.0	336.5	338.1	339.7	341.3	342.9	344.5	346.1	347.7	349.2	350.8	352.4	354.0
14	355.6	357.2	358.8	360.4	361.9	363.5	365.1	366.7	368.3	369.9	371.5	373.1	374.6	376.2	377.8	379.4
15	381.0	382.6	384.2	385.8	387.3	388.9	390.5	392.1	393.7	395.3	396.9	398.5	400.0	401.6	403.2	404.8
16	406.4	408.0	409.6	411.2	412.7	414.3	415.9	417.5	419.1	420.7	422.3	423.9	425.4	427.0	428.6	430.2
17	431.8	433.4	435.0	436.6	438.1	439.7	441.3	442.9	444.5	446.1	447.7	449.3	450.8	452.4	454.0	455.6
18	457.2	458.8	460.4	462.0	463.5	465.1	466.7	468.3	469.9	471.5	473.1	474.7	476.2	477.8	479.4	481.0
19	482.6	484.2	485.8	487.4	488.9	490.5	492.1	493.7	495.3	496.9	498.5	500.1	501.6	503.2	504.8	506.4
20	508.0	509.6	511.2	512.8	514.3	515.9	517.5	519.1	520.7	522.3	523.9	525.5	527.0	528.6	530.2	531.8
21	533.4	535.0	536.6	538.2	539.7	541.3	542.9	544.5	546.1	547.7	549.3	550.9	552.4	554.0	555.6	557.2
22	558.8	560.4	562.0	563.6	565.1	566.7	568.3	569.9	571.5	573.1	574.7	576.3	577.8	579.4	581.0	582.6
23	584.2	585.8	587.4	589.0	590.5	592.1	593.7	595.3	596.9	598.5	600.1	601.7	603.2	604.8	606.4	608.0
24	609.6	611.2	612.8	614.4	615.9	617.5	619.1	620.7	622.3	623.9	625.5	627.1	628.6	630.2	631.8	633.4
25	635.0	636.6	638.2	639.8	641.4	642.9	644.5	646.1	647.7	649.3	650.9	652.5	654.1	655.6	657.2	658.8
26	660.4	662.0	663.6	665.2	666.8	668.3	669.9	671.5	673.1	674.7	676.3	677.9	679.5	681.0	682.6	684.2

CONVERSION TABLE INCHES INTO MILLIMETRES
from 27 to 53 15/16 in

Pouces	0	1/16	1/8	3/16	1/4	5/16	3/8	7/16	1/2	9/16	5/8	11/16	3/4	13/16	7/8	15/16
27	685.8	687.4	689.0	690.6	692.2	693.7	695.3	696.9	698.5	700.1	701.7	703.3	704.9	706.4	708.0	709.6
28	711.2	712.8	714.4	716.0	717.6	719.1	720.7	722.3	723.9	725.5	727.1	728.7	730.3	731.8	733.4	735.0
29	736.6	738.2	739.8	741.4	743.0	744.5	746.1	747.7	749.3	750.9	752.5	754.1	755.7	757.2	758.8	760.4
30	762.0	763.6	765.2	766.8	768.4	769.9	771.5	773.1	774.7	776.3	777.9	779.5	781.1	782.6	784.2	785.8
31	787.4	789.0	790.6	792.2	793.8	795.3	796.9	798.5	800.1	801.7	803.3	804.9	806.5	808.0	809.6	811.2
32	812.8	814.4	816.0	817.6	819.2	820.7	822.3	823.9	825.5	827.1	828.7	830.3	831.9	833.4	835.0	836.6
33	838.2	839.8	841.4	843.0	844.6	846.1	847.7	849.3	850.9	852.5	854.1	855.7	857.3	858.8	860.4	862.0
34	863.6	865.2	866.8	868.4	870.0	871.5	873.1	874.7	876.3	877.9	879.5	881.1	882.7	884.2	885.8	887.4
35	889.0	890.6	892.2	893.8	895.4	896.9	898.5	900.1	901.7	903.3	904.9	906.5	908.1	909.6	911.2	912.8
36	914.4	916.0	917.6	919.2	920.8	922.3	923.9	925.5	927.1	928.7	930.3	931.9	933.5	935.0	936.6	938.2
37	939.8	941.4	943.0	944.6	946.2	947.7	949.3	950.9	952.5	954.1	955.7	957.3	958.9	960.4	962.0	963.6
38	965.2	966.8	968.4	970.0	971.6	973.2	974.7	976.3	977.9	979.5	981.1	982.7	984.3	985.9	987.5	989.0
39	990.6	992.2	993.8	995.4	997.0	998.5	1000.1	1001.7	1003.3	1004.9	1006.5	1008.1	1009.7	1011.2	1012.8	1014.4
40	1016.0	1017.6	1019.2	1020.8	1022.4	1024.0	1025.5	1027.1	1028.7	1030.3	1031.9	1033.5	1035.1	1036.6	1038.2	1039.8
41	1041.4	1043.0	1044.6	1046.2	1047.8	1049.3	1050.9	1052.5	1054.1	1055.7	1057.3	1058.9	1060.4	1062.0	1063.6	1065.2
42	1066.8	1068.4	1070.0	1071.6	1073.2	1074.7	1076.3	1077.9	1079.5	1081.1	1082.7	1084.3	1085.8	1087.4	1089.0	1090.6
43	1092.2	1093.8	1095.4	1097.0	1098.6	1100.1	1101.7	1103.3	1104.9	1106.5	1108.1	1109.7	1111.2	1112.8	1114.4	1116.0
44	1117.6	1119.2	1120.8	1122.4	1124.0	1125.5	1127.1	1128.7	1130.3	1131.9	1133.5	1135.1	1136.6	1138.2	1139.8	1141.4
45	1143.0	1144.6	1146.2	1147.8	1149.4	1150.9	1152.5	1154.1	1155.7	1157.3	1158.9	1160.5	1162.0	1163.6	1165.2	1166.8
46	1168.4	1170.0	1171.6	1173.2	1174.8	1176.3	1177.9	1179.5	1181.1	1182.7	1184.3	1185.9	1187.4	1189.0	1190.6	1192.2
47	1193.8	1195.4	1197.0	1198.6	1200.2	1201.7	1203.3	1204.9	1206.5	1208.1	1209.7	1211.3	1212.8	1214.4	1216.0	1217.6
48	1219.2	1220.8	1222.4	1224.0	1225.6	1227.1	1228.7	1230.3	1231.9	1233.5	1235.1	1236.7	1238.2	1239.8	1241.4	1243.0
49	1244.6	1246.2	1247.8	1249.4	1251.0	1252.5	1254.1	1255.7	1257.3	1258.9	1260.5	1262.1	1263.6	1265.2	1266.8	1268.4
50	1270.0	1271.6	1273.2	1274.8	1276.4	1277.9	1279.5	1281.1	1282.7	1284.3	1285.9	1287.5	1289.0	1290.6	1292.2	1293.8
51	1295.4	1297.0	1298.6	1300.2	1301.8	1303.3	1304.9	1306.5	1308.1	1309.7	1311.3	1312.9	1314.4	1316.0	1317.6	1319.2
52	1320.8	1322.4	1324.0	1325.6	1327.2	1328.7	1330.3	1331.9	1333.5	1335.1	1336.7	1338.3	1339.8	1341.4	1343.0	1344.6
53	1346.2	1347.8	1349.4	1351.0	1352.6	1354.1	1355.7	1357.3	1358.9	1360.5	1362.1	1363.7	1365.2	1366.8	1368.4	1370.0

CONVERSION TABLE FEET INTO METRES
from 1 to 100 ft

Feet	Metres	Feet	Metres	Feet	Metres	Feet	Metres
1	0,3048	26	7,9248	51	15,545	76	23,165
2	0,6096	27	8,2296	52	15,850	77	23,470
3	0,9144	28	8,5344	53	16,154	78	23,774
4	1,2192	29	8,8392	54	16,459	79	24,079
5	1,5240	30	9,1440	55	16,764	80	24,384
6	1,8288	31	9,4488	56	17,069	81	24,689
7	2,1336	32	9,7536	57	17,374	82	24,994
8	2,4384	33	10,058	58	17,678	83	25,298
9	2,7432	34	10,363	59	17,983	84	25,603
10	3,0480	35	10,668	60	18,288	85	25,908
11	3,3528	36	10,973	61	18,593	86	26,213
12	3,6576	37	11,278	62	18,898	87	26,518
13	3,9624	38	11,582	63	19,202	88	26,822
14	4,2672	39	11,887	64	19,507	89	27,127
15	4,5720	40	12,192	65	19,812	90	27,432
16	4,8768	41	12,497	66	20,117	91	27,737
17	5,1816	42	12,802	67	20,422	92	28,042
18	5,4864	43	13,106	68	20,726	93	28,346
19	5,7912	44	13,411	69	21,031	94	28,651
20	6,0960	45	13,716	70	21,336	95	28,956
21	6,4008	46	14,021	71	21,641	96	29,261
22	6,7056	47	14,326	72	21,946	97	29,566
23	7,0104	48	14,630	73	22,250	98	29,870
24	7,3152	49	14,935	74	22,555	99	30,175
25	7,6200	50	15,240	75	22,860	100	30,480

CONVERSION TABLE FEET INTO METRES
from 100 to 20 000 ft

Feet	Metres	Feet	Metres	Feet	Metres	Feet	Metres	Feet	Metres
100	30,48	4 100	1249,68	8 100	2468,88	12 100	3688,08	16 100	4907,28
200	60,96	4 200	1280,16	8 200	2499,36	12 200	3718,56	16 200	4937,76
300	91,44	4 300	1310,64	8 300	2529,84	12 300	3749,04	16 300	4968,24
400	121,92	4 400	1341,12	8 400	2560,32	12 400	3779,52	16 400	4998,72
500	152,40	4 500	1371,60	8 500	2590,80	12 500	3810,00	16 500	5029,20
600	182,88	4 600	1402,08	8 600	2621,28	12 600	3840,48	16 600	5059,68
700	213,36	4 700	1432,56	8 700	2651,76	12 700	3870,96	16 700	5090,16
800	243,84	4 800	1463,04	8 800	2682,24	12 800	3901,44	16 800	5120,64
900	274,32	4 900	1493,52	8 900	2712,72	12 900	3931,92	16 900	5151,12
1 000	304,80	5 000	1524,00	9 000	2743,20	13 000	3962,40	17 000	5181,60
1 100	335,28	5 100	1554,48	9 100	2773,68	13 100	3992,88	17 100	5212,08
1 200	365,76	5 200	1584,96	9 200	2804,16	13 200	4023,36	17 200	5242,56
1 300	396,24	5 300	1615,44	9 300	2834,64	13 300	4053,84	17 300	5273,04
1 400	426,72	5 400	1645,92	9 400	2865,12	13 400	4084,32	17 400	5303,52
1 500	457,20	5 500	1676,40	9 500	2895,60	13 500	4114,80	17 500	5335,00
1 600	487,68	5 600	1706,88	9 600	2926,08	13 600	4145,28	17 600	5364,48
1 700	518,16	5 700	1737,36	9 700	2956,56	13 700	4175,76	17 700	5394,96
1 800	548,64	5 800	1767,84	9 800	2987,04	13 800	4206,24	17 800	5425,44
1 900	579,12	5 900	1798,32	9 900	3017,52	13 900	4236,72	17 900	5455,92
2 000	609,60	6 000	1828,80	10 000	3048,00	14 000	4267,20	18 000	5486,40
2 100	640,08	6 100	1859,28	10 100	3078,48	14 100	4297,68	18 100	5516,88
2 200	670,56	6 200	1889,76	10 200	3108,96	14 200	4328,16	18 200	5547,36
2 300	701,04	6 300	1920,24	10 300	3139,44	14 300	4358,64	18 300	5577,84
2 400	731,52	6 400	1950,72	10 400	3169,92	14 400	4389,12	18 400	5608,32
2 500	762,00	6 500	1981,20	10 500	3200,40	14 500	4419,60	18 500	5638,80
2 600	792,48	6 600	2011,68	10 600	3230,88	14 600	4450,08	18 600	5669,28
2 700	822,96	6 700	2042,16	10 700	3261,36	14 700	4480,56	18 700	5699,76
2 800	853,44	6 800	2072,64	10 800	3291,84	14 800	4511,04	18 800	5730,24
2 900	883,92	6 900	2103,12	10 900	3322,32	14 900	4541,52	18 900	5760,72
3 000	914,40	7 000	2133,60	11 000	3352,80	15 000	4572,00	19 000	5791,20
3 100	944,88	7 100	2164,08	11 100	3383,28	15 100	4602,48	19 100	5821,68
3 200	975,36	7 200	2194,56	11 200	3413,76	15 200	4632,96	19 200	5852,16
3 300	1005,84	7 300	2225,04	11 300	3444,24	15 300	4663,44	19 300	5882,64
3 400	1036,32	7 400	2255,52	11 400	3474,72	15 400	4693,92	19 400	5913,12
3 500	1066,80	7 500	2286,00	11 500	3505,20	15 500	4724,40	19 500	5943,60
3 600	1097,28	7 600	2316,48	11 600	3535,68	15 600	4754,88	19 600	5974,08
3 700	1127,76	7 700	2346,96	11,700	3566,16	15 700	4785,36	19 700	6004,56
3 800	1158,24	7 800	2377,44	11 800	3596,64	15 800	4815,84	19 800	6035,04
3 900	1188,72	7 900	2407,92	11 900	3627,12	15 900	4846,32	19 900	6065,52
4 000	1219,20	8 000	2438,40	12 000	3657,60	16 000	4876,80	20 000	6096,00

CONVERSION TABLE
LITERS PER MINUTE INTO GALLONS (US) PER MINUTE AND VICE-VERSA

l/min into gal (US)/min

l/min	gal (US)/min	l/min	gal (US)/min	l/min	gal (US)/min
100	26.4	1 750	462	3 400	898
150	39.6	1 800	476	3 450	911
200	52.8	1 850	489	3 500	925
250	66.0	1 900	502	3 550	938
300	79.3	1 950	515	3 600	951
350	92.5	2 000	528	3 650	964
400	105.7	2 050	542	3 700	977
450	118.9	2 100	555	3 750	991
500	132.1	2 150	568	3 800	1 004
550	145.3	2 200	581	3 850	1 017
600	158.5	2 250	594	3 900	1 030
650	171.7	2 300	608	3 950	1 044
700	184.9	2 350	621	4 000	1 057
750	198.1	2 400	634	4 050	1 070
800	211.3	2 450	647	4 100	1 083
850	224.6	2 500	660	4 150	1 096
900	237.8	2 550	674	4 200	1 110
950	251.0	2 600	687	4 250	1 123
1 000	264.2	2 650	700	4 300	1 136
1 050	277.4	2 700	713	4 350	1 149
1 100	290.6	2 750	726	4 400	1 162
1 150	303.8	2 800	740	4 450	1 176
1 200	317.0	2 850	753	4 500	1 189
1 250	330.2	2 900	766	4 550	1 202
1 300	343.4	2 950	779	4 600	1 215
1 350	356.6	3 000	793	4 650	1 228
1 400	369.8	3 050	806	4 700	1 242
1 450	383.1	3 100	819	4 750	1 255
1 500	396.3	3 150	832	4 800	1 268
1 550	409.5	3 200	845	4 850	1 281
1 600	422.7	3 250	859	4 900	1 294
1 650	435.9	3 300	872	4 950	1 308
1 700	449.1	3 350	885	5 000	1 321

gal (US)/min into l/min

gal (US)/min	l/min	gal (US)/min	l/min	gal (US)/min	l/min	gal (US)/min	l/min
40	151	370	1 401	700	2 650	1 030	3 899
50	189	380	1 438	710	2 688	1 040	3 937
60	227	390	1 476	720	2 725	1 050	3 975
70	265	400	1 514	730	2 763	1 060	4 012
80	303	410	1 552	740	2 801	1 070	4 050
90	341	420	1 590	750	2 839	1 080	4 088
100	379	430	1 628	760	2 877	1 090	4 126
110	416	440	1 666	770	2 915	1 100	4 164
120	454	450	1 703	780	2 953	1 110	4 202
130	492	460	1 741	790	2 990	1 120	4 240
140	530	470	1 779	800	3 028	1 130	4 277
150	568	480	1 817	810	3 066	1 140	4 315
160	606	490	1 855	820	3 104	1 150	4 353
170	644	500	1 893	830	3 142	1 160	4 391
180	681	510	1 931	840	3 180	1 170	4 429
190	719	520	1 968	850	3 218	1 180	4 467
200	757	530	2 006	860	3 255	1 190	4 505
210	795	540	2 044	870	3 293	1 200	4 542
220	833	550	2 082	880	3 331	1 210	4 580
230	871	560	2 120	890	3 369	1 220	4 618
240	908	570	2 158	900	3 407	1 230	4 656
250	946	580	2 195	910	3 445	1 240	4 694
260	984	590	2 233	920	3 483	1 250	4 732
270	1 022	600	2 271	930	3 520	1 260	4 770
280	1 060	610	2 309	940	3 558	1 270	4 807
290	1 098	620	2 347	950	3 596	1 280	4 845
300	1 136	630	2 385	960	3 634	1 290	4 883
310	1 173	640	2 423	970	3 672	1 300	4 921
320	1 211	650	2 460	980	3 710	1 310	4 959
330	1 249	660	2 498	990	3 747	1 320	4 997
340	1 287	670	2 536	1 000	3 785	1 330	5 034
350	1 325	680	2 574	1 010	3 823	1 340	5 072
360	1 363	690	2 612	1 020	3 861	1 350	5 110

CONVERSION TABLE
BARRELS (US Oil) INTO CUBIC METRES

	0	1	2	3	4	5	6	7	8	9
0		0,159	0,318	0,477	0,636	0,795	0,954	1,113	1,272	1,431
10	1,590	1,749	1,908	2,067	2,226	2,385	2,544	2,703	2,862	3,021
20	3,180	3,339	3,498	3,657	3,816	3,975	4,134	4,293	4,452	4,611
30	4,770	4,929	5,088	5,247	5,406	5,565	5,724	5,883	6,042	6,201
40	6,360	6,519	6,678	6,837	6,996	7,155	7,314	7,473	7,632	7,791
50	7,950	8,109	8,268	8,427	8,586	8,745	8,904	9,063	9,222	9,381
60	9,540	9,699	9,858	10,017	10,176	10,335	10,494	10,653	10,812	10,971
70	11,130	11,289	11,448	11,607	11,766	11,925	12,084	12,243	12,402	12,561
80	12,720	12,879	13,038	13,197	13,356	13,515	13,674	13,833	13,992	14,151
90	14,310	14,469	14,628	14,787	14,946	15,105	15,264	15,423	15,582	15,741
100	15,900	16,059	16,218	16,377	16,536	16,695	16,854	17,013	17,172	17,331
110	17,490	17,649	17,808	17,967	18,126	18,285	18,444	18,603	18,762	18,921
120	19,080	19,239	19,398	19,557	19,716	19,875	20,034	20,193	20,352	20,511
130	20,670	20,829	20,988	21,147	21,306	21,465	21,624	21,783	21,942	22,101
140	22,260	22,419	22,578	22,737	22,896	23,055	23,214	23,373	23,532	23,691
150	23,850	24,009	24,168	24,327	24,486	24,645	24,804	24,963	25,122	25,281
160	25,440	25,599	25,758	25,917	26,076	26,235	26,394	26,553	26,712	26,871
170	27,030	27,189	27,348	27,507	27,666	27,825	27,984	28,143	28,302	28,461
180	28,620	28,779	28,938	29,097	29,256	29,415	29,574	29,733	29,892	30,051
190	30,210	30,369	30,528	30,687	30,846	31,005	31,164	31,323	31,482	31,641

CONVERSION TABLE
BARRELS (US Oil) INTO GALLONS (US)

	0	1	2	3	4	5	6	7	8	9
0		42	84	126	168	210	252	294	336	378
10	420	462	504	546	588	630	672	714	756	798
20	840	882	924	966	1008	1050	1092	1134	1176	1218
30	1260	1302	1344	1386	1428	1470	1512	1554	1596	1638
40	1680	1722	1764	1806	1848	1890	1932	1974	2016	2058
50	2100	2142	2184	2226	2268	2310	2352	2394	2436	2478
60	2520	2562	2604	2646	2688	2730	2772	2814	2856	2898
70	2940	2982	3024	3066	3108	3150	3192	3234	3276	3318
80	3360	3402	3444	3486	3528	3570	3612	3654	3696	3738
90	3780	3822	3864	3906	3948	3990	4032	4074	4116	4158
100	4200	4242	4284	4326	4368	4410	4452	4494	4536	4578
110	4620	4662	4704	4746	4788	4830	4872	4914	4956	4998
120	5040	5082	5124	5166	5208	5250	5292	5334	5376	5418
130	5460	5502	5544	5586	5628	5670	5712	5754	5796	5838
140	5880	5922	5964	6006	6048	6090	6132	6174	6216	6258
150	6300	6342	6384	6426	6468	6510	6552	6594	6636	6678
160	6720	6762	6804	6846	6888	6930	6972	7014	7056	7098
170	7140	7182	7224	7266	7308	7350	7392	7434	7476	7518
180	7560	7602	7644	7686	7728	7770	7812	7854	7896	7938
190	7980	8022	8064	8106	8148	8190	8232	8274	8316	8358

CONVERSION TABLE CUBIC FEET INTO CUBIC METRES

	0	1	2	3	4	5	6	7	8	9
0	0	0,028	0,057	0,085	0,113	0,142	0,170	0,198	0,227	0,255
10	0,283	0,311	0,340	0,368	0,396	0,425	0,453	0,481	0,510	0,538
20	0,566	0,595	0,623	0,651	0,680	0,708	0,736	0,765	0,793	0,821
30	0,850	0,878	0,906	0,934	0,963	0,991	1,019	1,048	1,076	1,104
40	1,133	1,161	1,189	1,218	1,246	1,274	1,303	1,331	1,359	1,388
50	1,416	1,444	1,473	1,501	1,529	1,557	1,586	1,614	1,642	1,671
60	1,699	1,727	1,755	1,784	1,812	1,841	1,869	1,897	1,925	1,953
70	1,982	2,011	2,039	2,067	2,095	2,124	2,152	2,180	2,209	2,237
80	2,265	2,294	2,322	2,350	2,379	2,407	2,435	2,464	2,492	2,520
90	2,549	2,577	2,605	2,633	2,662	2,690	2,718	2,747	2,775	2,803
100	2,832	2,860	2,888	2,917	2,945	2,973	3,002	3,030	3,058	3,087
110	3,115	3,143	3,172	3,200	3,228	3,256	3,285	3,313	3,341	3,370
120	3,398	3,426	3,455	3,483	3,511	3,540	3,568	3,596	3,625	3,653
130	3,681	3,710	3,738	3,766	3,794	3,823	3,851	3,879	3,908	3,936
140	3,964	3,993	4,021	4,049	4,078	4,106	4,134	4,163	4,191	4,219
150	4,248	4,276	4,304	4,333	4,361	4,389	4,417	4,446	4,474	4,502
160	4,531	4,559	4,587	4,616	4,644	4,672	4,701	4,729	4,757	4,786
170	4,814	4,842	4,871	4,899	4,927	4,955	4,984	5,012	5,040	5,069
180	5,097	5,125	5,154	5,182	5,210	5,239	5,267	5,295	5,324	5,352
190	5,380	5,409	5,437	5,465	5,493	5,522	5,550	5,578	5,607	5,635
200	5,663	5,692	5,720	5,748	5,777	5,805	5,833	5,862	5,890	5,918
210	5,947	5,975	6,003	6,032	6,060	6,088	6,116	6,145	6,173	6,201
220	6,230	6,258	6,286	6,315	6,343	6,371	6,400	6,428	6,456	6,485
230	6,513	6,541	6,570	6,598	6,626	6,654	6,683	6,711	6,739	6,768
240	6,796	6,824	6,853	6,881	6,909	6,938	6,966	6,994	7,023	7,051
250	7,079	7,108	7,136	7,164	7,193	7,221	7,249	7,277	7,306	7,334
260	7,362	7,380	7,419	7,447	7,475	7,504	7,532	7,560	7,589	7,617
270	7,646	7,674	7,702	7,731	7,759	7,787	7,815	7,844	7,872	7,900
280	7,929	7,957	7,985	8,014	8,042	8,070	8,098	8,127	8,155	8,183
290	8,212	8,240	8,268	8,297	8,325	8,353	8,381	8,410	8,438	8,466
300	8,494	8,523	8,551	8,580	8,608	8,636	8,664	8,693	8,721	8,749
310	8,777	8,806	8,834	8,863	8,891	8,919	8,947	8,976	9,004	9,032
320	9,060	9,089	9,117	9,146	9,174	9,202	9,230	9,259	9,287	9,315
330	9,343	9,372	9,400	9,429	9,457	9,485	9,513	9,542	9,570	9,598
340	9,626	9,655	9,683	9,712	9,740	9,768	9,796	9,825	9,853	9,881
350	9,909	9,938	9,966	9,995	10,023	10,051	10,079	10,108	10,136	10,164
360	10,192	10,221	10,249	10,278	10,306	10,334	10,362	10,391	10,419	10,447
370	10,475	10,504	10,532	10,561	10,589	10,617	10,645	10,674	10,702	10,730
380	10,758	10,787	10,815	10,843	10,872	10,900	10,928	10,957	10,985	11,013
390	11,041	11,070	11,098	11,126	11,155	11,183	11,211	11,240	11,268	11,296
400	11,324	11,353	11,381	11,409	11,438	11,466	11,494	11,523	11,551	11,579
410	11,607	11,636	11,664	11,692	11,721	11,749	11,777	11,806	11,834	11,862
420	11,890	11,919	11,947	11,975	12,004	12,032	12,060	12,089	12,117	12,145
430	12,173	12,202	12,230	12,258	12,287	12,315	12,343	12,372	12,400	12,428
440	12,456	12,485	12,513	12,541	12,570	12,598	12,626	12,655	12,683	12,711
450	12,739	12,768	12,796	12,824	12,853	12,881	12,909	12,938	12,966	12,994
460	13,022	13,051	13,079	13,107	13,136	13,164	13,192	13,221	13,249	13,277
470	13,305	13,334	13,362	13,390	13,419	13,447	13,475	13,504	13,532	13,560
480	13,588	13,617	13,645	13,673	13,702	13,730	13,758	13,787	13,815	13,843
490	13,871	13,900	13,928	13,956	13,985	14,013	14,041	14,070	14,098	14,126
—	0	1	2	3	4	5	6	7	8	9

CONVERSION TABLE
POUNDS PER SQUARE INCH INTO KILOGRAMMES-FORCE PER SQUARE CENTIMETRE AND BARS
from 1 to 100 psi

psi	kgf/cm²	bars	psi	kgf/cm²	bars	psi	kgf/cm²	bars	psi	kgf/cm²	bars
1	0.070	0.069	26	1.828	1.793	51	3.586	3.516	76	5.343	5.240
2	0.141	0.138	27	1.898	1.862	52	3.656	3.585	77	5.414	5.309
3	0.211	0.207	28	1.969	1.931	53	3.726	3.654	78	5.484	5.378
4	0.281	0.276	29	2.039	1.999	54	3.797	3.723	79	5.554	5.447
5	0.352	0.345	30	2.109	2.068	55	3.867	3.792	80	5.625	5.516
6	0.422	0.414	31	2.180	2.137	56	3.937	3.861	81	5.695	5.585
7	0.492	0.483	32	2.250	2.206	57	4.007	3.930	82	5.765	5.654
8	0.562	0.552	33	2.320	2.275	58	4.078	3.999	83	5.835	5.723
9	0.633	0.621	34	2.390	2.344	59	4.148	4.068	84	5.906	5.792
10	0.703	0.689	35	2.461	2.413	60	4.218	4.137	85	5.976	5.861
11	0.773	0.758	36	2.531	2.482	61	4.289	4.206	86	6.046	5.929
12	0.844	0.827	37	2.601	2.551	62	4.359	4.275	87	6.117	5.998
13	0.914	0.896	38	2.672	2.620	63	4.429	4.344	88	6.187	6.067
14	0.984	0.965	39	2.742	2.689	64	4.500	4.413	89	6.257	6.136
15	1.055	1.034	40	2.812	2.758	65	4.570	4.482	90	6.328	6.205
16	1.125	1.103	41	2.883	2.827	66	4.640	4.551	91	6.398	6.274
17	1.195	1.172	42	2.953	2.896	67	4.711	4.619	92	6.468	6.343
18	1.266	1.241	43	3.023	2.965	68	4.781	4.688	93	6.539	6.412
19	1.336	1.310	44	3.094	3.034	69	4.851	4.757	94	6.609	6.481
20	1.406	1.379	45	3.164	3.103	70	4.921	4.826	95	6.679	6.550
21	1.476	1.448	46	3.234	3.172	71	4.992	4.895	96	6.749	6.619
22	1.547	1.517	47	3.304	3.241	72	5.062	4.964	97	6.820	6.688
23	1.617	1.586	48	3.375	3.309	73	5.132	5.033	98	6.890	6.757
24	1.687	1.655	49	3.445	3.378	74	5.203	5.102	99	6.960	6.826
25	1.758	1.724	50	3.515	3.447	75	5.273	5.171	100	7.031	6.895

CONVERSION TABLE
POUNDS PER SQUARE INCH INTO KILOGRAMMES-FORCE PER SQUARE CENTIMETRE AND BARS
from 100 to 10 000 psi

psi	kgf/cm²	bars	psi	kgf/cm²	bars	psi	kgf/cm²	bars	psi	kgf/cm²	bars
100	7.031	6.895	2 600	182.798	179.264	5 100	358.566	351.633	7 600	534.333	524.002
200	14.061	13.790	2 700	189.829	186.159	5 200	365.596	358.528	7 700	541.364	530.897
300	21.092	20.684	2 800	196.860	193.053	5 300	372.627	365.422	7 800	548.394	537.791
400	28.123	27.579	2 900	203.890	199.948	5 400	379.658	372.317	7 900	555.425	544.686
500	35.153	34.474	3 000	210.921	206.843	5 500	386.688	379.212	8 000	562.456	551.581
600	42.184	41.369	3 100	217.952	213.738	5 600	393.719	386.107	8 100	569.487	558.476
700	49.215	48.263	3 200	224.982	220.632	5 700	400.750	393.001	8 200	576.517	565.370
800	56.246	55.158	3 300	232.013	227.527	5 800	407.781	399.896	8 300	583.548	572.265
900	63.276	62.053	3 400	239.044	234.422	5 900	414.811	406.791	8 400	590.579	579.160
1 000	70.307	68.948	3 500	246.074	241.317	6 000	421.842	413.686	8 500	597.609	586.055
1 100	77.338	75.842	3 600	253.105	248.211	6 100	428.873	420.580	8 600	604.640	592.949
1 200	84.368	82.737	3 700	260.136	255.106	6 200	435.903	427.475	8 700	611.671	599.844
1 300	91.399	89.632	3 800	267.167	262.001	6 300	442.934	434.370	8 800	618.701	606.739
1 400	98.430	96.527	3 900	274.197	268.896	6 400	449.965	441.265	8 900	625.732	613.634
1 500	105.460	103.421	4 000	281.228	275.790	6 500	456.995	448.159	9 000	632.763	620.528
1 600	112.491	110.316	4 100	288.259	282.685	6 600	464.026	455.054	9 100	639.794	627.423
1 700	119.522	117.211	4 200	295.289	289.580	6 700	471.057	461.949	9 200	646.824	634.318
1 800	126.553	124.106	4 300	302.320	296.475	6 800	478.088	468.844	9 300	653.855	641.213
1 900	133.583	131.000	4 400	309.351	303.369	6 900	485.118	475.738	9 400	660.886	648.107
2 000	140.614	137.895	4 500	316.381	310.264	7 000	492.149	482.633	9 500	667.916	655.002
2 100	147.645	144.790	4 600	323.412	317.159	7 100	499.180	489.528	9 600	674.947	661.897
2 200	154.675	151.685	4 700	330.443	324.054	7 200	506.210	496.423	9 700	681.978	668.792
2 300	161.706	158.579	4 800	337.474	330.948	7 300	513.241	503.317	9 800	689.008	675.686
2 400	168.737	165.474	4 900	344.504	337.843	7 400	520.272	510.212	9 900	696.039	682.581
2 500	175.767	172.369	5 000	351.535	344.738	7 500	527.302	517.107	10 000	703.070	689.476

CONVERSION TABLE
POUNDS PER FOOT INTO KILOGRAMMES PER METRE
(1 lb/ft = 1,4882 kg/m)

	0	1	2	3	4	5	6	7	8	9
0		1,49	2,98	4,46	5,95	7,44	8,92	10,41	11,90	13,39
10	14,88	16,37	17,85	19,34	20,83	22,32	23,81	25,29	26,78	28,27
20	29,76	31,25	32,74	34,22	35,71	37,20	38,69	40,18	41,66	43,15
30	44,64	46,13	47,62	49,10	50,59	52,08	53,57	55,06	56,55	58,03
40	59,52	61,01	62,50	63,99	65,47	66,96	68,45	69,94	71,43	72,92
50	74,40	75,89	77,38	78,87	80,36	81,84	83,33	84,82	86,31	87,80
60	89,29	90,77	92,26	93,75	95,24	96,73	98,21	99,70	101,19	102,68
70	104,17	105,65	107,14	108,63	110,12	111,61	113,10	114,58	116,07	117,56
80	119,05	120,54	122,02	123,51	125,00	126,49	127,98	129,47	130,95	132,44
90	133,93	135,42	136,91	138,39	139,88	141,37	142,86	144,35	145,84	147,32
100	148,81	150,30	151,79	153,28	154,77	156,25	157,74	159,23	160,72	162,21
110	163,70	165,18	166,67	168,16	169,65	171,14	172,62	174,11	175,60	177,09
120	178,58	180,07	181,55	183,04	184,53	186,02	187,51	188,99	190,48	191,97
130	193,46	194,95	196,43	197,92	199,41	200,90	202,39	203,88	205,36	206,85
140	208,34	209,83	211,32	212,80	214,29	215,78	217,27	218,76	220,25	221,73
150	223,22	224,71	226,20	227,69	229,17	230,66	232,15	233,64	235,13	236,62
160	238,10	239,59	241,08	242,57	244,06	245,54	247,03	248,52	250,01	251,50
170	252,99	254,48	255,97	257,46	258,95	260,44	261,92	263,42	264,90	266,39
180	267,88	269,36	270,85	272,34	273,83	275,32	276,81	278,29	279,78	281,27
190	282,76	284,25	285,73	287,22	288,71	290,20	291,69	293,18	294,66	296,15
200	297,64	299,13	300,62	302,10	303,59	305,08	306,57	308,06	309,55	311,03
210	312,52	314,01	315,50	316,99	318,47	319,96	321,45	322,94	324,43	325,92
220	327,40	328,89	330,38	331,87	333,36	334,85	336,33	337,82	339,31	340,80
230	342,29	343,77	345,26	346,75	348,24	349,73	351,22	352,70	354,19	355,68
240	357,17	358,66	360,14	361,63	363,12	364,61	366,10	367,59	369,07	370,56
250	372,05	373,54	375,03	376,51	378,00	379,49	380,98	382,47	383,96	385,44
260	386,93	388,42	389,91	391,40	392,88	394,37	395,86	397,35	398,84	400,33
270	401,81	403,30	404,79	406,28	407,77	409,26	410,74	412,23	413,72	415,21
280	416,70	418,18	419,67	421,16	422,65	424,14	425,63	427,11	428,60	430,09
290	431,58	433,07	434,55	436,04	437,53	439,02	440,51	442,00	443,48	444,97
300	446,46	447,95	449,44	450,92	452,41	453,90	455,39	456,88	458,37	459,85

CONVERSION TABLE
TON-MILES INTO 10³ DECANEWTON-KILOMETRES

	0	1	2	3	4	5	6	7	8	9
0	0	1,43	2,86	4,30	5,73	7,16	8,59	10,02	11,46	12,89
10	14,3	15,8	17,2	18,6	20,0	21,5	22,9	24,3	25,8	27,2
20	28,6	30,1	31,5	32,9	34,4	35,8	37,2	38,7	40,1	41,5
30	43,0	44,4	45,8	47,3	48,7	50,1	51,6	53,0	54,4	55,9
40	57,3	58,7	60,2	61,6	63,0	64,4	65,9	67,3	68,7	70,2
50	71,6	73,0	74,5	75,9	77,3	78,8	80,2	81,6	83,1	84,5
60	85,9	87,4	88,8	90,2	91,7	93,1	94,5	96,0	97,4	98,8
70	100,2	101,7	103,1	104,6	106,0	107,4	108,8	110,3	111,7	113,1
80	114,6	116,0	117,4	118,8	120,3	121,7	123,1	124,6	126,0	127,4
90	128,9	130,3	131,7	133,2	134,6	136,0	137,5	138,9	140,3	141,8
100	143,2	144,6	146,1	147,5	148,9	150,4	151,8	153,2	154,7	156,1
110	157,5	159,0	160,4	161,8	163,3	164,7	166,1	167,6	169,0	170,4
120	171,8	173,3	174,7	176,1	177,6	179,0	180,4	181,9	183,3	184,7
130	186,2	187,6	189,0	190,5	191,9	193,3	194,8	196,2	197,6	199,1
140	200,5	201,9	203,4	204,8	206,2	207,7	209,1	210,5	212,0	213,4
150	214,8	216,2	217,7	219,1	220,5	222,0	223,4	224,8	226,3	227,7
160	229,1	230,6	232,0	233,4	234,9	236,3	237,7	239,2	240,6	242,0
170	243,4	244,9	246,3	247,8	249,2	250,6	252,1	253,5	254,9	256,4
180	257,8	259,2	260,7	262,1	263,5	265,0	266,4	267,8	269,3	270,7
190	272,1	273,6	275,0	276,4	277,8	279,3	280,7	282,1	283,6	285,0

CONVERSION TABLE
TON-MILES PER FOOT INTO 10³ DECANEWTON-KILOMETRES
PER METRE

	0	1	2	3	4	5	6	7	8	9
0		4,7	9,4	14,1	18,8	23,5	28,2	32,9	37,6	42,3
10	47,0	51,7	56,4	61,1	65,8	70,5	75,2	79,9	84,6	89,3
20	94,0	98,7	103,4	108,1	112,8	117,5	122,2	126,9	131,6	136,3
30	141,0	145,7	150,4	155,1	159,8	164,5	169,2	173,9	178,6	183,3
40	188,0	192,7	197,4	202,1	206,8	211,5	217,2	221,9	226,6	231,3
50	235,0	239,7	244,4	249,1	253,8	258,5	263,2	267,9	272,6	277,3
60	282,0	286,7	291,4	296,0	300,8	305,5	310,2	314,9	319,6	324,3
70	329,0	333,7	338,4	343,1	347,8	352,5	357,2	361,9	366,6	371,3
80	376,0	380,7	385,4	390,1	394,8	399,5	404,2	408,9	413,6	418,3
90	423,0	427,7	432,4	437,1	441,8	446,5	451,2	455,9	460,6	465,3
100	470,0	474,7	479,4	484,1	488,8	493,5	498,2	502,9	507,6	512,3
110	517,0	521,7	526,4	531,1	535,8	540,5	545,2	549,9	554,6	559,3
120	564,0	568,7	573,4	578,1	582,8	587,5	592,2	596,9	601,6	606,3
130	611,0	615,7	620,4	625,1	629,8	634,5	639,2	643,9	648,6	653,3
140	658,0	662,7	667,4	672,1	676,8	681,5	686,2	690,9	695,6	700,3
150	705,0	709,7	714,4	719,1	723,8	728,5	733,2	737,9	742,6	747,3
160	752,0	756,7	761,4	766,1	770,8	775,5	780,2	784,9	789,6	794,3
170	799,0	803,7	808,4	813,1	817,8	822,5	827,2	831,9	836,6	841,3
180	846,0	850,7	855,4	860,1	864,8	869,5	874,2	878,9	883,6	888,3
190	893,0	897,7	902,4	907,1	911,8	916,5	921,2	925,9	930,6	935,3

CONVERSION TABLE
FOOT-POUNDS INTO METRE-DECANEWTONS

	0	1	2	3	4	5	6	7	8	9
0	—	0,14	0,27	0,41	0,54	0,68	0,81	0,95	1,08	1,22
10	1,36	1,49	1,63	1,76	1,90	2,03	2,17	2,30	2,44	2,58
20	2,71	2,85	2,98	3,12	3,26	3,39	3,53	3,66	3,80	3,93
30	4,07	4,20	4,34	4,47	4,61	4,74	4,88	5,02	5,15	5,29
40	5,42	5,56	5,69	5,83	5,97	6,11	6,24	6,37	6,51	6,64
50	6,78	6,91	7,05	7,19	7,32	7,46	7,59	7,73	7,86	8,00
60	8,14	8,27	8,41	8,54	8,68	8,81	8,95	9,08	9,22	9,36
70	9,49	9,63	9,71	9,90	10,03	10,17	10,30	10,44	10,58	10,71
80	10,84	10,98	11,12	11,25	11,39	11,52	11,66	11,80	11,93	12,07
90	12,20	12,34	12,47	12,61	12,74	12,88	13,02	13,15	13,29	13,42
100	13,55	13,69	13,83	13,96	14,10	14,24	14,37	14,51	14,64	14,78
110	14,92	15,05	15,19	15,32	15,46	15,59	15,73	15,86	16,00	16,13
120	16,28	16,41	16,54	16,68	16,81	16,95	17,08	17,22	17,35	17,49
130	17,63	17,76	17,90	18,03	18,17	18,30	18,44	18,57	18,71	18,85
140	18,98	19,12	19,25	19,39	19,52	19,66	19,79	19,93	20,07	20,20
150	20,33	20,47	20,61	20,74	20,88	21,02	21,15	21,29	21,42	21,56
160	21,68	21,83	21,96	22,10	22,24	22,37	22,50	22,64	22,78	22,91
170	23,04	23,18	23,32	23,46	23,59	23,73	23,86	24,00	24,14	24,27
180	24,40	24,54	24,68	24,81	24,95	25,08	25,22	25,35	25,49	25,62
190	25,75	25,90	26,03	26,17	26,31	26,41	26,58	26,72	26,85	26,98
200	27,10	27,25	27,39	27,53	27,66	27,79	27,93	28,07	28,20	28,34
210	28,47	28,61	28,74	28,88	29,01	29,15	29,29	29,42	29,56	29,69
220	29,84	29,96	30,10	30,23	30,37	30,51	30,64	30,78	30,91	31,05
230	31,20	31,32	31,46	31,60	31,73	31,86	32,00	32,14	32,27	32,40
240	32,56	32,68	32,82	32,96	33,10	33,23	33,39	33,53	33,67	33,80
250	33,91	34,03	34,16	34,30	34,43	34,57	34,70	34,84	34,98	35,11
260	35,26	35,39	35,53	35,67	35,80	35,94	36,07	36,20	36,34	36,48
270	36,61	36,74	36,88	37,01	37,15	37,28	37,42	37,55	37,68	37,82
280	37,96	38,10	38,24	38,38	38,38	38,65	38,79	38,92	39,06	39,20
290	39,31	39,45	39,59	39,72	39,85	39,99	40,13	40,26	40,40	40,54
300	40,66	40,81	40,95	41,08	41,21	41,35	41,49	41,62	41,76	41,90
310	42,01	42,17	42,30	42,44	42,58	42,71	42,84	42,98	43,11	43,24
320	43,36	43,52	43,65	43,79	43,93	44,07	44,21	44,34	44,47	44,60
330	44,72	44,88	45,02	45,15	45,29	45,43	45,57	45,71	45,85	45,98
340	46,08	46,23	46,36	46,50	46,63	46,76	46,90	47,03	47,16	47,40
350	47,44	47,59	47,73	47,86	48,00	48,13	48,26	48,40	48,53	48,66
360	48,80	48,95	49,09	49,22	49,35	49,49	49,63	49,76	49,90	50,03
370	50,15	50,30	50,43	50,57	50,70	50,83	50,97	51,10	51,23	51,37
380	51,50	51,66	51,80	51,93	52,07	52,20	52,33	52,47	52,60	52,73
390	52,86	53,01	53,14	53,27	53,40	53,53	53,67	53,80	53,93	54,07
400	54,20	54,37	54,50	54,63	54,77	54,90	55,03	55,16	55,30	55,43
410	55,57	55,72	55,86	56,00	56,13	56,26	56,40	56,53	56,66	56,80
420	56,94	57,08	57,21	57,34	57,47	57,60	57,73	57,86	58,00	58,14
430	58,31	58,44	58,57	58,70	58,83	58,96	59,10	59,23	59,36	59,50
440	59,66	59,79	59,93	60,06	60,20	60,33	60,46	60,60	60,73	60,86
450	61,01	61,15	61,29	61,43	61,56	61,70	61,83	61,96	62,10	62,24
460	62,40	62,50	62,63	62,76	62,90	63,04	63,17	63,30	63,44	63,58
470	63,74	63,86	64,00	64,13	64,26	64,40	64,53	64,66	64,80	64,96
480	65,10	65,21	65,34	65,47	65,60	65,74	65,87	66,00	66,14	66,28
490	66,44	66,57	66,70	66,83	66,97	67,10	67,23	67,36	67,50	67,64
—	0	1	2	3	4	5	6	7	8	9

TEMPERATURE CONVERSION TABLE

$$t\ °F = \frac{9}{5}\,t\ °C + 32 \qquad\qquad t\ °C = \frac{5}{9}(t\ °F - 32)$$

INSTRUCTIONS. The central figures refer to the temperatures either in degrees Celsius or degrees Fahrenheit which require conversion. The corresponding temperatures in degrees Fahrenheit or degrees Celsius will be found to the right or left respectively.

Example:

	C		F
	6.67	44	111.2

44° Celsius → 111.2° Fahrenheit
44° Fahrenheit → 6.67° Celsius

°C	central	°F
-56.7	-70	-94.0
-53.9	-65	-85.0
-51.2	-60	-76.0
-48.4	-55	-67.0
-45.6	-50	-58.0
-42.8	-45	-49.0
-40.0	-40	-40.0
-37.2	-35	-31.0
-34.4	-30	-22.0
-31.7	-25	-13.0
-28.9	-20	-4.0
-26.1	-15	5.0
-23.3	-10	14.0
-20.6	-5	23.0
-17.8	0	32.0
-16.7	2	35.6
-15.6	4	39.2
-14.4	6	42.8
-13.3	8	46.4
-12.2	10	50.0

°C	central	°F
-11.1	12	53.6
-10.0	14	57.2
-8.89	16	60.8
-7.78	18	64.4
-6.67	20	68.0
-5.55	22	71.6
-4.44	24	75.2
-3.33	26	78.8
-2.22	28	82.4
-1.11	30	86.0
0	32	89.6
1.11	34	93.2
2.22	36	96.8
3.33	38	100.4
4.44	40	104.0
5.55	42	107.6
6.67	**44**	**111.2**
7.78	46	114.8
8.89	48	118.4
10.0	50	122.0

°C	central	°F
11.1	52	125.6
12.2	54	129.2
13.3	56	132.8
14.4	58	136.4
15.6	60	140.0
16.7	62	143.6
17.8	64	147.2
18.9	66	150.8
20.0	68	154.4
21.1	70	158.0
22.2	72	161.6
23.3	74	165.2
24.4	76	168.8
25.6	78	172.4
26.7	80	176.0
27.8	82	179.6
28.9	84	183.2
30.0	86	186.8
31.1	88	190.4
32.2	90	194.0

°C	central	°F
33.3	92	197.6
34.4	94	201.2
35.6	96	204.8
36.7	98	208.4
37.8	100	212.0
48.9	120	248
60.0	140	284
71.1	160	320
82.2	180	356
93.3	200	392
104.4	220	428
115.6	240	464
126.7	260	500
137.8	280	536
148.9	300	572
160	320	608
171	340	644
182	360	680
193	380	716
204	400	752

°C	central	°F
216	420	788
227	440	824
238	460	860
249	480	896
260	500	932
271	520	968
282	540	1004
293	560	1040
304	580	1076
316	600	1112
327	620	1148
338	640	1184
349	660	1220
360	680	1256
371	700	1292
382	720	1328
393	740	1364
404	760	1400
416	780	1436
427	800	1472

INTERPOLATION TABLE

°C	0.56	1.11	1.67	2.22	2.78	3.33	3.89	4.44	5	5.56	6.11	6.67	7.22	7.78	8.33	8.89	9.44	10	10.56	11.11
	1	2	3	4	5	6	7	8	9	10	11	12	13	14	15	16	17	18	19	20
°F	1.8	3.6	5.4	7.2	9	10.8	12.6	14.4	16.2	18	19.8	21.6	23.4	25.2	27	28.8	30.6	32.4	34.2	36

CONVERSION TABLE

Specific gravity into degrees API at 15.56 °C in relation to water at 15.56 °C and 760 mm Hg

Specific gravity	Degrees API	Specific gravity	Degrees API	Specific gravity	Degrees API	Specific gravity	Degrees API	Specific gravity	Degrees API	Specific gravity	Degrees API	Specific gravity	Degrees API	Specific gravity	Degrees API	Specific gravity	Degrees API
0.600	104.3	0.650	86.2	0.700	70.6	0.750	57.2	0.800	45.4	0.850	35.0	0.900	25.7	0.950	17.5	1.000	10.0
0.602	103.5	0.652	85.5	0.702	70.1	0.752	56.7	0.802	44.9	0.852	34.6	0.902	25.4	0.952	17.1	1.002	9.7
0.604	102.8	0.654	84.9	0.704	69.5	0.754	56.2	0.804	44.5	0.854	34.2	0.904	25.0	0.954	16.8	1.004	9.4
0.606	102.0	0.656	84.2	0.706	68.9	0.756	55.7	0.806	44.1	0.856	33.8	0.906	24.7	0.956	16.5	1.006	9.2
0.608	101.2	0.658	83.6	0.708	68.4	0.758	55.2	0.808	43.6	0.858	33.4	0.908	24.3	0.958	16.2	1.008	8.9
0.610	100.5	0.660	82.9	0.710	67.8	0.760	54.7	0.810	43.2	0.860	33.0	0.910	24.0	0.960	15.9	1.010	8.6
0.612	99.7	0.662	82.2	0.712	67.2	0.762	54.2	0.812	42.8	0.862	32.7	0.912	23.7	0.962	15.6	1.012	8.3
0.614	99.0	0.664	81.6	0.714	66.7	0.764	53.7	0.814	42.3	0.864	32.3	0.914	23.3	0.964	15.3	1.014	8.1
0.616	98.2	0.666	81.0	0.716	66.1	0.766	53.2	0.816	41.9	0.866	31.9	0.916	23.0	0.966	15.0	1.016	7.8
0.618	97.5	0.668	80.3	0.718	65.6	0.768	52.7	0.818	41.5	0.868	31.5	0.918	22.6	0.968	14.7	1.018	7.5
0.620	96.7	0.670	79.7	0.720	65.0	0.770	52.3	0.820	41.1	0.870	31.1	0.920	22.3	0.970	14.4	1.020	7.2
0.622	96.0	0.672	79.1	0.722	64.5	0.772	51.8	0.822	40.6	0.872	30.8	0.922	22.0	0.972	14.1	1.022	7.0
0.624	95.3	0.674	78.4	0.724	63.9	0.774	51.3	0.824	40.2	0.874	30.4	0.924	21.6	0.974	13.8	1.024	6.7
0.626	94.5	0.676	77.8	0.726	63.4	0.776	50.9	0.826	39.8	0.876	30.0	0.926	21.3	0.976	13.5	1.026	6.4
0.628	93.8	0.678	77.2	0.728	62.9	0.778	50.4	0.828	39.4	0.878	29.7	0.928	21.0	0.978	13.2	1.028	6.2
0.630	93.1	0.680	76.6	0.730	62.3	0.780	49.9	0.830	39.0	0.880	29.3	0.930	20.7	0.980	12.9	1.030	5.9
0.632	92.4	0.682	76.0	0.732	61.8	0.782	49.5	0.832	38.6	0.882	28.9	0.932	20.3	0.982	12.6	1.032	5.6
0.634	91.7	0.684	75.4	0.734	61.3	0.784	49.0	0.834	38.2	0.884	28.6	0.934	20.0	0.984	12.3	1.034	5.4
0.636	91.0	0.686	74.8	0.736	60.8	0.786	48.5	0.836	37.8	0.886	28.2	0.936	19.7	0.986	12.0	1.036	5.1
0.638	90.3	0.688	74.2	0.738	60.2	0.788	48.1	0.838	37.4	0.888	27.9	0.938	19.4	0.988	11.7	1.038	4.8
0.640	89.6	0.690	73.6	0.740	59.7	0.790	47.6	0.840	37.0	0.890	27.5	0.940	19.0	0.990	11.4	1.040	4.6
0.642	88.9	0.692	73.0	0.742	59.2	0.792	47.2	0.842	36.6	0.892	27.1	0.942	18.7	0.992	11.1	1.042	4.3
0.644	88.2	0.694	72.4	0.744	58.7	0.794	46.7	0.844	36.2	0.894	26.8	0.944	18.4	0.994	10.9	1.044	4.0
0.646	87.5	0.696	71.8	0.746	58.2	0.796	46.3	0.846	35.8	0.896	26.4	0.946	18.1	0.996	10.6	1.046	3.8
0.648	86.9	0.698	71.2	0.748	57.7	0.798	45.8	0.848	35.4	0.898	26.0	0.948	17.8	0.998	10.3	1.048	3.5

$$\text{Degree API} = \frac{141.5}{d\,(15.56°C/15.56°C)} - 131.5$$

$$d\,(15.56°C/15.56°C) = \text{specific gravity } (60°F/60°F)$$

Approximate temperature correction to obtain temperatures at 15 °C

Specific gravity	Correction for 1 °C
de 0.600 à 0.700	0.0009
de 0.700 à 0.800	0.0008
de 0.800 à 0.840	0.00075
de 0.840 à 0.880	0.0007
de 0.880 à 0.920	0.00065
de 0.920 à 1.000	0.0006

add   if t > 15°C
subtract   if t < 15°C

NUMERICAL CONSTANTS
AND MATHEMATICAL FORMULAE

π	3,1415927	$\dfrac{1}{\pi}$	0,3183099	$\dfrac{\pi}{2}$	1,5707963	$\dfrac{\pi}{180}$	0,0174533
π^2	9,8696044	$\dfrac{1}{\pi^2}$	0,1013212	$\dfrac{\pi}{3}$	1,0471976	$\dfrac{\pi}{200}$	0,0157080
π^3	31,0062767	$\dfrac{1}{\pi^3}$	0,0322515	$\dfrac{\pi}{4}$	0,7853982	$\dfrac{180}{\pi}$	57,2957795
$\sqrt{\pi}$	1,7724539	$\dfrac{1}{\sqrt{\pi}}$	0,5641896	$\dfrac{4\pi}{3}$	4,1887902	$\dfrac{200}{\pi}$	63,6619763
$\sqrt[3]{\pi}$	1,4645919	$\dfrac{1}{\sqrt[3]{\pi}}$	0,6827840				

$\sqrt{2}$	1,414214	$\sqrt{3}$	1,732051	$\sqrt{5}$	2,236068	$\sqrt{10}$	3,162278
$\dfrac{1}{\sqrt{2}}$	0,70711	$\dfrac{1}{\sqrt{3}}$	0,57735	$\dfrac{1}{\sqrt{5}}$	0,44721	$\dfrac{1}{\sqrt{10}}$	0,31623

e	2,7182818	$\dfrac{1}{e}$	0,3678794	$\log_{10} e\ =\ 0,4342945$		$g\ =\ 9,80665\ \mathrm{m\ s^2}$	

$$\frac{1}{\log_{10} e} = \mathrm{colog}\ e = \log_e 10 = 2,3025851 \qquad \log_e x = 2,3025851\ \log_{10} x \qquad \log_{10} x = 0,4342945\ \log_e x$$

Arithmetic progression

$$a \quad a + r \quad a + 2r \quad a + 3r\ldots \quad a + (n-1)\,r$$

$$S_n = \left(\frac{a + l}{2}\right)n = \frac{n}{2}\,[2a + (n-1)\,r]$$

a = first term,
r = common difference,
n = number of terms,
l = last term.

Geometric progression

$$a \quad aq \quad aq^2 \quad aq^3 \quad \ldots \quad aq^{n-1}$$

$$\text{Si } q \neq 1 \quad S_n = \frac{lq - a}{q - 1} = \frac{a\,(q^n - 1)}{q - 1}$$

a = first term,
q = common ratio.
n = number of terms,
l = last term aq^{n-1}

PROPERTIES OF NUMBERS FROM 1 TO 25
Powers, roots, reciprocals, circumference, area and common log

n	n^2	n^3	$\sqrt{n}$	$\sqrt[3]{n}$	$\dfrac{1}{n}$	πn	$\dfrac{1}{4}\pi n^2$	log n
1	1	1	1,0000	1,0000	1,0000	3,1416	0,7854	0,00000
2	4	8	1,4142	1,2599	0,5000	6,2832	3,1416	0,30103
3	9	27	1,7321	1,4422	0,3333	9,425	7,0686	0,47712
4	16	64	2,0000	1,5874	0,2500	12,57	12,566	0,60206
5	25	125	2,2361	1,7100	0,2000	15,71	19,635	0,69897
6	36	216	2,4495	1,8171	0,1667	18,85	28,274	0,77815
7	49	343	2,6458	1,9129	0,1429	21,99	38,485	0,84510
8	64	512	2,8284	2,0000	0,1250	25,13	50,266	0,90309
9	81	729	3,0000	2,0801	0,1111	28,27	63,617	0,95424
10	100	1000	3,1623	2,1544	0,1000	31,42	78,540	1,00000
11	121	1331	3,3166	2,2240	0,0909	34,56	95,033	1,04139
12	144	1728	3,4641	2,2894	0,0833	37,70	113,10	1,07918
13	169	2197	3,6056	2,3513	0,0769	40,84	132,73	1,11394
14	196	2744	3,7417	2,4101	0,0714	43,98	153,94	1,14613
15	225	3375	3,8730	2,4662	0,0667	47,12	176,72	1,17609
16	256	4096	4,0000	2,5198	0,0625	50,27	201,06	1,20412
17	289	4913	4,1231	2,5713	0,0588	53,41	226,98	1,23045
18	324	5832	4,2426	2,6207	0,0556	56,55	254,47	1,25527
19	361	6859	4,3589	2,6684	0,0526	59,69	283,53	1,27875
20	400	8000	4,4721	2,7144	0,0500	62,83	314,16	1,30103
21	441	9261	4,5826	2,7589	0,0476	65,97	346,36	1,32222
22	484	10648	4,6904	2,8020	0,0454	69,12	380,13	1,34242
23	529	12167	4,7958	2,8439	0,0435	72,26	415,48	1,36173
24	576	13824	4,8990	2,8845	0,0416	75,40	452,39	1,38021
25	625	15625	5,0000	2,9240	0,0400	78,54	490,87	1,39794

PROPERTIES OF NUMBERS FROM 26 TO 50
Powers, roots, reciprocals, circumference, area and common log

n	n^2	n^3	$\sqrt{n}$	$\sqrt[3]{n}$	$\dfrac{1}{n}$	πn	$\dfrac{1}{4}\pi n^2$	log n
26	676	17576	5,0990	2,9625	0,0385	81,68	530,93	1,41497
27	729	19683	5,1962	3,0000	0,0370	84,82	572,56	1,43136
28	784	21952	5,2915	3,0366	0,0357	87,96	615,75	1,44716
29	841	24389	5,3852	3,0723	0,0345	91,11	660,52	1,46240
30	900	27000	5,4772	3,1072	0,0333	94,25	706,86	1,47712
31	961	29791	5,5678	3,1414	0,0323	97,39	754,77	1,49136
32	1024	32768	5,6569	3,1748	0,0312	100,53	804,25	1,50515
33	1089	35937	5,7446	3,2075	0,0303	103,7	855,30	1,51851
34	1156	39304	5,8310	3,2396	0,0294	106,8	907,92	1,53148
35	1225	42875	5,9161	3,2711	0,0286	110,0	962,11	1,54407
36	1296	46656	6,0000	3,3019	0,0278	113,1	1017,9	1,55630
37	1369	50653	6,0828	3,3322	0,0270	116,2	1075,2	1,56820
38	1444	54872	6,1644	3,3620	0,0263	119,4	1134,1	1,57978
39	1521	59319	6,2450	3,3912	0,0256	122,5	1194,6	1,59106
40	1600	64000	6,3246	3,4200	0,0250	125,7	1256,6	1,60206
41	1681	68921	6,4031	3,4482	0,0244	128,8	1320,3	1,61278
42	1764	74088	6,4807	3,4760	0,0238	131,9	1385,4	1,62325
43	1849	79507	6,5574	3,5034	0,0233	135,1	1452,2	1,63347
44	1936	85184	6,6332	3,5303	0,0227	138,2	1520,5	1,64345
45	2025	91125	6,7082	3,5569	0,0222	141,4	1590,4	1,65321
46	2116	97336	6,7823	3,5830	0,0217	144,5	1661,9	1,66276
47	2209	103823	6,8557	3,6088	0,0213	147,7	1734,9	1,67210
48	2304	110592	6,9282	3,6342	0,0208	150,8	1809,6	1,68124
49	2401	117649	7,0000	3,6593	0,0204	153,9	1885,7	1,69020
50	2500	125000	7,0711	3,6840	0,0200	157,1	1963,5	1,69897

PROPERTIES OF NUMBERS FROM 51 TO 75
Powers, roots, reciprocals, circumference, area and common log

n	n^2	n^3	$\sqrt{n}$	$\sqrt[3]{n}$	$\dfrac{1}{n}$	πn	$\dfrac{1}{4}\pi n$	log n
51	2601	132651	7,1414	3,7084	0,0196	160,2	2042,8	1,70757
52	2704	140608	7,2111	3,7325	0,0192	163,4	2123,7	1,71600
53	2809	148877	7,2801	3,7563	0,0189	166,5	2206,2	1,72428
54	2916	157464	7,3485	3,7798	0,0185	169,6	2290,2	1,73239
55	3025	166375	7,4162	3,8030	0,0182	172,8	2375,8	1,74036
56	3136	175616	7,4833	3,8259	0,0179	175,9	2463,0	1,74819
57	3249	185193	7,5498	3,8485	0,0175	179,1	2551,8	1,75587
58	3364	195112	7,6158	3,8709	0,0172	182,2	2642,1	1,76343
59	3481	205379	7,6811	3,8930	0,0169	185,4	2734,0	1,77085
60	3600	216000	7,7460	3,9149	0,0167	188,5	2827,4	1,77815
61	3721	226981	7,8102	3,9365	0,0164	191,6	2922,5	1,78533
62	3844	238328	7,8740	3,9579	0,0161	194,8	3019,1	1,79239
63	3969	250047	3,9373	3,9791	0,0159	197,9	3117,3	1,79934
64	4096	262144	8,0000	4,0000	0,0156	201,1	3217,0	1,80618
65	4225	274625	8,0623	4,0207	0,0154	204,2	3318,3	1,81291
66	4356	287496	8,1240	4,0412	0,0151	207,3	3241,2	1,81954
67	4489	300763	8,1854	4,0615	0,0149	210,5	3525,7	1,82607
68	4624	314432	8,2462	4,0817	0,0147	213,6	3631,7	1,83251
69	4761	328509	8,3066	4,1016	0,0145	216,8	3739,3	1,83885
70	4900	343000	8,3666	4,1213	0,0143	219,9	3848,5	1,84510
71	5041	357911	8,4261	4,1408	0,0141	223,1	3959,2	1,85126
72	5184	372248	8,4853	4,1602	0,0139	226,2	4071,5	1,85733
73	5329	389017	8,5440	4,1793	0,0137	229,3	4185,4	1,86332
74	5476	405224	8,6023	4,1983	0,0135	232,5	4300,8	1,86923
75	5625	421875	8,6603	4,2172	0,0133	235,6	4417,9	1,87506

PROPERTIES OF NUMBERS FROM 76 TO 100
Powers, roots, reciprocals, circumference, area and common log

n	n^2	n^3	$\sqrt{n}$	$\sqrt[3]{n}$	$\dfrac{1}{n}$	πn	$\dfrac{1}{4}\pi n^2$	log n
76	5776	438976	8,7178	4,2358	0,0132	238,8	4536,5	1,88081
77	5929	456533	8,7750	4,2543	0,0130	241,9	4656,6	1,88649
78	6084	474552	8,8318	4,2727	0,0128	245,0	4778,4	1,89209
79	6241	493039	8,8882	4,2908	0,0127	248,2	4901,7	1,89763
80	6400	512000	8,9443	4,3089	0,0125	251,3	5026,6	1,90309
81	6561	531441	9,0000	4,3267	0,0123	254,5	5153,0	1,90849
82	6724	551368	9,0554	4,3445	0,0122	257,6	5281,0	1,91381
83	6889	571787	9,1104	4,3621	0,0120	260,8	5410,6	1,91908
84	7056	592704	9,1652	4,3795	0,0119	263,9	5541,8	1,92428
85	7225	614125	9,2195	4,3968	0,0118	267,0	5674,5	1,92942
86	7396	636056	9,2736	4,4140	0,0116	270,2	5808,8	1,93450
87	7569	658503	9,3274	4,4310	0,0115	273,3	5944,7	1,93952
88	7744	681472	9,3808	4,4480	0,0114	276,5	6082,1	1,94448
89	7921	704969	9,4340	4,4647	0,0112	279,6	6221,1	1,94939
90	8100	729000	9,4868	4,4814	0,0111	282,7	6361,7	1,95424
91	8281	753571	9,5394	4,4979	0,0110	285,9	6503,9	1,95904
92	8464	778688	9,5917	4,5144	0,0109	289,0	6647,6	1,96379
93	8649	804357	9,6437	4,5307	0,0107	292,2	6792,9	1,96848
94	8836	830584	9,6954	4,5468	0,0106	295,3	6939,8	1,97313
95	9025	857375	9,7468	4,5629	0,0105	298,5	7088,2	1,97772
96	9216	884736	9,7980	4,5789	0,0104	301,6	7238,2	1,98227
97	9409	912673	9,8489	4,5947	0,0103	304,7	7389,8	1,98677
98	9604	941192	9,8995	4,6104	0,0102	307,9	7543,0	1,99123
99	9801	970299	9,9499	4,6261	0,0101	311,0	7697,7	1,99564
100	10000	1000000	10,0000	4,6416	0,0100	314,2	7854,0	2,00000

NATURAL TRIGONOMETRIC FUNCTIONS FROM 0° TO 23°45'

Angles	Cosine	Sine	Tangent	Angles	Cosine	Sine	Tangent
0°00'	1.0000	0.0000	0.0000	12°00'	0.9781	0.2079	0.2126
15	1.0000	.0044	.0041	15	.9772	.2122	.2171
30	1.0000	.0087	.0087	30	.9763	.2164	.2217
45	0.9999	0.0131	0.0131	45	0.9753	0.2207	0.2263
1°00'	0.9998	0.0175	0.0175	13°00'	0.9744	0.2250	0.2309
15	.9998	.0218	.0218	15	.9734	.2292	.2355
30	.9997	.0262	.0262	30	.9724	.2334	.2401
45	0.9995	0.0305	0.0306	45	0.9713	0.2377	0.2447
2°00'	0.9994	0.0349	0.0349	14°00'	0.9703	0.2419	0.2493
15	.9992	.0393	.0393	15	.9692	.2462	.2540
30	.9990	.0436	.0437	30	.9681	.2504	.2586
45	0.9988	0.0480	0.0480	45	0.9670	.2546	0.2633
3°00'	0.9986	0.0523	0.0524	15°00'	0.9659	0.2588	0.2679
15	.9984	.0567	.0568	15	.9648	.2630	.2726
30	.9981	.0610	.0612	30	.9636	.2672	.2773
45	0.9979	0.0654	0.0655	45	0.9625	0.2714	0.2820
4°00'	0.9976	0.0698	0.0699	16°00'	0.9613	0.2756	0.2867
15	.9973	.0741	.0743	15	.9600	.2798	.2915
30	.9969	.0785	.0787	30	.9588	.2840	.2962
45	0.9966	0.0828	0.0831	45	0.9576	0.2882	0.3010
5°00'	0.9962	0.0872	0.0875	17°00'	0.9563	0.2924	0.3057
15	.9958	.0915	.0919	15	.9550	.2965	.3105
30	.9954	.0958	.0963	30	.9537	.3007	.3153
45	0.9950	0.1002	0.1007	45	0.9524	0.3049	0.3201
6°00'	0.9945	0.1045	0.1051	18°00'	0.9511	0.3090	0.3249
15	.9941	.1089	.1095	15	.9497	.3132	.3298
30	.9936	.1132	.1139	30	.9483	.3173	.3346
45	0.9931	0.1175	0.1184	45	0.9469	0.3214	0.3395
7°00'	0.9925	0.1219	0.1228	19°00'	0.9455	0.3256	0.3443
15	.9920	.1262	.1272	15	.9441	.3297	.3492
30	.9914	.1305	.1317	30	.9426	.3338	.3541
45	0.9909	0.1349	0.1361	45	0.9412	0.3379	0.3590
8°00'	0.9903	0.1392	0.1405	20°00'	0.9397	0.3420	0.3640
15	.9897	.1435	.1450	15	.9382	.3461	.3689
30	.9890	.1478	.1495	30	.9367	.3502	.3739
45	0.9884	0.1521	0.1539	45	0.9351	0.3543	0.3789
9°00	0.9877	0.1564	0.1584	21°00'	0.9336	0.3584	0.3839
15	.9870	.1607	.1629	15	.9320	.3624	.3889
30	.9863	.1650	.1673	30	.9304	.3665	.3939
45	0.9856	0.1693	0.1718	45	0.9288	0.3706	0.3990
10°00'	0.9848	0.1736	0.1763	22°00'	0.9272	0.3746	0.4040
15	.9840	.1779	.1808	15	.9255	.3786	.4091
30	.9833	.1822	.1853	30	.9239	.3827	.4142
45	0.9825	0.1865	0.1899	45	0.9222	0.3867	0.4193
11°00'	0.9816	0.1908	0.1944	23°00'	0.9205	0.3907	0.4245
15	.9808	.1951	.1989	15	.9188	.3947	.4296
30	.9799	.1994	.2035	30	.9171	.3987	.4348
45	0.9790	0.2036	0.2080	45	0.9153	0.4027	0.4400

NATURAL TRIGONOMETRIC FUNCTIONS FROM 24° TO 47°45'

Angles	Cosine	Sine	Tangent	Angles	Cosine	Sine	Tangent
24°00'	0.9135	0.4067	0.4452	**36°00'**	0.8090	0.5878	0.7266
15	.9118	.4107	.4505	15	.8064	.5913	.7332
30	.9100	.4147	.4557	30	.8030	.5948	.7400
45	0.9081	0.4187	0.4610	45	0.8013	0.5983	0.7467
25°00'	0.9063	0.4226	0.4663	**37°00'**	0.7986	0.6018	0.7536
15	.9045	.4266	.4716	15	.7960	.6063	.7604
30	.9026	.4305	.4770	30	.7934	.6088	.7673
45	0.9007	0.4344	0.4823	45	0.7907	0.6122	0.7743
26°00'	0.8988	0.4384	0.4877	**38°00'**	0.7880	0.6157	0.7813
15	.8969	.4423	.4931	15	.7853	.6191	.7883
30	.8949	.4462	.4986	30	.7826	.6225	.7954
45	0.8930	0.4501	0.5040	45	0.7799	0.6259	0.8026
27°00'	0.8910	0.4540	0.5095	**39°00'**	0.7771	0.6293	0.8098
15	.8890	.4579	.5150	15	.7744	.6327	.8170
30	.8870	.4617	.5206	30	.7716	.6361	.8243
45	0.8850	0.4656	0.5261	45	0.7688	0.6394	0.8317
28°00'	0.8829	0.4695	0.5317	**40°00'**	0.7660	0.6428	0.8391
15	.8809	.4733	.5373	15	.7632	.6461	.8466
30	.8788	.4772	.5430	30	.7604	.6494	.8541
45	0.8767	0.4810	0.5486	45	0.7576	0.6528	0.8617
29°00'	0.8746	0.4848	0.5543	**41°00'**	0.7547	0.6561	0.8693
15	.8725	.4886	.5600	15	.7518	.6593	.8770
30	.8704	.4924	.5658	30	.7490	.6626	.8847
45	0.8682	0.4962	0.4715	45	0.7461	0.6659	0.8925
30°00'	0.8660	0.5000	0.5774	**42°00'**	0.7431	0.6691	0.9004
15	.8638	.5038	.5832	15	.7402	.6724	.9083
30	.8617	.5075	.5890	30	.7373	.6756	.9163
45	0.8594	0.5113	0.5950	45	0.7343	0.6788	0.9244
31°00'	0.8572	0.5150	0.6009	**43°00'**	0.7314	0.6820	0.9325
15	.8549	.5188	.6068	15	.7284	.6852	.9407
30	.8526	.5225	.6128	30	.7254	.6884	.9490
45	0.8504	0.5262	0.6188	45	0.7224	0.6915	0.9573
32°00'	0.8480	0.5299	0.6249	**44°00'**	0.7193	0.6947	0.9657
15	.8457	.5336	.6310	15	.7163	.6978	.9742
30	.8434	.5373	.6371	30	.7133	.7009	.9827
45	0.8410	0.5410	0.6432	45	0.7102	0.7040	0.9913
33°00'	0.8387	0.5446	0.6494	**45°00'**	0.7071	0.7071	cot 1.0000
15	.8363	.5483	.6556	15	.7040	.7102	.9913
30	.8339	.5519	.6619	30	.7009	.7133	.9827
45	0.8315	0.5556	0.6682	45	0.6978	0.7163	0.9742
34°00'	0.8290	0.5592	0.6745	**46°00'»**	0.6947	0.7193	0.9657
15	.8266	.5628	.6809	15	.6915	.7224	.9573
30	.8241	.5664	.6873	30	.6884	.7254	.9490
45	0.8216	0.5700	0.6937	45	0.6852	0.7284	0.9407
35°00'	0.8192	0.5736	0.7002	**47°00'**	0.6820	0.7314	0.9325
15	.8166	.5771	.7067	15	.6788	.7343	.9244
30	.8141	.5807	.7133	30	.6756	.7373	.9163
45	0.8116	0.5842	0.7199	45	0.6724	0.7402	0.9083

NATURAL TRIGONOMETRIC FUNCTIONS FROM 48° TO 71°45′

Angles	Cosine	Sine	Cotangent	Angles	Cosine	Sine	Cotangent
48°00′	0.6691	0.7431	0.9004	**60°00′**	0.5000	0.8660	0.5774
15	.6659	.7461	.8925	15	.4962	.8682	.5715
30	.6626	.7490	.8847	30	.4924	.8704	.5658
45	0.6593	0.7518	0.8770	45	0.4886	0.8725	0.5600
49°00′	0.6561	0.7547	0.8693	**61°00′**	0.4848	0.8746	0.5543
15	.6528	.7576	.8617	15	.4810	.8767	.5486
30	.6494	.7604	.8541	30	.4772	.8788	.5430
45	0.6461	0.7632	0.8466	45	0.4733	0.8809	0.5373
50°00′	0.6428	0.7660	0.8391	**62°00′**	0.4695	0.8829	0.5317
15	.6394	.7688	.8317	15	.4656	.8850	.5261
30	.6361	.7716	.8243	30	.4617	.8870	.5206
45	0.9327	0.7744	0.8170	45	0.4579	0.8890	0.5150
51°00′	0.6293	0.7771	0.8098	**63°00′**	0.4540	0.8910	0.5095
15	.6259	.7799	.8026	15	.4501	.8930	.5040
30	.6225	.7826	.7954	30	.4462	.8949	.4986
45	0.6191	0.7853	0.7883	45	0.4423	0.8969	0.4931
52°00′	0.6157	0.7880	0.7813	**64°00′**	0.4384	0.8988	0.4877
15	.6122	.7907	.7743	15	.4344	.9007	.4823
30	.6088	.7934	.7673	30	.4305	.9026	.4770
45	0.6053	0.7960	0.7604	45	0.4266	0.9045	0.4716
53°00′	0.6018	0.7986	0.7536	**65°00′**	0.4226	0.9063	0.4663
15	.5983	.8013	.7467	15	.4187	.9081	.4610
30	.5948	.8039	.7400	30	.4147	.9100	.4557
45	0.5913	0.8064	0.7332	45	0.4107	0.9118	0.4505
54°00′	0.5878	0.8090	0.7265	**66°00′**	0.4067	0.9135	0.4452
15	.5842	.8116	.7199	15	.4027	.9153	.4400
30	.5807	.8141	.7133	30	.3987	.9171	.4348
45	0.5771	0.8166	0.7067	45	0.3947	0.9188	0.4296
55°00′	0.5736	0.8192	0.7002	**67°00′**	0.3907	0.9205	0.4245
15	.5700	.8216	.6937	15	.3867	.9222	.4193
30	.5664	.8241	.6873	30	.3827	.9239	.4142
45	0.5628	0.8266	0.6809	45	0.3786	0.9255	0.4091
56°00′	0.5592	0.8290	0.6745	**68°00′**	0.3746	0.9272	0.4040
15	.5556	.8315	.6682	15	.3706	.9288	.3990
30	.5519	.8339	.6619	30	.3665	.9304	.3939
45	0.5483	0.8363	0.6556	45	0.3624	0.9320	0.3889
57°00′	0.5446	0.8387	0.6494	**69°00′**	0.3584	0.9336	0.3839
15	.5410	.8410	.6432	15	.3543	.9351	.3789
30	.5373	.8434	.6371	30	.3502	.9367	.3739
45	0.5336	0.8457	0.6310	45	0.3461	0.9382	0.3689
58°00′	0.5299	0.8480	0.6249	**70°00′**	0.3420	0.9397	0.3640
15	.5262	.8504	.6188	15	.3379	.9412	.3590
30	.5225	.8526	.6128	30	.3338	.9426	.3541
45	0.5188	0.8549	0.6068	45	0.3297	0.9441	0.3492
59°00′	0.5150	0.8572	0.6009	**71°00′**	0.3256	0.9455	0.3443
15	.5113	.8594	.5949	15	.3214	.9469	.3395
30	.5075	.8616	.5890	30	.3173	.9483	.3346
45	0.5038	0.8638	0.5832	45	0.3132	0.9497	0.3298

NATURAL TRIGONOMETRIC FUNCTIONS FROM 72° TO 90°

Angles	Cosine	Sine	Cotangent	Angles	Cosine	Sine	Cotangent
72°00'	0.3090	0.9511	0.3249	**81°00'**	0.1564	0.9877	0.1584
15	.3049	.9524	.3201	15	.1521	.9884	.1539
30	.3007	.9537	.3153	30	.1478	.9890	.1495
45	0.2965	0.9550	0.3105	45	0.1435	0.9897	0.1450
73°00'	0.2924	0.9563	0.3057	**82°00'**	0.1392	0.9903	0.1405
15	.2882	.9576	.3010	15	.1349	.9909	.1361
30	.2840	.9588	.2962	30	.1305	.9914	.1317
45	0.2798	0.9600	0.2915	45	0.1262	0.9920	0.1272
74°00'	0.2756	0.9613	0.2867	**83°00'**	0.1219	0.9925	0.1228
15	.2714	.9625	.2820	15	.1175	.9931	.1184
30	.2672	.9636	.2773	30	.1132	.9936	.1139
45	0.2630	0.9648	0.2726	45	0.1089	0.9941	0.1095
75°00'	0.2588	0.9659	0.2679	**84°00'**	0.1045	0.9945	0.1051
15	.2546	.9670	.2633	15	.1002	.9950	.1007
30	.2504	.9681	.2586	30	.0958	.9954	.0963
45	0.2462	0.9692	0.2540	45	0.0915	0.9958	0.0919
76°00'	0.2419	0.9703	0.2493	**85°00'**	0.0872	0.9962	0.0875
15	.2377	.9713	.2447	15	.0828	.9966	.0831
30	.2334	.9724	.2401	30	.0785	.9969	.0787
45	0.2292	0.9734	0.2355	45	0.0741	0.9973	0.0743
77°00'	0.2250	0.9744	0.2309	**86°00'**	0.0698	0.9976	0.0699
15	.2207	.9753	.2263	15	.0654	.9979	.0655
30	.2164	.9763	.2217	30	.0610	.9981	.0612
45	0.2122	0.9772	0.2171	45	0.0567	0.9984	0.0568
78°00'	0.2079	0.9781	0.2126	**87°00'**	0.0523	0.9986	0.0524
15	.2036	.9790	.2080	15	.0480	.9988	.0480
30	.1994	.9799	.2035	30	.0436	.9990	.0437
45	0.1951	0.9808	0.1989	45	0.0393	0.9992	0.0393
79°00'	0.1908	0.9816	0.1944	**88°00'**	0.0349	0.9994	0.0349
15	.1865	.9825	.1899	15	.0305	.9995	.0306
30	.1822	.9833	.1853	30	.0262	.9997	.0262
45	0.1779	0.9840	0.1808	45	0.0218	.9998	.0218
80°00'	0.1736	0.9848	0.1763	**89°00'**	0.0175	0.9998	0.0175
15	.1693	.9856	.1718	15	.0131	.9999	.0131
30	.1650	.9863	.1673	30	.0087	1.0000	.0087
45	0.1607	0.9870	0.1629	45	0.0044	1.0000	0.0044
				90°00'	0.0000	1.0000	0.0000

TRIGONOMETRIC FORMULAE

DEFINITION

$$\cos \alpha = \frac{OP}{OM}$$

$$\sin \alpha = \frac{PM}{OM} \qquad \cotan \alpha = \frac{1}{\tan \alpha} = \frac{OP}{PM}$$

$$\tan \alpha = \frac{PM}{OP}$$

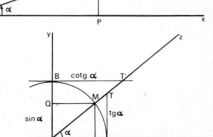

GEOMETRIC INTERPRETATION

$$\overline{OA} = \overline{OM} = R = 1$$

$$\overline{OQ} = \sin \alpha$$

$$\overline{OP} = \cos \alpha$$

$$\overline{AT} = \tan \alpha$$

$$\overline{BT'} = \cotan \alpha$$

TRIGONOMETRIC RELATIONS

$$\cos^2 \alpha + \sin^2 \alpha = 1$$

$$\tan \alpha = \frac{\sin \alpha}{\cos \alpha}$$

$$\cotan \alpha = \frac{\cos \alpha}{\sin \alpha} = \frac{1}{\tan \alpha}$$

$$\sin 2\alpha = 2 \sin \alpha \cos \alpha$$

$$\cos 2\alpha = \cos^2\alpha - \sin^2\alpha$$

$$= 1 - 2 \sin^2\alpha$$

$$\tan 2\alpha = \frac{2 \tan \alpha}{1 - \tan^2\alpha}$$

$$\sin (\alpha + \beta) = \sin \alpha \cos \beta + \cos \alpha \sin \beta$$
$$\cos (\alpha + \beta) = \cos \alpha \cos \beta - \sin \alpha \sin \beta$$
$$\sin (\alpha - \beta) = \sin \alpha \cos \beta - \cos \alpha \sin \beta$$
$$\cos (\alpha - \beta) = \cos \alpha \cos \beta + \sin \alpha \sin \beta$$

$$tg (\alpha + \beta) = \frac{\tan \alpha + \tan \beta}{1 - \tan \alpha \tan \beta}$$

$$tg (\alpha - \beta) = \frac{\tan \alpha - \tan \beta}{1 + \tan \alpha \tan \beta}$$

VALUES OF TRIGONOMETRIC FUNCTIONS RELATED TO HALF-ANGLE TANGENTS

$$\tan \frac{\alpha}{2} = t$$

$$\cos \alpha = \frac{1 - t^2}{1 + t^2}$$

$$\sin \alpha = \frac{2t}{1 + t^2}$$

$$\tan \alpha = \frac{2t}{1 - t^2}$$

RELATIONS BETWEEN SIDES AND ANGLE OF ANY TRIANGLE

$$\hat{A} + \hat{B} + \hat{C} = \pi$$

$$\frac{a}{\sin \hat{A}} = \frac{b}{\sin \hat{B}} = \frac{c}{\sin \hat{C}} = 2R$$

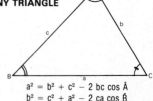

$$a^2 = b^2 + c^2 - 2bc \cos \hat{A}$$
$$b^2 = c^2 + a^2 - 2ca \cos \hat{B}$$
$$c^2 = a^2 + b^2 - 2ab \cos \hat{C}$$

MENSURATION FORMULAE

A R E A	V O L U M E

AREA

Triangle

$$p = \frac{a + b + c}{2}$$

$$S = \frac{ah}{2} = \frac{abc}{4R} = pr$$

Parallelogram

$$S = b\,h$$

Square : $S = a^2$ Rectangle $S = ab$

Trapezoid

$$S = \frac{AB + CD}{2}\,h = MN \cdot h$$

Circle

$$C = 2\pi R = \pi D$$

$$S = \pi R^2 = \frac{\pi D^2}{4} = \frac{C^2}{4\pi}$$

Sector of a circle

$$S = \frac{arc\ ACB \cdot R}{2} = \frac{\pi R^2 \alpha}{360}$$

(α is the angle between the radii in degrees)

Segment of a circle

$$S = \frac{\pi R^2 \beta}{360} - \frac{DF}{2}\,(R - f)$$

Annulars

$$S = \frac{\pi}{4}(D^2 - d^2) = \pi(R^2 - r^2)$$

$$= \frac{\pi}{4}(D + d)\,(D - d)$$

$$= \pi(R + r)\,(R - r)$$

Ellipse

$\mathbf{a}$ = semimajor axis

$\mathbf{b}$ = semiminor axis

$$S = \pi\,ab$$

VOLUME

Regular or oblique prism

$$V = B\,h$$

Right cylinder

$$V = \pi R^2 h = Bh$$

Hollow cylinder

$$V = \pi\,(R^2 - r^2)\,h = \pi\,(R + r)\,eh$$

Right cone

$$V = \frac{\pi R^2 h}{3}$$

Frustrum of a right cone

$$V = \frac{\pi h}{3}\,(R^2 + r^2 + Rr)$$

Pyramid

$$V = \frac{1}{3}\,Bh$$

Frustrum of a pyramid with parallel bases

$$V = \frac{1}{3}\,h\,(B + b + \sqrt{Bb})$$

Sphere $S = 4\pi R^2 = \pi D^2$

$$V = \frac{4}{3}\,\pi R^3 = 4,189\ R^3$$

Hollow sphere

$$V = \frac{4}{3}\,\pi\,(R^3 - r^3)$$

Spherical segment with one base

$$1°)\ V = \frac{1}{6}\,\pi h\,(h^2 + 3\ \overline{AI}^2)$$

$$2°)\ V = \frac{1}{3}\,\pi h^2\,(3R - h)$$

Spherical segment with two bases

$$V = \frac{1}{6}\,\pi h\,(3a^2 + 3b^2 + h^2)$$

CAPACITY OF HORIZONTAL CYLINDRICAL TANKS
Mensuration of tank : capacity = V, height = H

Fraction of H	Fraction of V	Fraction of H	Fraction of V	Fraction of H	Fraction of V
0,01	0,0017	0,34	0,2998	0,67	0,7122
0,02	0,0047	0,35	0,3119	0,68	0,7241
0,03	0,0087	0,36	0,3241	0,69	0,7360
0,04	0,0134	0,37	0,3364	0,70	0,7477
0,05	0,0187	0,38	0,3487	0,71	0,7593
0,06	0,0245	0,39	0,3611	0,72	0,7708
0,07	0,0308	0,40	0,3736	0,73	0,7821
0,08	0,0375	0,41	0,3860	0,74	0,7934
0,09	0,0446	0,42	0,3986	0,75	0,8045
0,10	0,0520	0,43	0,4111	0,76	0,8155
0,11	0,0599	0,44	0,4237	0,77	0,8263
0,12	0,0680	0,45	0,4364	0,78	0,8369
0,13	0,0764	0,46	0,4490	0,79	0,8474
0,14	0,0851	0,47	0,4617	0,80	0,8576
0,15	0,0941	0,48	0,4745	0,81	0,8677
0,16	0,1033	0,49	0,4872	0,82	0,8776
0,17	0,1127	0,50	0,5000	0,83	0,8873
0,18	0,1223	0,51	0,5128	0,84	0,8967
0,19	0,1323	0,52	0,5255	0,85	0,9059
0,20	0,1424	0,53	0,5383	0,86	0,9149
0,21	0,1526	0,54	0,5510	0,87	0,9236
0,22	0,1631	0,55	0,5636	0,88	0,9320
0,23	0,1737	0,56	0,5763	0,89	0,9401
0,24	0,1845	0,57	0,5889	0,90	0,9480
0,25	0,1955	0,58	0,6014	0,91	0,9554
0,26	0,2066	0,59	0,6140	0,92	0,9625
0,27	0,2179	0,60	0,6264	0,93	0,9692
0,28	0,2292	0,61	0,6389	0,94	0,9755
0,29	0,2407	0,62	0,6513	0,95	0,9813
0,30	0,2523	0,63	0,6636	0,96	0,9866
0,31	0,2640	0,64	0,6759	0,97	0,9913
0,32	0,2759	0,65	0,6881	0,98	0,9952
0,33	0,2878	0,66	0,7002	0,99	0,9983

Example : Let us take a tank with capacity V = 12 000 l and height H = 2 m. Measurements show a liquid height of 0,20 m in the tank. How much liquid does the tank contain.

Answer : Fraction of height 0,20/2 = 0,10, corresponding to a volume fraction of 0,0520 in the table. The contenance is thus : 0,0520 × 12 000 = 624 l.

MECHANICS AND STRENGTH OF MATERIALS

MOMENT OF A FORCE ABOUT A CENTRE   **MOMENT OF A TORQUE**

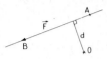

$$M_o^! \vec{F} = Fd$$

$$M_o^! C = Fd$$

M⁄ in metre-newton, F in newton and d in metres

UNIFORM STRAIGHT LINE MOTION

$$\ell = \ell_o + vt$$

ℓ = distance covered (m),
ℓ_o = initial distance (m),
v = velocity (m/s),
t = time (s).

UNIFORMLY ACCELERATED MOTION

$$\ell = \ell_o + v_o t + \frac{\gamma t^2}{2}$$

ℓ = distance covered (m),
ℓ_o = initial distance (m),
v_o = initial velocity (m/s),
t = time (s),
γ = acceleration (m/s²).

UNIFORM CIRCULAR MOTION

Angular velocity $\omega = \dfrac{\alpha}{t}$ a = angle of rotation

or $\alpha = \omega t$

Angular velocity versus revolutions per minute

$$\omega = \frac{2\pi N}{60}$$

(ω in radians per second and N in revolutions per minute)

Circumferential velocity

$$v \text{ m/min} = 2\pi RN$$

or

$$v \text{ m/s} = \omega R = \frac{2\pi RN}{60}$$

(ω in radians per second, R in metres and N in rotations per minute)

Centripetal acceleration γ_c

$$\gamma_c = \omega^2 R \qquad \text{ou} \qquad \gamma_c = \frac{V^2}{R}$$

(γ_c in metres per second per second, ω in radians per second, R in metres and V in metres per second).

FUNDAMENTAL FORMULA OF DYNAMICS

$$F = m\gamma \qquad m = \text{mass} \qquad \gamma = \text{acceleration}$$

(F in newtons, m in kilogrammes and γ in metres per second per second)

Special case for gravity P = m$\vec{g}$

$\vec{g}$ gravity acceleration,
$\vec{g}$ about 9,81 m/s² in Paris.

MECHANICS AND STRENGTH OF MATERIALS
(continued)

CENTRIFUGAL FORCE

$$f_c = m \ \omega^2 \ R \qquad \text{or} \qquad f_c = m \ \frac{V^2}{R}$$

(F_C in newtons, m in kilogrammes, ω in radians per second,
R in metres and V in metres per second).

WORK OF A FORCE

Constant force in quantity and direction displacing its point of application :

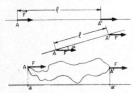

(1) On its acting line $\qquad\qquad T = F \ \ell$;

(2) On an oblique line to its acting line $\qquad T = F \ \ell \cos \alpha$;

(3) On a curve in its plane $\qquad\qquad T = F \ aa'$.

(T in joules, F in newtons and ℓ in metres).

Constant force displacing itself tangentially to a circle :

$$T = F R \alpha = M\!\flat \ F \ \alpha$$

for one rotation $T = 2\pi \ RF$

((T in joules, F in newtons, R in metres,
α in radians and M$\flat$ in metres-newtons).

WORK OF A TORQUE

Torque producing rotation about an axis perpendicular to its plane :

$$T = F d \alpha = M\!\flat \ C \ \alpha$$

for one rotation $T = 2\pi \ m\!\flat \ C = F \ d$

(T in joules, F in newtons, d in metres, α in radians,
M$\flat$ in metres-newtons)

POWER

Work product per unit $P = \dfrac{T}{t}$

(P in watts, T in joules, t in seconds)

Power of a torque rotating at a constant speed ω :

$$P = M\!_o^! \ C \ \omega \qquad \text{or} \qquad P = F \ d \ \omega = Fd \ \frac{2\pi \ N}{60}$$

(P in watts, M$\flat$ in metres-newtons, ω in radians per second,
F in newtons, d in metres and N in rotations per minute).

MECHANICS AND STRENGTH OF MATERIALS
(continued)

KINETIC ENERGY

$$W = \frac{1}{2}m \, v^2$$

(W in joules, m in kilogrammes and v in metres per second)

STRENGTH OF MATERIALS

Tensile and compressive strength

Stress : $n = \frac{N}{S} \, 10^{-7}$

n = stress (hbar),
N = compressive or tensile strength (N),
S = area (m²).

Hooke's law : $n = E\frac{\Delta \ell}{\ell}$

E = Young's modulus or longitudinal elasticity modulus 20 000 to 22 000 hbars for steel,

$\Delta \boldsymbol{\ell}$ = elongation
$\boldsymbol{\ell}$ = length
} expressed in the same unit.

Torsion

Torsional moment M_t = FR
(M_t metres-newtons, F in newtons, R in metres)

Basic torsion : $\theta = \frac{\alpha}{\ell}$

θ = basic torsion (rd/m),
α = rotation angle (rd),
Hooke's law : $t = G \, r \, \theta$ $\boldsymbol{\ell}$ = length (m),
t = tangential twist or shear strength (hbar),
G = transversal elasticity modulus : $G = 0,4 \, E =$ about 8 000 hbars for steel,
r = radius of cylinder (m).

ELECTRICITY
Direct current

CURRENT : I
Unit : Ampere (A)
The ampere is defined as that current which, if maintained in two straight parallel conductors of infinite length, of negligible circular cross-section, and placed 1 metre apart in a vacuum, would produce between these conductors a force of 2×10^{-7} newton per metre of length.

QUANTITY OF ELECTRICITY : Q

Unit : Coulomb (C)
Quantity of electricity passing a given point in a circuit when a current of 1 ampere is maintained for 1 second.
Practical unit : Ampere-hour (Ah)
Quantity of electricity passing in 1 hour in a circuit carrying a current of 1 ampere (1 Ah = 3 600 C)

$$Q = I \quad t$$
$$(Ah) \quad (A) \quad (h)$$

POTENTIAL DIFFERENCE : (Voltage) : U

Unit : volt (V)
Namely the difference of potential between two points of a conducting wire carrying a current of 1 ampere when the power dissipated between these points is equal to 1 watt.

RESISTANCE : R
Unit : ohm (Ω)
Namely the resistance between two points of a conductor when a potential difference of 1 volt, applied between these points, produces in this conductor a current of 1 ampere, the conductor not beeing a source of any electromotive force.

Resistivity : ρ ($\Omega/m/mm^2$) at 15 °C
Resistance of a conductor of 1 metre length and a cross-sectional area of 1 square millimetre.

copper	ρ = 0,017 - 0,0175	Ω/mm^2	-	iron	ρ = 0,11	Ω/mm^2
silver	ρ = 0,016 - 0,018	Ω/mm^2	-	steel	ρ = 0,10 - 0,25	$\Omega/m/mm^2$
aluminum	ρ = 0,029	$\Omega/m/mm^2$	-	nickel-silver	ρ = 0,36 - 0,39	$\Omega/m/mm^2$
				(Cu 60%, Zn 20%, Ni 20%)		

$$R = \rho \frac{1 \ (m)}{s \ (mm^2)}$$
$$(\Omega) \quad (\Omega/m/mm^2)$$

TEMPERATURE COEFFICIENT OF RESISTANCE AND RESISTIVITY

$$R_t = R_0 \ (1 + \alpha t) \qquad\qquad \rho_t = \rho_0 \ (1 + \alpha t)$$
$$(\Omega) \quad (\Omega) \quad (°C) \qquad\qquad (\Omega/m/mm^2) \ (\Omega/m/mm^2) \ (°C)$$

R_t, ρ_t = resistance, resistivity at t degrees Celsius.
R_0, ρ_0 = resistance, resistivity at 0 degree Celsius.
α = temperature coefficient at 15 °C.

copper	$\alpha = 3,93.10^{-3}$	iron	: $4,7.10^{-3}$
silver	$\alpha = 3,6 \ .10^{-3}$	steel	: 5.10^{-3}
aluminum	$\alpha = 3,9 \ .10^{-3}$	nickel-silver (Cu 60%, Zn 20%, Ni 20%)	: 3.10^{-4}

ELECTRICITY
Direct current (continued)

RESISTANCE CONNECTIONS

1) Connection in series

$$R = R_1 + R_2 + R_3....$$
$$U = U_1 + U_2 + U_3....$$

I constant

(2) Connection in parallel

$$\frac{1}{R} = \frac{1}{R_1} + \frac{1}{R_2} + \frac{1}{R_3}....$$
$$I = I_1 + I_2 + I_3....$$

U constant

For two resistances in parallel

$$R = \frac{R_1 R_2}{R_1 + R_2}$$

$$I_1 = I \frac{R_2}{R_1 + R_2}$$

$$I_2 = I \frac{R_1}{R_1 + R_2}$$

OHM'S LAW

$$U = R\ I$$
(V) (Ω) (A)

$$I = \frac{U}{R} \frac{(V)}{(Ω)}$$
(A)

$$R = \frac{U}{I} \frac{(V)}{(A)}$$
(Ω)

ELECTRICAL ENERGY (W) OR HEAT QUANTITY (Q)

Unit : Joule (J)

Electrical energy expended per second by a current of one ampere flowing through a one ohm resistance

$$W = R\ I^2\ t$$
(J) (Ω) (A) (s)

$$W = U\ I\ t$$
(J) (V) (A) (s)

Non SI units :

(1) Watt-hour (Wh)

Energy expended in 1 hour by a power of 1 watt

$$W = R\ I^2\ t$$
(Wh) = (Ω) (A) (h)

1 Wh = 3 600 J

(2) Calorie (cal)

$$Q = 0,24\ R\ I^2\ t$$
(cal) (Ω) (A) (s)

1 cal = 4,1855 J   1 J = 0,2389 cal

ELECTRICAL POWER (P)

Unit : Watt (W)

Power of 1 joule per second

$$P = R\ I^2$$
(W) (Ω) (A)

$$P = U\ I$$
(W) (V) (A)

$$P = \frac{U^2}{R} \frac{(V)}{(Ω)}$$
(W)

ELECTRICITY
Alternating current

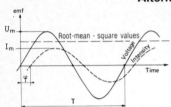

Period $T = \dfrac{1}{F}$

Frequency $F = \dfrac{1}{T}$ (Hz)

Angular frequency $\omega = 2\pi F$ (rd/s)

INSTANTANEOUS VALUES

$u = U_m \cos \omega t$
$i = I_m \cos(\omega t - \varphi)$
φ = angle of phase difference between current and voltage

ROOT-MEAN-SQUARE VALUES (r. m. s. values)

$$U = \frac{U_m}{\sqrt{2}} \qquad I = \frac{I_m}{\sqrt{2}}$$

POWER :

(1) Applied power : $S = UI$ in volt-amperes (VA),
(2) Active power : $P = UI \cos \varphi$ in watts (W),
(3) Reactive power : $Q = UI \sin \varphi$ in reactive volt-amperes (VAR).

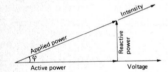

$S^2 = P^2 + Q^2$

$\tan \varphi = \dfrac{Q}{P}$

$\cos \varphi = \dfrac{P}{S}$ (power factor)

Three-phase system

PHASE WINDINGS
(formulas valid with same load for all 3 phases)

Star connection
$U_v = 1{,}73 \, U_p$
$I = I_p$

Mesh or Delta connection
$U_v = U_p$
$I = 1{,}73 \, I_p$

(1) Applied power $S = UI$ (VA)
(2) Active power $P = 1{,}73 \, U_v \, I \cos \varphi$ (W)
       $= 3 \quad U_p \, I \cos \varphi$ (W)
(3) Reactive power $Q = \sqrt{S^2 - P^2}$
       $= 1{,}73 \, U_v \, I \sin \varphi$ (VAR)
       $= 3 \quad U_p \, I_p \sin \varphi$ (VAR)

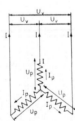
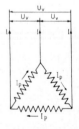

with :
U_v : emf in volt between 2 conductors of the three-phase winding,
U_p : emf for each phase,
I : intensity in amperes through each conductor of the three-phase winding,
I_p : intensity for each phase,
φ : angle of phase difference between current and voltage.

ELECTRICITY
Alternating current. Three-phase system
(continued)

CAPACITANCE : C

Unit : farad (F). A capacitance of one farad requires one coulomb of electricity to raise its potential one volt.

$$1 \text{ farad} = \frac{1 \text{ coulomb}}{1 \text{ volt}} \qquad\qquad C = \frac{Q}{U}$$

CONNECTIONS OF CAPACITORS (OR CONDENSERS)

Capacitors in parallel :

$$C = C_1 + C_2 + C_3 + \ldots$$

Capacitors in series :

$$\frac{1}{C} = \frac{1}{C_1} + \frac{1}{C_2} \quad \frac{1}{C_3} + \ldots \qquad \text{for 2 capacitors :} \quad C = \frac{C_1 \times C_2}{C_1 + C_2}$$

ADMISSIBLE INTENSITIES THROUGH CONDUCTORS

	Nominal area	INTENSITY				
		Temperature rise = 45° C			Temperature rise ≠ 45° C	
Number of conductors		2	3	4		
	(mm²)	A	A	A	multiply the intensities read in the opposite columns by the following coefficients	
	2	20	17	15	temperature rise	coefficient
	3	27	22,5	21	20°	0,67
	5	35	31	28	25°	0,75
	10	53	47	44	30°	0,82
	16	66	60	55	35°	0,88
	25	88	81	70	40°	0,94
	40	110	103	88	45°	1
	50	130	123	105	50°	1,05
	75	167	154	132	55°	1,10
	95	192	184	155	60°	1,15

SOME CHEMICAL SYMBOLS
ATOMIC NUMBERS AND WEIGHTS

Name	Symbol	Atomic number	Atomic weigt	Name	Symbol	Atomic number	Atomic weight
Aluminum	Al	13	27	Mercury	Hg	80	200,6
Antimony	Sb	51	122	Molybdenum	Mo	42	96
Argon	A	18	40	Neon	Ne	10	20
Arsenic	As	33	75	Nickel	Ni	28	58,7
Barium	Ba	56	137	Nitrogen	N	7	14
Bismuth	Bi	83	209	Oxygen	O	8	16
Boron	B	5	11	Phosphorus	P	15	31
Bromine	Br	35	80	Platinum	Pt	78	195
Cadmium	Cd	48	112	Plutonium	Pu	94	242
Calcium	Ca	20	40	Potassium	K	19	39
Carbon	C	6	12	Radium	Ra	88	226
Chlorine	Cl	17	35,5	Selenium	Se	34	79
Chromium	Cr	24	52	Silicon	Si	14	28
Cobalt	Co	27	59	Silver	Ag	47	108
Copper	Cu	29	63,5	Sodium	Na	11	23
Fluorine	F	9	19	Strontium	Sr	38	87,6
Gold	Au	79	197	Sulphur	S	16	32
Helium	He	2	4	Tin	Sn	50	119
Hydrogen	H	1	1	Titanium	Ti	22	48
Iodine	I	53	127	Tungsten	W	74	184
Iron	Fe	26	56	Uranium	U	92	238
Lead	Pb	82	207	Vanadium	V	23	51
Lithium	Li	3	7	Xenon	Xe	54	131,3
Magnesium	Mg	12	24	Zinc	Zn	30	65,4
Manganese	Mn	25	55	Zirconium	Zr	40	91

SPECIFIC GRAVITY OF VARIOUS MATERIALS AND LIQUIDS

Name	Specific gravity	Name	Specific gravity
ROCK		MISCELLANEOUS MATERIALS	
Granite	2,4 to 3,0	Baryte (Baryum sulfate)	4,5
Gypsum	1,2	Brick (common and hard)	2,2
Limestone hard	2,4 to 2,7	Cement (Portland, powder)	3 to 3,3
Limestone medium hard	1,9 to 2,3	Cement (Portland, slurry)	1,8 to 2,0
Marble	2,5 to 2,9	Clay (compact)	2,1
Quartzite	2,2 to 2,8	Concrete	2,25
Salt	2,16	Glass	2,53
Sand dry	2,6	Walnut shells	1,3
Sandstone	1,9 to 2,6	GAS (at 10° C and 760 mm Hg to air)	
LIQUID	(at 25° C)		
Acetone	0,791	Air	1
Alcohol ethyl	0,816	Butane - iso	2,067
Alcohol methyl	0,792	Butane - n	2,0854
Benzene	0,878	Carbon dioxide	1,529
Carbon tetrachloride	1,595	Carbon monoxide	0,9671
Chloroform	1,482	Ethane	1,0493
Ether	0,714	Ethylene	0,9749
Glycerin	1,260	Hydrogen	0,06952
Trichloroethylene	1,455	Methane	0,5544
Water at 4° C	1	Oxygen	1,10527
		Propane	1,554

PHYSICAL PROPERTIES OF METALS

Name	Symbol	Specific gravity	Melting point (° C)	Brinell hardness	Mohs' scale
Aluminum	Al	2,7	660	16 (annealed)	2 to 2,9
Antimony	Sb	6,7	631	—	3 to 3,3
Bismuth	Bi	9,75	271	—	2,5
Cadmium	Cd	8,65	321	23 (cast)	2
Chromium	Cr	7,19	1890	70-130	9
Cobalt	Co	8,9	1495	124 (cast)	
Copper	Cu	8,94	1083	—	2,5 to 3
Gold	Au	19,32	1063	—	2,5 to 3
Iron	Fe	7,88	1535	77 (annealed)	4 to 5
Lead	Pb	11,34	327	7 (cast)	1,5
Magnesium	Mg	1,74	651	29 (annealed)	2
Manganese	Mn	7,2	1260	—	5
Mercury	Hg	13,55	-39	—	—
Molybdenum	Mo	10,2	2620	150-200 (drawn)	—
Nickel	Ni	8,9	1455	110-300 (cold rolled)	—
Platinum	Pt	21,45	1774	64 (drawn)	4,3
Silver	Ag	10,5	961	—	2,5 to 7
Tin	Sn	7,3	232	—	1,5 to 1,8
Titanium	Ti	4,5	1800	—	—
Tungsten	W	19,3	3370	350 (drawn)	—
Vanadium	V	5,96	1710	—	—
Zinc	Zn	7,14	419	—	2,5

RELATION BETWEEN NaCl CONTENT
AND SPECIFIC GRAVITIES OF SOLUTIONS

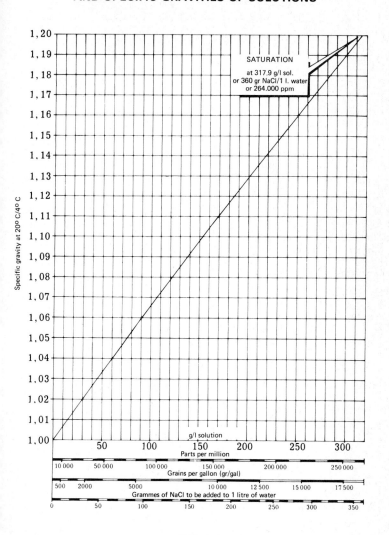

SATURATION

at 317.9 g/l sol.
or 360 gr NaCl/1 l. water
or 264.000 ppm

Specific gravity at 20° C/4° C

g/l solution

Parts per million

Grains per gallon (gr/gal)

Grammes of NaCl to be added to 1 litre of water

NaCl content

STRATIGRAPHIC SCALE

ERA	PERIOD	FORMATIONS
QUATERNARY (Psychozoic)	Holocene (Neolithic)(Paleolithic) Pleistocene	Flandrian, Tyrrhenian, Sicilian
TERTIARY (Cenozoic)	Pliocene	Calabrian (Villafranchian), Astian, Plaisancian
	Miocene	Sahelian (Pontian), Vindobonian, Burdigalian
	Oligocene	Aquitanian, Chattian, Stampian, Samoisian
	Eocene	Ludian, Bartonian, Lutetian, Ypresian, Sparnacian, Thanetian, Montian

ERA	PERIOD	FORMATIONS
SECONDARY (Mesozoic)	Upper Cretaceous	Danian, Senonian, Turonian, Cenomanian
	Lower Cretaceous	Albian, Aptian, Barremian (Urgonian), Hauterivian, Valanginian
	Upper Jurassic (Malm)	(Purbekian), Portlandian (Tithonic), Kimmeridgian, Sequanian, Rauracian, Argovian, Oxfordian, Callovian
	Middle Jurassic (Dogger)	Bathonian, Bajocian
	Lower Jurassic (Lias)	Aalenian, Toarcian, Charmoutian, Sinemurian, Hettangian, Rhetian
	Trias	Keuper, Muschelkalk, Bunter

ERA	PERIOD	FORMATIONS
PRIMARY (Paleozoic)	Permian	Zechstein or Thuringian, Saxonian, Autunian
	Carboniferous	Coal form. (Stephanian)(Westphalian), Dinantian (Culm)
	Devonian	Fammenian, Frasnian, Givetian, Eifelian, Coblencian, Gedinnian, Downtonian
	Silurian	Gothlandian, Ordovician
	Cambrian	Potsdamian, Acadian, Georgian
	Precambrian (algonkian)	
	Archean	

MAP AND DRAWING SCALES

Scales are axpressed :
for the French in a decimal scale,
for the Anglo-American in :
miles to 1 inch for maps and surveys,
feet to 1 inch for drawing and site plans.

Decimal scale	Imperial equivalent	
	miles to 1 inch	feet to 1 inch
I : 1 000 000	15,78	
I : 950 700	15,00	
I : 633 600	10,00	
I : 500 000	7,89	
I : 316 800	5,00	
I : 253 440	4,00	
I : 200 000	3,16	
I : 126 720	2,00	
I : 100 000	1,58	
I : 80 000	1,26	
I : 63 360	1,00	
I : 50 000	0,79	
I : 40 000	0,63	
I : 31 680	0,50	
I : 25 000	0,40	
I : 24 000	0,38	
I : 20 000	0,32	
I : 12 000	0,19	1 000,0
I : 5 000	0,08	416,7
I : 2 500		208,3
I : 1 200		100,0
I : 600		50,0
I : 300		25,0
I : 150		12,5
I : 30		2,5
I : 12		1,0
I : 6		0,5
I : 3		0,25
I : 2		0,167

GREEK ALPHABET

English equivalent	Greek letter		Greek name	English equivalent	Greek letter		Greek name
	Capital	Small			Capital	Small	
a	A	α	alpha	n	N	ν	nu
b	B	β	beta	x	Ξ	ξ	xi
g	Γ	γ	gamma	o	O	o	omicron
d	Δ	δ	delta	p	Π	π	pi
e	E	e	epsilon	r, rh	P	ρ	rho
z	Z	ζ	zeta	s	Σ	σ	sigma
$\bar{e}$	H	η	eta	t	T	τ	tau
th	Θ	θ	theta	y, u	Υ	υ	upsilon
i	I	ι	iota	ph	Φ	φ	phi
k	K	κ	kappa	ch	X	χ	chi
l	Λ	λ	lambda	ps	Ψ	ψ	psi
m	M	μ	mu	$\bar{o}$	Ω	ω	omega

B PHYSICAL PROPERTIES OF DRILL PIPE, DRILL COLLAR, KELLY

Summary

DRILL PIPE
STEEL GRADES AND TENSILE REQUIREMENTS

TENSILE PROPERTIES	UNIT	GRADES				
		D	E	X95	G105	S135
Minimum yield strength	hbar	37.9	51.7	65.5	72.4	93.0
	psi	55.000	75.000	95.000	105.000	135.000
Maximum yield strength	hbar	—	72.4	86.1	93.0	113.8
	psi	—	105.000	125.000	135.000	165.000
Minimum tensile strength	hbar	65.5	68.9	75.8	79.3	100.0
	psi	95.000	100.000	110.000	115.000	145.000
Elongation, mini, per cent $L = 2''$ cross section area = 0.75 sq in (1)		19.5	18.5	17.0	16.0	13.5
Average yield strength (API RP 7G)	hbar	44.8	58.6	75.8	82.7	100.0
	psi	65.000	85.000	110.000	120.000	145.000

(1) Formula given in API Std 5A : $\quad e = 625\,000\,\dfrac{A^{0.2}}{U^{0.9}}$

where :

A = cross section area (sq. in),
U = specified tensile strength (psi),
e = minimum elongation in 2 inches length (%).

API DRILL PIPES LIST
(API Std 5A and API RP 7G)

Outside diameter		Nominal weight (lb/ft)	Wall thickness (mm)	Inside diameter		Outside diameter upset (mm) (4)	Cross section area (mm²)	Polar sectional modulus (mm³)
in	mm			body (mm)	upset (mm) (3)			
INTERNAL UPSET DRILL PIPE (IU)								
2 7/8 (1)	73,0	6,85	5,51	62,0	47,6	—	1169	36790
2 7/8	73,0	10,40	9,19	54,6	33,3	—	1843	52504
3 1/2	88,9	9,50	6,45	76,0	57,2	—	1671	64270
3 1/2	88,9	13,30	9,35	70,2	49,2	—	2337	84295
3 1/2	88,9	15,50	11,40	66,1	49,2	—	2776	95799
4 (2)	101,6	11,85	6,65	88,3	74,6	—	1984	88490
4	101,6	14,00	8,38	84,8	69,8	—	2454	105828
4 1/2 (2)	114,3	13,75	6,88	100,5	85,7	—	2322	117725
5 (2)	127,0	16,25	7,52	112,0	95,2	—	2823	159250
EXTERNAL UPSET DRILL PIPE (EU)								
2 3/8 (1)	60,3	4,85	4,83	50,7	—	67,5	841,4	21631
2 3/8	60,3	6,65	7,11	46,1	—	67,5	1189	28415
2 7/8 (1)	73,0	6,85	5,51	62,0	—	81,8	1169	36790
2 7/8	73,0	10,40	9,19	54,6	—	81,8	1843	52504
3 1/2	88,9	9,50	6,45	76,0	—	97,1	1671	64270
3 1/2	88,9	13,30	9,35	70,2	—	97,1	2337	84295
3 1/2	88,9	15,50	11,40	66,1	—	97,1	2776	95799
4 (2)	101,6	11,85	6,65	88,3	—	114,3	1984	88490
4	101,6	14,00	8,38	84,8	—	114,3	2454	105828
4 1/2 (2)	114,3	13,75	6,88	100,5	—	127,0	2322	117725
4 1/2	114,3	16,60	8,56	97,2	—	127,0	2844	139979
4 1/2	114,3	20,00	10,92	92,5	—	127,0	3547	167673
INTERNAL EXTERNAL UPSET DRILL PIPE (IEU)								
4 1/2	114,3	16,60	8,56	97,2	80,2	118,3	2844	139979
4 1/2	114,3	20,00	10,92	92,5	76,2	121,4	3547	167673
5	127,0	19,50	9,19	108,6	93,7	131,8	3403	187075
5 (6)	127,0	25,60	12,70	101,6	87,3	131,8	4560	237449
5 1/2	139,7	21,90	9,17	121,4	101,6	141,3	3760	230445
5 1/2	139,7	24,70	10,54	118,6	101,6	141,3	4277	257081
6 5/8	168,3	25,20	8,38	151,5	(5)	(5)	4210	320728

(1) These sizes are non API.
(2) These sizes are classified as tentative.
(3) Tolerances on inside diameter of the upset : ± 1,59 mm.
(4) Tolerances on outside diameter of the upset : + 3,18 mm — 0,79 mm.
(5) Upset requirements for 6 5/8'' not established.
(6) This size has been API standard since April 1972.

UPSET TUBING FOR SMALL DIAMETER WORK STRINGS (1)
(API Std 5A and API Spec 7)
(Grade N80)

| Outside diameter | | Nominal weight (lb/ft) | Wall thickness (mm) | Weight with tool-joint (kg/m) | Inside diameter | | Upset outside diameter (mm) | Cross section area (mm²) | Inside capacity (l/m) | Volume of steel (l/m) | Tensile yield strength (10³ daN) | Torsional yield strength (m·daN) | Internal pressure (bar) | Collapse resistance (bar) |
in	mm				body (mm)	upset (mm)								
1.050	26.7	1,55	3,91	2,20	18,9	17,5	36,5	280	0,280	0,280	15,8	89	1410	1390
1.315	33.4	2,30	4,55	3,36	24,3	21,4	42,8	413	0,464	0,428	23,4	168	1310	1310
1.660	42.2	3,29	5,03	4,77	32,1	22,2	47,6	588	0,810	0,607	33,4	310	1150	1165
1.900	48.3	4,19	5,56	6,10	37,2	23,8	55,5	746	1,085	0,777	42,3	455	1110	1115

(1) The specified upset dimensions were chosen to accomodate the various bore of tool joints with API numbered rotary shouldered connections NC 10, NC 12, NC 13, NC 16 and to maintain a satisfactory cross section in the weld zone after final machining of the assembly.

CLASSIFICATION OF DRILL PIPES
(API RP 7G)

CLASS I : with a white band. Nominal dimensions of new drill-pipes.

PREMIUM CLASS : with two white bands :
Remaining wall not less than 80 % due to uniform wear,
Remaining wall not less than 80 % eccentric wear,
Calculated area with not more than 20 % uniform wall reduction.

CLASS II : with a yellow band :
Remaining wall not less than 80 % due to uniform wear,
Remaining wall not less than 65 % due to eccentric wear,
Calculated area with not more than 20 % uniform wall reduction.

CLASS III : with a blue band :
Remaining wall not less than 62,5 % due to uniform wear,
Remaining wall not less than 55 % due to eccentric wear,
Calculated area with not more than 37,5 % uniform wall reduction.

CLASS IV : with a green band :
Remaining wall less than 62,5 % due to uniform wear,
Remaining wall less than 55 % due to eccentric wear,
Calculated area with greater than 37,5 % uniform wall reduction.

Notes :

(1) In any classification where fatigue cracks appear, the pipe will be identified by a red band and considered unfit for further drilling service.

(2) Eccentric wear supposes that all wear occurs on one side, the outside section remaining circular but with its centre offset from the centre of the circular inside section.

(3) The inside diameter is assumed constant and at the nominal inside diameter of the pipe throughout its life.

(4) For more details refer to Table 10-1 of API RP 7G.

MECHANICAL PROPERTIES OF DRILL PIPES
Class I, Premium Class (S), Class II, Class III
TENSILE AND TORSIONAL DATA (API RP 7G)

Size outside diameter (in)	Nominal weight (lb/ft)	Class	Load at minimum yield strength (2) (10^3daN)					Torsional yield strength (1) (m-daN)				
			D	E	95	105	135	D	E	95	105	135
2 3/8	4,85	I	33	44	56	62	80	473	645	816	903	1159
		S	26	35	44	48	62	370	504	639	706	907
		II	26	35	44	48	62	310	430	540	600	770
		III	19	26	33	37	47	265	365	460	510	655
	6,65	I	45	61	78	86	110	620	845	1069	1183	1523
		S	35	48	61	67	81	477	651	824	911	1172
		II	35	48	61	67	81	410	560	710	790	1010
		III	27	37	46	51	66	350	480	605	670	865
2 7/8	6,85	I	44	61	77	85	109	803	1093	1385	1532	1970
		S	35	48	60	67	86	629	857	1085	1200	1542
		II	35	48	60	67	86	530	725	920	1015	1305
		III	27	37	46	51	66	450	615	780	865	1110
	10,40	I	70	95	120	133	171	1145	1560	1980	2190	2811
		S	54	74	94	104	133	878	1200	1519	1678	2158
		II	54	74	94	104	133	760	1040	1320	1460	1875
		III	41	57	71	79	102	650	890	1125	1245	1595
3 1/2	9,50	I	63	86	110	121	156	1403	1915	2422	2676	3 441
		S	50	68	86	95	122	1101	1501	1902	2102	2702
		II	50	68	86	95	122	930	1270	1605	1775	2280
		III	38	52	67	73	94	790	1080	1370	1510	1945
	13,30	I	89	120	153	169	218	1838	2510	3180	3510	4515
		S	69	95	120	132	170	1425	1944	2462	2710	3498
		II	69	95	120	132	170	1225	1670	2115	2335	3005
		III	53	72	92	101	130	1045	1420	1800	1990	2560
	15,50	I	105	144	182	201	259	2091	2850	3610	4000	5130
		S	82	112	141	156	201	1603	2186	2769	3059	3932
		II	82	112	141	156	201	1395	1900	2405	2660	3420
		III	62	85	107	119	153	1190	1620	2055	2270	2915
4	11,85	I	75	103	130	144	185	1933	2640	3333	3685	4737
		S	59	81	103	114	146	1520	2145	2625	2901	3729
		II	59	81	103	114	146	1280	1745	2210	2440	3140
		III	46	64	79	87	113	1090	1485	1880	2080	2675
	14,00	I	93	127	160	178	228	2308	3150	3990	4415	5670
		S	74	101	127	141	181	1806	2462	3120	3447	4434
		II	74	101	127	141	181	1530	2090	2650	2925	3760
		III	56	77	97	108	138	1305	1780	2255	2495	3205
	15,70	I	106	144	183	202	259	2557	3490	4420	4895	6290
		S	83	113	143	158	203	2025	2762	3498	3865	4970
		II	83	113	143	158	203	1700	2320	2940	3250	4175
		III	64	87	110	121	156	1450	1975	2505	2770	3560

(1) Based on shear strength equal to 57,7 % of minimum yield strength and :
    20 % uniform wear for Class S,
    35 % eccentric wear for Class II,
    45 % eccentric wear for Class III.
(2) Based on : 20 % uniform wear for Class S and Class II,
           37,5 % uniform wear for Class III.

MECHANICAL PROPERTIES OF DRILL PIPES
Class I, Premium Class (S), Class II, Class III
TENSILE AND TORSIONAL DATA (API RP 7G)
(continued)

Size outside diameter (in)	Nominal weight (lb/ft)	Class	Load at minimum yield strength (2) (10³ daN)					Torsional yield strength (1) (m-daN)				
			D	E	95	105	135	D	E	95	105	135
4 1/2	13,75	I	88	120	152	168	216	2575	3505	4435	4902	6302
		S	70	95	120	133	171	2030	2770	3498	3865	4970
		II	70	95	120	133	171	1700	2320	2935	3245	4170
		III	54	73	93	103	132	1450	1975	2500	2765	3555
	16,60	I	108	147	186	206	265	3053	4170	5280	5835	7505
		S	84	116	147	162	209	2390	3268	4135	4575	5880
		II	84	116	147	162	209	2025	2760	3500	3870	4960
		III	65	89	113	125	160	1725	2355	2980	3295	4235
	20,00	I	135	183	232	257	330	3656	4990	6325	6990	8990
		S	105	144	182	201	259	2847	3880	4920	5420	6970
		II	105	144	182	201	259	2430	3320	4200	4640	5970
		III	81	110	140	154	199	2075	2825	3580	3960	5090
5	16,25	I	107	146	185	205	263	3447	4735	5999	6630	8525
		S	84	115	146	161	208	2740	3740	4740	5230	6725
		II	84	115	146	161	208	2300	3130	3970	4390	5640
		III	65	89	113	125	160	1960	2670	3380	3740	4805
	19,50	I	129	176	223	246	317	4079	5570	7050	7800	10030
		S	102	139	176	194	250	3205	4370	5335	6115	7860
		II	102	139	176	194	250	2700	3690	4670	5160	6630
		III	78	107	136	150	193	2305	3145	3980	4400	5660
	25,60	I	173	236	298	330	425	5179	7080	8950	9900	12712
		S	136	186	238	261	334	4025	5490	6950	7670	9880
		II	136	186	238	261	334	3445	4700	5950	6580	8460
		III	104	141	179	198	255	2940	4005	5075	5610	7210
5 1/2	19,20	I		166	210	232	298		5981	7575	8373	10765
		S										
		II										
		III										
	21,90	I	143	194	246	272	350	5026	6860	8690	9600	12340
		S	112	154	194	215	279	3960	5390	6883	7550	9710
		II	112	154	194	215	279	3330	4540	5750	6355	8170
		III	87	118	150	166	213	2835	3870	4900	5415	6960
	24,70	I	162	221	280	310	399	5606	7630	9700	10700	13761
		S	135	174	234	259	332	4400	6000	7590	8410	10750
		II	135	174	234	259	332	3720	5070	6420	7100	9130
		III	98	134	170	188	242	3170	4320	5475	6050	7780
6 5/8	25,20	I	160	218	276	305	392	7004	9560	12096	13371	17220
		S	126	172	218	242	310	5970	8200	10440	11520	14800
		II	126	172	218	242	310					
		III	98	133	169	187	240					

(1) Based on shear strength equal to 57,7 % of minimum yield strength and :
    20 % uniform wear for Class S,
    35 % eccentric wear for Class II,
    45 % eccentric wear for Class III.
(2) Based on : 20 % uniform wear for Class S and Class II,
         37,5 % uniform wear for Class III.

MECHANICAL PROPERTIES OF DRILL PIPES
Class I, Premium Class (S), Class II, Class III
INTERNAL PRESSURE - COLLAPSE PRESSURE

Size outside diameter (in)	Nominal weight (lb/ft)	Class	Internal pressure at minimum yield strength (1) (bar)					Collapse pressure based on minimum values (1) (bar)				
			D	E	95	105	135	D	E	95	105	135
2 3/8	4,85	I	531	723	916	1013	1303	558	761	964	1067	1314
		S	485	662	838	927	1191	461	590	700	752	891
		II	394	538	681	753	968	336	415	474	499	554
		III	334	455	576	637	819	250	294	316	332	369
	6,65	I	782	1066	1309	1492	1924	789	1076	1361	1505	1935
		S	715	976	1236	1366	1756	676	923	1169	1291	1660
		II	581	792	1003	1109	1426	580	791	1002	1108	1424
		III	492	670	849	939	1207	510	691	831	899	1087
2 7/8	6,85	I	501	683	865	956	1229	530	722	892	966	1176
		S	458	625	791	874	1124	418	529	621	663	773
		II	325	444	562	621	798	299	363	407	424	456
		III	275	375	476	525	676	217	248	276	289	312
	10,40	I	855	1139	1442	1594	2050	835	1138	1442	1593	2049
		S	764	1042	1320	1458	1875	719	980	1242	1373	1765
		II	620	845	1071	1183	1521	619	844	1069	1182	1519
		III	525	715	907	1002	1288	546	744	943	1025	1257
3 1/2	9,50	I	483	656	832	919	1182	510	692	832	900	1160
		S	441	601	761	841	1081	390	490	570	607	698
		II	358	488	618	683	878	275	330	363	375	414
		III	302	413	550	578	743	197	223	253	263	276
	13,30	I	697	951	1204	1331	1711	714	973	1233	1362	1752
		S	638	870	1102	1218	1566	607	829	1049	1160	1491
		II	518	706	894	986	1271	518	706	855	926	1125
		III	438	597	756	835	1073	435	553	651	698	823
	15,50	I	851	1160	1470	1624	2088	848	1156	1465	1619	2082
		S	778	1061	1344	1486	1911	732	998	1264	1397	1796
		II	632	862	1092	1207	1552	631	861	1091	1205	1550
		III	535	730	924	1021	1313	557	759	962	1063	1307
4	11,85	I	434	593	750	830	1046	454	580	687	738	950
		S	397	542	687	758	976	322	395	447	470	515
		II	323	440	560	615	795	217	248	276	289	314
		III	275	375	470	520	670	152	177	192	195	197
	14,00	I	547	746	945	1040	1343	574	783	991	1096	1409
		S	501	683	865	956	1229	483	623	743	801	956
		II	405	553	700	775	995	358	446	513	544	610
		III	345	470	595	659	845	269	335	352	363	401
	15,70	I	630	859	1088	1199	1546	652	978	1128	1245	1600
		S	576	786	996	1100	1415	552	753	953	1050	1282
		II	468	638	809	894	1149	462	590	700	753	892
		III	396	541	685	756	973	360	448	516	546	616

(1) Based on : minimum wall 80 % of nominal wall for Premium Class,
minimum wall 65 % of nominal wall for Class II,
minimum wall 55 % of nominal wall for Class III.

MECHANICAL PROPERTIES OF DRILL PIPES
Class I, Premium Class (S), Class II, Class III
INTERNAL PRESSURE - COLLAPSE PRESSURE (continued)

Size outside diameter (in)	Nominal weight (lb/ft)	Class	Internal pressure at minimum yield strength (1) (bar)					Collapse pressure based on minimum values (1) (bar)				
			D	E	95	105	135	D	E	95	105	135
4 1/2	13,75	I	399	544	690	763	978	394	496	579	617	792
		S	365	499	631	698	897	272	325	356	368	407
		II	297	405	513	567	729	175	204	227	234	240
		III	251	343	434	480	616	128	144	148	149	150
	16,60	I	497	677	858	944	1213	525	716	879	953	1225
		S	454	620	785	867	1116	412	521	610	652	758
		II	369	503	637	704	905	294	356	397	414	447
		III	311	425	538	595	765	212	241	271	282	305
	20,00	I	634	864	1095	1206	1516	656	894	1132	1251	1609
		S	580	791	1001	1107	1423	555	757	958	1058	1299
		II	472	643	814	901	1160	468	599	712	760	903
		III	398	542	686	759	976	365	455	525	557	627
5	16,25	I	392	535	678	748	964	383	481	558	594	697
		S	359	490	621	686	882	262	311	339	349	391
		II	292	398	504	557	716	166	195	216	222	228
		III	247	336	427	472	606	123	138	141	142	142
	19,50	I	480	655	830	916	1179	510	690	844	896	1152
		S	439	599	758	838	1078	388	487	567	604	693
		II	358	488	618	683	878	274	328	361	373	412
		III	302	411	521	576	741	195	221	250	260	270
	25,60	I	663	904	1124	1266	1627	683	931	1179	1303	1676
		S	607	827	1048	1158	1489	579	790	1000	1106	1416
		II	493	672	851	940	1209	493	650	777	838	1006
		III	417	568	720	796	989	397	500	583	622	718
5 1/2	19,20	I		500	633	700	900	338	418	478	504	650
		S	335	457	579	641	823	225	259	285	299	325
		II		372	471	520	669	144	168	180	182	183
		III		352	445	492	633	105	112	113	113	113
	21,90	I	435	594	752	832	1069	456	582	690	741	952
		S	399	543	688	761	978	323	397	450	473	519
		II	323	441	559	618	794	219	252	279	292	316
		III	274	373	473	523	672	154	180	196	199	199
	24,70	I	500	682	864	955	1228	529	721	891	965	1241
		S	458	624	792	874	1124	418	529	621	663	772
		II	372	507	643	710	913	299	363	407	424	456
		III	314	429	543	600	771	216	247	275	288	312
6 5/8	25,20	I	331	451	570	630	810	277	332	365	379	487
		S	291	396	502	555	714	173	202	224	231	236
		II	245	335	424	469	603	117	129	131	131	131
		III	208	283	359	396	510	81	81	82	82	82

(1) Based on : minimum wall 80 % of minimal wall for Premium Class,
           minimum wall 65 % of nominal wall for Class II,
           minimum wall 55 % of nominal wall for Class III.

API THREAD FORMS
(API Spec 7)

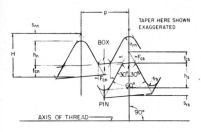

V - 0,038 R

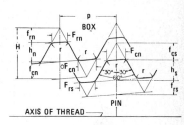

V - 0,055

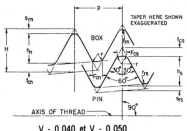

V - 0,040 et V - 0,050

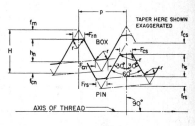

V - 0,065 (1)

Thread form	Taper (%) (3)	H (mm)	$h_n = h_s$ (mm)	$S_{rn} = S_{rs}$ $f_{rn} = f_{rs}$ (mm)	$f_{cn} = f_{cs}$ (mm)	$F_{cn} = F_{cs}$ (mm)	$F_{cn} = F_{cs}$ (in)	$F_{rn} = F_{rs}$ (mm)	$r_{rn} = r_{rs}$ (mm)	r (mm)
V.0,038 R	16,66	5,49	3,09	0,97	1,43	1,65	0,065	—	0,97 (2)	0,38
V.0,038 R	25	5,47	3,08	0,97	1,42	1,65	0,065	—	0,97	0,38
V.0,040	25	4,38	2,99	0,51	0,88	1,02	0,040	—	0,51	0,38
V.0,050	25	5,47	3,74	0,63	1,09	1,27	0,050	—	0,63	0,38
V.0,050	16,66	5,49	3,75	0,63	1,10	1,27	0,050	—	0,63	0,38
V.0,055 (4)	12,50	3,70	1,4	1	1,2	1,4	0,055	1,2	—	0,38
V.0,065 (1)	16,66	5,49	2,83	1,23	1,43	1,65	0,065	1,42	—	0,38

(1) Obsolescent thread form,
(2) 0,97 mm = 0,038 in which gives the name of this thread form,
(3) Taper in % = 8,33 × taper in inches per foot,
(4) Thread form V-0,055 flat for small diameter rotary shouldered connection (NC 10, NC 12, NC 13, NC 16).

API ROTARY SHOULDERED CONNECTION DIMENSIONS
(API Spec 7)

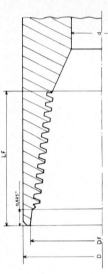

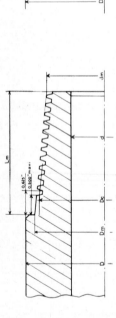

Connection number or size	Drill pipe Diameter-Upset-Weight (in) - (lb/ft)	Tool joint Outside diameter (in)	(mm)	Inside diameter (in)	(mm)	Pin Dm (mm)	dm (mm)	Lm (mm)	Box Dc (mm)	Df (mm)	Lf (mm)	Taper (%)	Threads per inch	Thread form
				NUMBER (NC) STYLE (1) (2)										
NC 10*	1.050- 1.55	1.375	34.9	0.719	18.3	30.2	25.5	38.1	27.0	30.6	54.0	12.5	6	V-0.055
NC 12*	1.315- 2.30	1.625	41.3	0.906	23.0	35.4	29.8	44.4	32.1	35.7	60.3	12.5	6	V-0.055
NC 13*	1.660- 3.29	1.812	46.0	0.937	23.8	38.6	33.0	44.4	35.3	38.9	60.3	12.5	6	V-0.055
NC 16*	1.900- 4.19	2.125	54.0	1.000	25.4	44.1	38.5	44.4	40.9	44.5	60.3	12.5	6	V-0.055
NC 23*	—	—	—			65.1	52.4	76.2	59.8	66.7	92.1	16.66	4	V-0.038R
NC 26	2 3/8-EU- 6.65	3 3/8	85.7	1 3/4	44.4	73.1	60.4	76.2	67.8	74.6	92.1	16.66	4	V-0.038R

(1) The number of the connection in the number style is the pitch diameter of the pin thread at the gage point, rounded to units and tenths of inch.

(2) Connections in the number (NC) style are interchangeable with connections having the same pitch diameter in the FH and IF styles as follows :

NC 26 and 2 3/8 IF   NC 40 and 4   FH
NC 31 and 2 7/8 IF   NC 46 and 4   IF
NC 38 and 3 1/2 IF   NC 50 and 4 1/2 IF

* These connections are tentative :

NC 10, 12, 13, 16, 23 are tentative for small diameter work strings. An " 0 " ring is optional at the base of the pin shoulder for high pressure use.

API ROTARY SHOULDERED CONNECTION DIMENSIONS (continued)
(API Spec 7)

Connection number or size	Drill pipe Diameter-Upset-Weight (in) - (lb/ft)	Tool joint Outside diameter (in)	(mm)	Inside diameter (in)	(mm)	Pin Dm (mm)	dm (mm)	Lm (mm)	Dc (mm)	Box Df (mm)	Lt (mm)	Taper (%)	Threads per inch	Thread form
				NUMBER (NC) STYLE (1) (2)										
NC 31	2 7/8-EU-10.40	4 1/8	104.8	2 1/8	54.0	86.1	71.3	88.9	80.8	87.7	104.8	16.66	4	V-0.038R
NC 35						95.0	79.1	95.2	89.7	96.8	111.1	16.66	4	V-0.038R
NC 38	3 1/2-EU-13.30	4 3/4	120.6	2 11/16	68.3	102.0	85.1	101.6	96.7	103.6	117.5	16.66	4	V-0.038R
NC 38	3 1/2-EU-15.50	5	127.0	2 9/16	65.1	102.0	85.1	101.6	96.7	103.6	117.5	16.66	4	V-0.038R
NC 40	4-IU-14.00	Std 5 1/4	133.3	2 13/16	71.4	108.7	89.7	114.3	103.4	110.3	130.2	16.66	4	V-0.038R
NC 40	4-IU-14.00	Opt 5 1/2	139.7	2 13/16	71.4	108.7	89.7	114.3	103.4	110.3	130.2	16.66	4	V-0.038R
NC 44						117.5	98.4	114.3	112.2	119.1	130.2	16.66	4	V-0.038R
NC 46	4-EU-14.00	5 3/4	146.0	3 1/4	82.5	122.8	103.7	114.3	117.5	124.6	130.2	16.66	4	V-0.038R
NC 46	4 1/2-IU-16.60	Std 6	152.4	3 1/4	82.5	122.8	103.7	114.3	117.5	124.6	130.2	16.66	4	V-0.038R
NC 46	4 1/2-IU-16.60	Opt 6 1/4	158.7	3 1/4	82.5	122.8	103.7	114.3	117.5	124.6	130.2	16.66	4	V-0.038R
NC 46	4 1/2-IEU-20.00	Std 6	152.4	3	76.2	122.8	103.7	114.3	117.5	124.6	130.2	16.66	4	V-0.038R
NC 46	4 1/2-IEU-20.00	Opt 6 1/4	158.7	3	76.2	122.8	103.7	114.3	117.5	124.6	130.2	16.66	4	V-0.038R
NC 50	4 1/2-EU-16.60	Std 6 1/8	155.6	3 3/4	95.2	133.3	114.3	114.3	128.1	134.9	130.2	16.66	4	V-0.038R
NC 50	4 1/2-EU-16.60	Opt 6 1/4	158.7	3 3/4	95.2	133.3	114.3	114.3	128.1	134.9	130.2	16.66	4	V-0.038R
NC 50	5-IEU-19.50	Std 6 3/8	161.9	3 3/4	95.2	133.3	114.3	114.3	128.1	134.9	130.2	16.66	4	V-0.038R
NC 50	5-IEU-19.50	Opt 6 1/2	165.1	3 3/4	95.2	133.3	114.3	114.3	128.1	134.9	130.2	16.66	4	V-0.038R
NC 56	5 1/2-IEU-21.90	7	177.8	3 1/2	88.9	149.3	117.5	127.0	142.0	150.8	142.9	16.66	4	V-0.038R
NC 61						163.5	128.6	139.7	156.9	165.1	155.6	25	4	V-0.038R
NC 70						185.8	147.7	152.4	187.3	187.3	168.3	25	4	V-0.038R
NC 77*						203.2	162.0	165.1	196.6	204.8	181.0	25	4	V-0.038R

(1) The number of the connection in the number style is the pitch diameter of the pin thread at the gage point, rounded to units and tenths of inch.

(2) Connections in the number (NC) style are interchangeable with connections having the same pitch diameter in the FH and IF styles as follows :

NC 26 and 2 3/8 IF		
NC 31 and 2 7/8 IF	NC 40 and 4	FH
NC 38 and 3 1/2 IF	NC 46 and 4	IF
	NC 50 and 4 1/2 IF	

These connections are tentative.

API ROTARY SHOULDERED CONNECTION DIMENSIONS (continued)
(API Spec 7)

Connection number or size	Drill pipe Diameter-Upset-Weight (in) - (lb/ft)	Tool joint Outside diameter (in)	(mm)	Tool joint Inside diameter (in)	(mm)	Pin Dm (mm)	Pin dm (mm)	Pin Lm (mm)	Pin Dc (mm)	Box Df (mm)	Box Lf (mm)	Box Taper (%)	Threads per inch	Thread form
REGULAR (REG) STYLE														
2 3/8 REG	2 7/8-IU-10.40	3 3/4	95.2	1 1/4	31.7	66.7	47.6	76.2	60.1	68.3	92.1	25	5	V-0.040
2 7/8 REG	3 1/2-IU-13.30	4 1/4	107.9	1 1/2	38.1	76.2	54.0	88.9	69.6	77.8	104.8	25	5	V-0.040
3 1/2 REG	3 1/2-IU-16.60	5 1/2	139.7	2 1/4	57.1	88.9	65.1	95.2	82.3	90.5	111.1	25	5	V-0.040
4 1/2 REG	4 1/2-IEU-20.00	5 1/2	139.7	2 1/4	57.1	117.5	90.5	107.9	110.9	119.1	123.8	25	5	V-0.040
4 1/2 REG	5 1/2-IEU-21.90	6 3/4	171.4	2 3/4	69.8	117.5	90.5	107.9	110.9	119.1	123.8	25	5	V-0.040
5 1/2 REG						140.2	110.1	120.6	132.9	141.7	136.5	16.66	4	V-0.050
6 5/8 REG						152.0	131.0	127.0	146.2	154.0	142.9	25	4	V-0.050
7 5/8 REG						177.8	144.5	133.3	170.5	180.2	149.2	25	4	V-0.050
8 5/8 REG						202.0	167.8	136.5	194.7	204.4	152.4	25	4	V-0.050
FULL HOLE (FH) STYLE (4)														
3 1/2 FH	3 1/2-IU-13.30	4 5/8	117.5	2 7/16	61.9	101.4	77.6	95.2	94.8	102.8	111.1	25	5	V-0.040
3 1/2 FH	3 1/2-IU-15.50	4 5/8	117.5	2 7/16	61.9	101.4	77.6	95.2	94.8	102.8	111.1	25	5	V-0.040
4 FH	4-IU-14.00	5 1/4	133.3	2 13/16	71.4	108.7	89.7	101.6	103.4	110.3	130.2	16.66	5	V-0.065
4 1/2 FH	4 1/2-IU-16.60	5 3/4	146.0	3	76.2	121.7	96.3	101.6	115.1	123.8	117.5	25	5	V-0.040
4 1/2 FH	4 1/2-IEU-20.00	5 3/4	146.0	3	76.2	121.7	96.3	101.6	115.1	123.8	117.5	25	4	V-0.050
5 1/2 FH	5 1/2-IEU-21.90	7	177.8	4	101.6	148.0	126.8	127.0	142.0	150.0	142.9	16.66	4	V-0.050
5 1/2 FH	5 1/2-IEU-24.70	7	177.8	4	101.6	148.0	126.8	127.0	142.0	150.0	142.9	16.66	4	V-0.050
6 5/8 FH						171.5	150.4	127.0	165.6	173.8	142.9	16.66	4	V-0.050
INTERNAL FLUSH (IF) STYLE (4)														
2 3/8 IF	2 3/8-EU- 6.65	3 3/8	85.7	1 3/4	44.4	73.1	60.4	76.2	67.8	74.6	92.1	16.66	4	V-0.065
2 7/8 IF	2 7/8-EU-10.40	4 1/8	104.8	2 1/8	54.0	86.1	71.3	88.9	80.8	87.7	104.8	16.66	4	V-0.065
3 1/2 IF	3 1/2-EU-13.30	4 3/4	120.6	2 11/16	68.3	102.0	85.1	101.6	96.7	103.6	117.5	16.66	4	V-0.065
3 1/2 IF	3 1/2-EU-15.50	5	127.0	2 11/16	68.3	102.0	85.1	101.6	96.7	103.6	117.5	16.66	4	V-0.065
4 IF	4-EU-14.00	5 3/4	146.0	3 1/4	82.5	122.8	103.7	101.6	117.5	124.6	124.6	16.66	4	V-0.065
4 1/2 IF	4 1/2-EU-16.60	Std 6 1/8	155.6	3 3/4	95.2	133.3	114.3	114.3	128.1	134.9	130.2	16.66	4	V-0.065
4 1/2 IF	4 1/2-EU-16.60	Opt 6 1/4	158.7	3 3/4	95.2	133.3	114.3	114.3	128.1	134.9	130.2	16.66	4	V-0.065
4 1/2 IF*	5-IEU-19.50	Std 6 3/8	161.9	3 3/4	95.2	133.3	114.3	114.3	128.1	134.9	130.2	16.66	4	V-0.065
4 1/2 IF*	5-IEU-19.50	Opt 6 1/2	165.1	3 1/2	88.9	133.3	114.3	114.3	128.1	134.9	130.2	16.66	4	V-0.065
5 IF						162.5	141.3	127.0	157.2	163.9	142.9	16.66	4	V-0.065

(3) For tool joint thread, the use of a thread compound containing 40 to 60 % by weight of finely powdered metallic zinc is recommended.

(4) These styles are obsolescent.

* NC 50 (4 1/2 IF) joint with 6 3/8 in and 6 1/2 in outside diameter is used on 5 in IEU drill pipe to produce an assembly variously known as 5 in Extra-Hole and 5 in Semi-Internal-Flush.

CHARACTERISTICS OF SOME
NON API TOOL JOINT THREADS

The dimensions in the tables are given solely to identify the style of the thread.

Size (in)	Pin		Box Df * (mm)	Taper (%)	Threads per inch	Thread form	Make-up torque (m-daN)
	Dm * (mm)	Dc * (mm)					
EXTRA HOLE (XH) STYLE							
2 7/8	84,5	79,2	85,3	16,66	4	V 0,065	760-950
3 1/2	96,8	91,5	98,4	16,66	4	V 0,065	975-1220
4 1/2	122,8	117,5	124,6	16,66	4	V 0,065	1950-2440
5	133,3	128,1	134,9	16,66	4	V 0,065	2140-2515
DOUBLE STREAMLINE STYLE							
3 1/2	84,5	79,2	85,3	16,66	4	V 0,065	570-705
4	98,7	93,4	99,6	16,66	4	V 0,065	870-1080
4 1/2	108,7	103,4	110,3	16,66	4	V 0,065	1060-1330
5 1/2	133,3	128,1	134,9	16,66	4	V 0,065	1900-2370
SLIM HOLE (SH) STYLE							
2 7/8	73,1	67,8	74,6	16,66	4	V 0,065	390-490
3 1/2	86,1	80,8	87,7	16,66	4	V 0,065	650-800
4	96,8	91,5	98,4	16,66	4	V 0,065	870-1080
4 1/2	102,0	96,7	103,6	16,66	4	V 0,065	1060-1330
HUGUES (H 90) STYLE							
3 1/2	104,8	99,8	106,4	16,66	3 1/2	H 90	1300-1630
4	114,3	109,3	115,9	16,66	3 1/2	H 90	2000-2450
4 1/2	122,8	117,8	124,2	16,66	3 1/2	H 90	2200-2700
REED WIDE OPEN (WO) STYLE							
2 3/8	71,5	66,2	72,6	16,66	4	V 0,065	245-300
2 7/8	84,5	79,3	85,7	16,66	4	V 0,065	405-515
3 1/2	102,0	96,7	103,6	16,66	4	V 0,065	730-920
4	122,8	117,5	124,6	16,66	4	V 0,065	670-2050
4 1/2	133,3	128,1	134,9	16,66	4	V 0,065	900-2370

* See page 66 (B 10) for identification.

CHARACTERISTICS OF SOME
NON API TOOL JOINT THREADS (continued)

Size (in)	Pin		Box Df * (mm)	Taper (%)	Threads per inch	Thread form	Make-up torque (m-daN)
	Dm * (mm)	Dc * (mm)					
AMERICAN OPEN HOLE (OH) STYLE							
2 3/8	69.8	65.7	71.4	12.5	4		260-325
2 7/8	79.9	75.8	81.8	12.5	4	Special	490-610
3 1/2	98.8	94.7	100.4	12.5	4		650-810
4	116.3	112.2	117.9	12.5	4	American	1520-1900
4 1/2	124.8	120.7	126.6	12.5	4		1170-1460

Size (in)	Pin		Taper (%)	Threads per inch	Hydril special thread	Make-up torque (m-daN)
	Dm * (mm)	Dc * (mm)				
HYDRIL IF STYLE						
2 3/8	71.3	100.0	4.17	3	2 steps	515
2 7/8	80.8	100.0	4.17	3	2 steps	730
3 1/2	97.5	100.0	4.17	3	2 steps	895
4 1/2	132.1	101.6	4.47	3	2 steps	1550
HYDRIL IEU STYLE						
3 1/2	95.0	107.9	4.17	3	2 steps	895
4	118.4	109.6	4.17	3	2 steps	1550
4 1/2	120.4	112.7	4.17	3	2 steps	1550
5 1/2	148.2	139.7	4.17	2	2 steps	2060
HYDRIL SH STYLE						
2 7/8	71.3	100.0	4.17	3	2 steps	580
3 1/2	80.8	100.0	4.17	3	2 steps	730
4	97.5	100.0	4.17	3	2 steps	895
4 1/2	106.5	104.8	4.17	3	2 steps	1180
5 1/2	132.1	101.6	4.17	3	2 steps	1550
HYDRIL F STYLE						
2 3/8	48.9	65.1	4.17	4	1 step	215
2 7/8	60.1	90.5	4.17	4	2 steps	365
3 1/2	71.3	101.6	4.17	3	2 steps	580
4	84.8	100.0	4.17	3	2 steps	730
4 1/2	97.5	100.0	4.17	3	2 steps	895
5 1/2	118.4	106.4	4.17	3	2 steps	1550

* See page 66 (B 10) for identification.

INTERCHANGEABILITY OF TOOL JOINT THREADS

STYLE	INTERCHANGEABLE WITH
2 3/8 IF 2 3/8 IF Hydril	2 7/8 SH - n° 26 2 7/8 SH Hydril
2 7/8 IF 2 7/8 XH 2 7/8 SH 2 7/8 SH Hydril 2 7/8 IF Hydril	3 1/2 SH - n° 31 3 1/2 Dble Streamline 2 3/8 IF - n° 26 2 3/8 IF Hydril 3 1/2 SH Hydril
3 1/2 IF 3 1/2 XH 3 1/2 Dble Streamline 3 1/2 SH 3 1/2 WO 3 1/2 SH Hydril 3 1/2 IF Hydril	4 1/2 SH - 3 1/2 WO - n° 38 4 SH 2 7/8 XH 2 7/8 IF - n° 31 3 1/2 IF - 4 1/2 SH - n° 38 2 7/8 IF Hydril 4 SH Hydril
4 IF 4 FH 4 SH 4 WO 4 SH Hydril	4 1/2 XH - 4 WO - n° 46 4 1/2 Dble Streamline - n° 40 3 1/2 XH 4 IF - 4 1/2 XH - n° 46 3 1/2 IF Hydril
4 1/2 IF 4 1/2 XH 4 1/2 Dble Streamline 4 1/2 SH 4 1/2 WO 4 1/2 IF Hydril	5 XH - 5 1/2 Dble Streamline - 4 1/2 WO - n° 50 4 IF - 4 WO - n° 46 4 FH - n° 40 3 1/2 IF - 3 1/2 WO - n° 38 4 1/2 IF - 5 XH - 5 1/2 Dble Streamline - n° 50 5 1/2 SH Hydril
5 XH	4 1/2 IF - 5 1/2 Dble Streamline - 4 1/2 WO - n° 50
5 1/2 Dble Streamline 5 1/2 SH Hydril	4 1/2 IF - 5 XH - 4 1/2 WO - n° 50 4 1/2 IF Hydril
n° 26 n° 31 n° 38 n° 40 n° 46 n° 50	2 3/8 IF - 2 7/8 SH 2 7/8 IF - 3 1/2 SH 3 1/2 IF - 4 1/2 SH - 3 1/2 WO 4 FH - 4 1/2 Dble Streamline 4 IF - 4 1/2 XH - 4 WO 4 1/2 IF - 5 XH - 5 1/2 Dble Streamline - 4 1/2 WO

PROPERTIES OF DRILL PIPES AND TOOL JOINTS
(API RP 7G)

			1	2 3/8 · 6,65 #			2 7/8 · 10,40 #					
Nominal size			1									
Grade			2	E	95	105	E					95
Type (1)			3	EU	EU	EU	EU	EU	EU	IU	IU	EU
Tool joint data	Size and style		4	2 3/8 IF NC26	2 3/8 IF NC26	2 3/8 IF NC26	2 7/8 IF NC31	2 7/8 OH	2 7/8 SL-H90	2 7/8 XH	2 7/8 SH	2 7/8 IF NC31
	Outside diameter	in	5	3 3/8	3 3/8	3 3/8	4 1/8	3 7/8	3 7/8	4 1/4	3 3/8	4 1/8
		mm	6	85,7	85,7	85,7	104,8	98,4	98,4	108,0	85,7	104,8
	Inside diameter	in	7	1 3/4	1 3/4	1 3/4	2 1/8	2 5/32	2 5/32	1 7/8	1 3/4	2
		mm	8	44,4	44,4	44,4	54,0	54,8	54,8	47,6	44,4	50,8
	Bevel diameter	in	9	3 17/64	3 17/64	3 17/64	3 61/64		3 5/8	4	3 1/4	3 61/64
		mm	10	83,0	83,0	83,0	101,4		92,1	101,6	79,4	101,4
Weight pipe + tool joint		lb/ft	11	7,00	7,08	7,08	10,82	10,57	10,53	11,20	10,37	10,89
		kg/m	12	10,42	10,54	10,54	16,10	15,73	15,67	16,67	15,43	16,21
Capacity		gal/ft	13	0,135	0,135	0,135	0,190	0,190	0,190	0,185	0,185	0,190
		l/m	14	1,68	1,68	1,68	2,36	2,36	2,36	2,30	2,30	2,36
open end displacement		gal/ft	15	0,107	0,108	0,108	0,165	0,161	0,161	0,171	0,159	0,166
		l/m	16	1,33	1,34	1,34	2,05	2,00	2,00	2,12	1,97	2,06
closed end displacement		gal/ft	17	0,242	0,243	0,243	0,355	0,351	0,351	0,356	0,344	0,356
		l/m	18	3,01	3,02	3,02	4,41	4,36	4,36	4,42	4,27	4,42
Tensile yield	Pipe	10³lb	19	138,2	175,1	193,5	214,3	214,3	214,3	214,3	214,3	271,5
		10³daN	20	61,5	77,9	86,1	95,3	95,3	95,3	95,3	95,3	120,8
	Tool joint	10³lb	21	313,7	313,7	313,7	447,1	345,4	382,6	505,1	313,7	495,7
		10³daN	22	139,5	139,5	139,5	198,9	153,6	170,2	224,7	139,7	220,5
Torsional yield	Pipe (2)	ft-lb	23	6250	7920	8750	11550	11550	11550	11550	11550	14640
		m-daN	24	845	1075	1185	1565	1565	1565	1565	1565	1985
	Tool joint (3)	ft-lb	25	6800	6800	6800	11800	8900	11300	13400	6800	13200
		m-daN	26	920	920	920	1600	1210	1530	1820	920	1790
Class I	Outside diam. tool joint	in	27	3 3/8	3 3/8	3 3/8	4 1/8	3 7/8	3 7/8	4 1/4	3 3/8	4 1/8
		mm	28	85,7	85,7	85,7	104,8	98,4	98,4	108,0	85,7	104,8
	Make-up torque	ft-lb	29	3500	3500	3500	5900	4500	5700	6700	3500	6600
		m-daN	30	475	475	475	800	610	770	910	475	895
Premium	Outside diam. tool joint	in	31	3 3/16	3 1/4	3 9/32	3 13/16	3 19/32	3 19/32	3 23/32	3 3/8	3 29/32
		mm	32	81,0	82,6	83,3	96,8	91,3	91,3	94,5	85,7	99,2
	Make-up torque	ft-lb	33	2500	3000	3300	4600	4300	4600	4400	3500	5700
		m-daN	34	340	405	450	620	580	620	600	475	770
Class II	Outside diam. tool joint	in	35	3 1/8	3 3/16	3 7/32	3 23/32	3 17/32	3 17/32	3 21/32	3 5/16	3 13/16
		mm	36	79,4	81,0	81,8	94,5	89,7	89,7	92,9	84,1	96,8
	Make-up torque	ft-lb	37	2000	2500	2700	3500	3600	3800	3700	3100	4600
		m-daN	38	270	340	370	475	490	515	500	420	620
Class III	Outside diam. tool joint	in	39	3 1/8	3 5/32	3 5/32	3 11/16	3 15/32	3 15/32	3 5/8	3 1/4	3 3/4
		mm	40	79,4	80,2	80,2	93,7	88,1	88,1	92,1	82,6	95,3
	Make-up torque	ft-lb	41	2000	2200	2200	3200	3000	3000	3100	3000	3900
		m-daN	42	270	300	300	430	410	410	420	410	530

(1) EU = external upset. IU = internal upset. IEU = internal external upset.

(2) The torsional yield strength is based on a shear strength of 57,7 % of the minimum yield strength.

(3) The torsional yield strengths for tool-joints are calculated on the pin and box. The smallest of the two torsional strengths thus obtained is the theoretical torsional strength of the tool joint.

PROPERTIES OF DRILL PIPES AND TOOL JOINTS (continued)
(API RP 7G)

1	2 7/8 · 10,40 #			3 1/2 · 9,50 #				3 1/2 · 13,30 #			
2	95	105	135	E				E			
3	EU	EU	EU	EU	EU	EU	EU	EU	EU	IU	IU
4	2 7/8 SL-H90	2 7/8IF NC31	2 7/8 IF NC31	3 1/2 WO	3 1/2IF NC38	3 1/2 OH	3 1/2 SL-H90	3 1/2IF NC38	3 1/2 OH	3 1/2 XH	3 1/2SH NC31
5/6	3 7/8 / 98,4	4 1/8 / 104,8	4 3/8 / 111,1	4 3/4 / 120,6	4 3/4 / 120,6	4 1/2 / 114,3	4 5/8 / 117,5	4 3/4 / 120,6	4 3/4 / 120,6	4 3/4 / 120,6	4 1/8 / 104,8
7/8	2 5/32 / 54,8	2 / 50,8	1 5/8 / 41,3	3 / 76,2	2 11/16 / 68,3	3 / 76,2	3 / 76,2	2 11/16 / 68,3	2 11/16 / 68,3	2 7/16 / 61,9	2 1/8 / 54,0
9/10	3 5/8 / 92,1	3 61/64 / 101,4	3 61/64 / 101,4	4 37/64 / 116,3	4 37/64 / 116,3		4 17/32 / 115,1	4 37/64 / 116,3	4 37/64 / 116,3	4 1/2 / 114,3	4 / 101,6
11/12	10,83 / 16,12	10,89 / 16,21	11,20 / 16,67	10,25 / 15,25	10,39 / 15,46	9,95 / 14,81	10,12 / 15,06	13,86 / 20,63	13,86 / 20,63	14,06 / 20,92	13,51 / 20,11
13/14	0,190 / 2,36	0,190 / 2,36	0,190 / 2,36	0,366 / 4,54	0,366 / 4,54	0,366 / 4,54	0,366 / 4,54	0,312 / 3,88	0,312 / 3,88	0,307 / 3,81	0,307 / 3,81
15/16	0,165 / 2,05	0,166 / 2,06	0,171 / 2,12	0,156 / 1,94	0,159 / 1,97	0,152 / 1,89	0,155 / 1,92	0,212 / 2,63	0,212 / 2,63	0,214 / 2,66	0,206 / 2,56
17/18	0,355 / 4,41	0,356 / 4,42	0,361 / 4,48	0,522 / 6,48	0,525 / 6,51	0,518 / 6,43	0,521 / 6,46	0,524 / 6,51	0,524 / 6,51	0,521 / 6,47	0,513 / 6,37
19/20	271,5 / 120,8	300,8 / 133,8	385,8 / 171,6	194,3 / 86,4	194,3 / 86,4	194,3 / 86,4	194,3 / 86,4	271,6 / 120,8	271,6 / 120,8	271,6 / 120,8	271,6 / 120,8
21/22	443,8 / 197,4	495,7 / 220,5	623,8 / 277,5	419,8 / 186,7	587,3 / 261,2	392,0 / 174,4	366,5 / 163,0	587,3 / 261,2	559,6 / 248,9	570,9 / 253,9	447,1 / 198,9
23/24	14640 / 1985	16180 / 2190	20800 / 2820	14150 / 1965	14150 / 1965	14150 / 1965	14150 / 1965	18550 / 2515	18550 / 2515	18550 / 2515	18550 / 2515
25/26	13200 / 1790	13200 / 1790	17000 / 2300	12800 / 1735	18100 / 2450	12100 / 1640	12600 / 1710	18100 / 2450	17400 / 2360	17100 / 2320	11800 / 1600
27/28	3 7/8 / 98,4	4 1/8 / 104,8	4 3/8 / 11,1	4 3/4 / 120,6	4 3/4 / 120,6	4 1/2 / 114,3	4 5/8 / 117,5	4 3/4 / 120,6	4 3/4 / 120,6	4 3/4 / 120,6	4 1/8 / 104,8
29/30	5700 / 770	6600 / 890	8500 / 1150	6400 / 870	9100 / 1230	6100 / 830	6300 / 850	9100 / 1230	8700 / 1180	8500 / 1150	6300 / 850
31/32	3 11/16 / 93,7	3 15/16 / 100,0	4 1/16 / 103,2	4 3/8 / 111,1	4 1/2 / 114,3	4 9/32 / 108,7	4 3/16 / 106,4	4 1/2 / 114,3	4 3/8 / 111,1		4 / 101,6
33/34	5700 / 770	6100 / 830	7700 / 1040	5300 / 720	7300 / 990	5500 / 745	5500 / 745	7300 / 990	7000 / 950		6000 / 810
35/36	3 19/32 / 91,3	3 27/32 / 97,6	3 31/32 / 100,8	4 11/32 / 110,3	4 13/32 / 111,9	4 7/32 / 107,2	4 1/8 / 104,8	4 13/32 / 111,9	4 5/16 / 109,5		3 29/32 / 99,2
37/38	4600 / 620	5000 / 670	6500 / 880	4800 / 650	5800 / 790	4600 / 620	4500 / 610	5800 / 790	6000 / 810		5500 / 745
39/40	3 17/32 / 89,7	3 25/32 / 96,0	3 7/8 / 98,4	4 9/32 / 108,7	4 11/32 / 110,3	4 5/32 / 105,6	4 3/32 / 104,0	4 11/32 / 110,3	4 1/4 / 108,0		3 27/32 / 97,6
41/42	3800 / 515	4200 / 570	5400 / 730	3800 / 515	4800 / 650	3700 / 500	4000 / 540	4800 / 650	5100 / 690		5000 / 680

(4) Calculations for recommended tool joint make-up torque are based on a tensile stress of 50 % of the minimum tensile yield for new joints and 60 % for used joints.

PROPERTIES OF DRILL PIPES AND TOOL JOINTS (continued)
(API RP 7G)

		#	3 1/2 - 13,30 #				3 1/2 - 15,50 #			4 - 11,85 #	
Nominal size		1									
Grade		2	95	105	135		E	95	105	E	
Type (1)		3	EU	EU	EU	EU	EU	EU	EU	EU	EU
Tool joint data Size and style		4	3 1/2 IF NC38	3 1/2 SL-H90	3 1/2 IF NC38	4 FH NC40	3 1/2 IF NC38	3 1/2 IF NC38	3 1/2 IF NC38	4 IF NC46	4 WO NC46
Outside diameter	in	5	5	4 5/8	5	5 3/8	5	5	5	5 3/4	5 3/4
	mm	6	127,0	117,5	127,0	136,5	127,0	127,0	127,0	146,1	146,1
Inside diameter	in	7	2 9/16	2 11/16	2 7/16	2 7/16	2 9/16	2 7/16	2 1/8	3 1/4	3 7/16
	mm	8	65,1	68,3	61,9	61,9	65,1	61,9	54,0	82,6	87,3
Bevel diameter	in	9	4 37/64	4 17/32	4 37/64	5 1/8	4 37/64	4 37/64	4 37/64	5 17/32	5 17/32
	mm	10	116,3	115,1	116,3	130,2	116,3	116,3	116,3	140,5	140,5
Weight pipe + tool joint	lb/ft	11	14,32	13,74	14,38	14,95	16,42	16,54	16,61	13,14	13,04
	kg/m	12	21,31	20,44	21,40	22,25	24,43	24,61	14,72	19,56	19,41
Capacity	gal/ft	13	0,319	0,314	0,312	0,312	0,279	0,279	0,279	0,493	0,493
	l/m	14	3,96	3,90	3,87	3,87	3,46	3,46	3,46	6,12	6,12
open end displacement	gal/ft	15	0,218	0,209	0,220	0,228	0,250	0,253	0,254	0,200	0,199
	l/m	16	2,71	2,60	2,73	2,83	3,11	3,14	3,15	2,49	2,47
closed end displacement	gal/ft	17	0,537	0,523	0,532	0,540	0,529	0,532	0,533	0,693	0,692
	l/m	18	6,67	6,50	6,60	6,70	6,57	6,60	6,61	8,61	8,59
Tensile yield Pipe	10^3lb	19	344,0	344,0	380,2	488,8	322,8	408,9	451,9	231,6	231,6
	10^3daN	20	153,0	153,0	169,1	217,4	143,6	181,9	201,0	103,0	103,0
Tool joint	10^3lb	21	649,2	537,3	708,1	897,2	649,2	708,1	836,4	921,7	800,3
	10^3daN	22	288,8	239,0	315,0	399,1	288,8	315,0	372,0	410,0	356,0
Torsional yield Pipe (2)	ft-lb	23	23500	23500	25970	33390	21090	26710	29520	19470	19470
	m-daN	24	3180	3180	3520	4530	2860	3620	4000	2640	2640
Tool joint (3)	ft-lb	25	20300	19180	22200	30000	20300	22200	26400	34080	29280
	m-daN	26	2750	2600	3000	4070	2750	3000	3580	4620	3970
Class I Outside diam. tool joint	in	27	5	4 5/8	5	5 3/8	5	5	5	5 3/4	5 3/4
	mm	28	127,0	117,5	127,0	136,5	127,0	127,0	127,0	146,1	146,1
Make-up torque	ft-lb	29	10200	9300	11100	15000	10200	11100	13300	16700	14400
	m-daN	30	1380	1315	1500	2030	1380	1500	1800	2260	1950
Premium Outside diam. tool joint	in	31	4 19/32	4 3/8	4 21/32	5	4 17/32	4 21/32	4 23/32	5 7/32	5 7/32
	mm	32	116,7	111,1	118,3	127,0	115,1	118,3	119,9	132,6	132,6
Make-up torque	ft-lb	33	8800	8800	9900	12600	7800	9900	10900	7900	7900
	m-daN	34	1190	1190	1340	1710	1060	1340	1480	1070	1070
Class II Outside diam. tool joint	in	35	4 1/2	4 9/32	4 9/16	4 7/8	4 15/32	4 9/16	4 5/8	5 5/32	5 5/32
	mm	36	114,3	108,7	115,9	123,8	113,5	115,9	123,8	131,0	131,0
Make-up torque	ft-lb	37	7300	7100	8300	10200	6800	8300	9300	6500	6500
	m-daN	38	990	960	1125	1380	920	1125	1260	880	880
Class III Outside diam. tool joint	in	39	4 7/16	4 7/32	4 15/32	4 25/32	4 3/8	4 15/32	4 17/32	5 5/32	5 5/32
	mm	40	112,7	107,2	113,5	121,4	111,1	113,5	115,1	131,0	131,0
Make-up torque	ft-lb	41	6300	6100	6800	8400	5300	6800	7800	5100	5100
	m-daN	42	850	830	920	1140	720	920	1060	690	690

(1) EU = external upset. IU = internal upset. IEU = internal external upset.

(2) The torsional yield strength is based on a shear strength of 57,7 % of the minimum yield strength.

(3) The torsional yield strengths for tool joints are calculated on the pin and box. The smallest of the two torsional strengths thus obtained is the theoretical torsional strength of the tool joint.

PROPERTIES OF DRILL PIPES AND TOOL JOINTS (continued)
(API RP 7G)

1	4 · 14,00 #									4 · 15,70 #	
2	E			95			105		135	E	
3	IU	IU	EU	IU	IU	EU	IU	IU	EU	IU	IU
4	4FH NC40	4 SH	4IF NC46	4FH NC40	4 H90	4IF NC46	4FH NC40	4 H90	4IF NC46	4FH NC40	4 H90
5	5 1/4	4 5/8	6	5 1/4	5 1/2	6	5 1/2	5 1/2	6	5 1/4	5 1/2
6	133,4	117,5	152,4	133,4	139,7	152,4	139,7	139,7	152,4	133,4	139,7
7	2 13/16	2 9/16	3 1/4	2 11/16	2 13/16	3 1/4	2 7/16	2 13/16	3	2 11/16	2 13/16
8	71,4	65,1	82,6	68,3	71,4	82,6	61,9	71,4	76,2	68,3	71,4
9	5 1/64	4 3/8	5 23/32	5 1/64	5 5/16	5 23/32	5 1/64	5 1/64	5 23/32	5 1/64	5 5/16
10	127,4	111,2	145,3	127,4	134,9	145,3	127,4	127,4	145,3	127,4	134,9
11	15,13	14,29	15,98	15,16	15,57	15,92	15,65	15,57	16,06	16,99	17,30
12	22,52	21,27	23,78	22,56	23,17	23,69	23,29	23,17	23,90	25,28	25,75
13	0,454	0,454	0,457	0,455	0,455	0,457	0,455	0,455	0,456	0,428	0,428
14	5,64	5,64	5,67	5,65	5,65	5,67	5,65	5,65	5,66	5,32	5,32
15	0,231	0,218	0,244	0,231	0,238	0,243	0,239	0,238	0,245	0,259	0,264
16	2,87	2,71	3,03	2,87	2,95	3,02	2,97	2,95	3,04	3,22	3,28
17	0,685	0,672	0,701	0,686	0,693	0,700	0,694	0,693	0,701	0,687	0,692
18	8,51	8,35	8,70	8,52	8,60	8,69	8,62	8,60	8,70	8,54	8,60
19	285,4	285,4	285,4	361,5	361,5	361,5	399,5	399,5	513,7	324,1	324,1
20	127,9	127,9	127,9	162,0	162,0	162,0	179,1	179,1	230,3	145,3	145,3
21	711,6	512,0	901,2	776,4	913,5	901,2	897,2	913,5	1048,4	776,4	913,5
22	319,0	229,5	404,0	348,0	409,5	404,0	402,1	409,5	469,9	348,0	409,5
23	23290	23290	23290	29500	29500	29500	32600	32600	41920	25810	25810
24	3160	3160	3160	4000	4000	4000	4420	4420	5680	3500	3500
25	23500	15000	33600	25400	35400	33600	30000	35400	39200	25400	35400
26	3190	2030	4560	3440	4800	4560	4070	4800	5310	3440	4800
27	5 1/4	4 5/8	6	5 1/4	5 1/2	6	5 1/2	5 1/2	6	5 1/4	5 1/2
28	133,4	117,5	152,4	133,4	139,7	152,4	139,7	139,7	152,4	133,4	139,7
29	11800	8000	16900	12700	17700	16900	15100	17700	19600	12800	17700
30	1600	1080	2290	1720	2400	2290	2050	2400	2660	1740	2400
31	4 13/16	4 7/16	5 9/32	4 15/16	5 1/32	5 3/8	5	5 3/32	5 9/16	4 7/8	4 31/32
32	122,2	112,7	134,1	125,4	127,8	136,5	127,0	129,4	134,1	123,8	126,2
33	9000	7400	9200	11400	11100	11400	12600	12500	15800	10200	9700
34	1220	1000	1250	1550	1500	1550	1710	1690	2140	1380	1315
35	4 23/32	4 11/32	5 3/16	4 13/16	4 31/32	5 9/32	4 7/8	5	5 7/16	4 3/4	4 29/32
36	119,9	110,3	131,8	122,2	126,2	134,1	123,8	127,0	138,1	120,7	124,6
37	7300	6500	7200	9000	9700	9200	10200	10400	12800	7900	8300
38	990	880	980	1220	1315	1250	1380	1410	1735	1070	1125
39	4 21/32	4 9/32	5 5/32	4 3/4	4 29/32	5 7/32	4 25/32	4 15/16	5 3/8	4 11/16	4 27/32
40	118,3	108,7	131,0	120,7	124,6	132,6	121,4	125,4	136,5	119,1	123,0
41	6200	6000	6500	7900	8300	7900	8500	9000	11400	6800	7000
42	840	810	880	1070	1125	1070	1150	1220	1550	920	950

(4) Calculations for recommended tool joint make-up torque are based on a tensile stress of 50 % of the minimum tensile yield for new joints and 60 % for used joints.

PROPERTIES OF DRILL PIPES AND TOOL JOINTS (continued)
(API RP 7G)

			1	4 · 15,70 #			4 1/2 · 13,75 #					
Nominal size			1	4 · 15,70 #			4 1/2 · 13,75 #					
Grade			2	95		105	E					
Type (1)			3	IU	IU	IU	EU	EU	EU	IU		
Tool joint data	Size and style		4	4FH NC40	4 H90	4 H90	4 1/2IF NC50	4 1/2WO NC50	4 1/2 OH	4 1/2 H90		
	Outside diameter	in	5	5 3/8	5 1/2	5 1/2	6 3/8	6 1/8	5 3/4	6		
		mm	6	136,5	139,7	139,7	161,9	155,6	146,1	152,4		
	Inside diameter	in	7	2 7/16	2 13/16	2 13/16	3 3/4	3 7/8	3 31/32	3 1/4		
		mm	8	61,9	71,4	71,4	95,2	98,4	100,8	82,6		
	Bevel diameter	in	9	5 1/8	5 5/16	5 5/16	5 59/64	5 59/64	5 17/32	5 3/4		
		mm	10	130,2	134,9	134,9	150,4	150,4	140,5	146,1		
Weight pipe + tool joint		lb/ft	11	17,41	17,30	17,30	15,40	14,90	14,10	15,24		
		kg/m	12	25,91	25,75	25,75	22,92	22,17	20,98	22,68		
Capacity		gal/ft	13	0,428	0,428	0,428	0,639	0,639	0,639	0,636		
		l/m	14	5,32	5,32	5,32	7,94	7,94	7,94	7,90		
open end displacement		gal/ft	15	0,266	0,264	0,264	0,235	0,227	0,215	0,232		
		l/m	16	3,30	3,28	3,28	2,92	2,82	2,67	2,89		
closed end displacement		gal/ft	17	0,694	0,692	0,692	0,874	0,863	0,854	0,868		
		l/m	18	8,62	8,60	8,60	10,86	10,76	10,61	10,69		
Tensile yield	Pipe	10³lb	19	410,6	410,6	453,8	270,0	270,0	270,0	270,0		
		10³daN	20	182,6	182,6	201,9	120,1	120,1	120,1	120,1		
	Tool joint	10³lb	21	897,2	913,5	913,5	944,0	849,6	554,8	938,2		
		10³daN	22	399,1	406,3	406,3	419,9	377,9	246,8	417,3		
Torsional yield	Pipe (2)	ft-lb	23	32690	32690	36130	25910	25910	25910	25910		
		m-daN	24	4430	4430	4890	3510	3510	3510	3510		
	Tool joint (3)	ft-lb	25	30000	35400	35400	34100	37700	21300	38900		
		m-daN	26	4070	4800	4800	4620	5110	2890	5270		
Class I	Outside diam. tool joint	in	27	5 3/8	5 1/2	5 1/2	6 3/8	6 1/8	5 3/4	6		
		mm	28	136,5	139,7	139,7	161,9	155,6	146,1	152,4		
	Make-up torque	ft-lb	29	15000	17700	17700	18900	17480	10770	19500		
		m-daN	30	2030	2400	2400	2560	2370	1460	2640		
Premium	Outside diam. tool joint	in	31	5	5 3/32	5 5/32	5 23/32	5 23/32	5 1/2	5 11/32		
		mm	32	127,0	129,4	131,0	145,3	145,3	139,7	135,7		
	Make-up torque	ft-lb	33	12600	12600	14000	11600	12100	10770	11900		
		m-daN	34	1710	1710	1900	1570	1640	1640	1610		
Class II	Outside diam. tool joint	in	35	4 7/8	5	5 1/32	5 21/32	5 21/32	5 7/16	5 1/4		
		mm	36	123,8	127,0	127,8	143,7	143,7	138,1	133,3		
	Make-up torque	ft-lb	37	10200	10400	11100	10000	10400	10400	9500		
		m-daN	38	1380	1410	1500	1360	1410	1410	1290		
Class III	Outside diam. tool joint	in	39	4 25/32	4 29/32	4 31/32	5 19/32	5 19/32	5 3/8	5 3/16		
		mm	40	121,4	124,6	126,2	142,1	142,1	136,5	131,8		
	Make-up torque	ft-lb	41	8500	8400	9700	8400	8780	8480	8000		
		m-daN	42	1150	1140	1320	1140	1190	1150	1080		

(1) EU = external upset. IU = internal upset. IEU = internal external upset.

(2) The torsional yield strength is based on a shear strength of 57,7 % of the minimum yield strength.

(3) The torsional yield strengths for tool-joints are calculated on the pin and box. The smallest of the two torsional strengths thus obtained is the theoretical torsional strength of the tool joint.

PROPERTIES OF DRILL PIPES AND TOOL JOINTS (continued)
(API RP 7G)

1	4 1/2 · 16,60 #										
2	E					95				105	
3	IU	IU	IU	EU	EU	IU	IU	IU	EU	IU	IU
4	4 1/2XH NC46	4 1/2 FH	4 1/2 H90	4 1/2IF NC50	4 1/2 OH	4 1/2 FH	4 1/2XH NX46	4 1/2 H90	4 1/2 IF NC50	4 1/2 FH	4 1/2XH NC46
5 6	6 1/4 158,8	6 152,4	6 152,4	6 3/8 161,9	5 7/8 149,2	6 152,4	6 1/4 158,8	6 152,4	6 3/8 161,9	6 152,4	6 1/4 158,8
7 8	3 1/4 82,6	3 76,2	3 1/4 82,6	3 3/4 95,2	3 3/4 95,2	3 76,2	3 76,2	3 1/4 82,6	3 3/4 95,2	3 76,2	3 76,2
9 10	5 23/32 145,3	5 23/32 145,3	5 3/4 146,0	5 59/64 150,4		5 23/32 145,3	5 23/32 145,3	5 3/4 146,0	5 59/64 150,4	5 23/32 145,3	5 23/32 145,3
11 12	18,40 27,38	18,23 27,13	17,94 26,70	18,10 26,94	17,22 25,63	18,23 27,13	18,51 27,55	18,09 26,92	18,10 26,94	18,23 27,13	18,51 27,55
13 14	0,584 7,25	0,584 7,25	0,584 7,25	0,599 7,44	0,599 7,44	0,584 7,25	0,584 7,25	0,584 7,25	0,599 7,44	0,584 7,25	0,584 7,25
15 16	0,281 3,49	0,279 3,46	0,274 3,40	0,276 3,43	0,262 3,26	0,279 3,46	0,283 3,51	0,276 3,43	0,276 3,43	0,279 3,46	0,283 3,51
17 18	0,865 10,74	0,863 10,71	0,858 10,65	0,875 10,87	0,861 10,70	0,863 10,71	0,867 10,76	0,860 10,68	0,875 10,87	0,863 10,71	0,867 10,76
19 20	330,6 147,1	330,6 147,1	330,6 147,1	330,6 147,1	330,6 147,1	418,7 186,2	418,7 186,2	418,7 186,2	418,7 186,2	462,8 205,9	462,8 205,9
21 22	901,2 400,9	976,2 434,2	938,2 417,3	944,0 419,9	713,9 317,6	976,2 434,2	1048,4 466,4	938,2 417,3	944,0 419,9	976,2 434,2	1048,4 466,4
23 24	30810 4180	30810 4180	30810 4180	30810 4180	30810 4180	39020 5290	39020 5290	39020 5290	39020 5290	43130 5850	43130 5850
25 26	33900 4600	34800 4720	38900 5270	37700 5100	27500 3730	34800 4720	39600 5370	38900 5270	37700 5100	34800 4720	39600 5370
27 28	6 1/4 158,8	6 152,4	6 152,4	6 3/8 161,9	5 7/8 149,2	6 152,4	6 1/4 158,8	6 152,4	6 3/8 161,9	6 152,4	6 1/4 158,8
29 30	17000 2300	17400 2360	19500 2640	18900 2560	13800 1870	17400 2360	19800 2680	19500 2640	18900 2560	17400 2360	19800 2680
31 32	5 13/32 137,3	5 3/8 136,5	5 11/32 135,7	5 23/32 145,3	5 15/32 138,9	5 3/8 136,5	5 17/32 140,5	5 11/32 135,7	5 27/32 148,4	5 3/8 136,5	5 19/32 142,1
33 34	12100 1640	12100 1640	11900 1610	11600 1570	12300 1670	12100 1640	15000 2030	11900 1610	15000 2030	12100 1640	16500 2230
35 36	5 5/16 134,9	5 9/32 134,1	5 1/4 133,3	5 21/32 143,7	5 3/8 136,5	5 9/32 134,1	5 15/32 138,9	5 1/4 133,3	5 3/4 146,1	5 9/32 134,1	5 15/32 138,9
37 38	9900 1340	10100 1370	9500 1290	10000 1360	10000 1360	10100 1370	12100 1640	9500 1290	12400 1680	10100 1370	13500 1830
39 40	5 1/4 133,3	5 3/16 131,8	5 3/16 131,8	5 19/32 142,1	5 5/16 134,9	5 3/16 131,8	5 11/32 135,7	5 3/16 131,8	5 11/16 144,5	5 3/16 131,8	5 3/8 136,5
41 42	8500 1150	8100 1100	8000 1080	8400 1140	8600 1170	8100 1100	10600 1440	8000 1080	10800 1460	8100 1100	11400 1550

(4) Calculations for recommended tool joint make-up torque are based on a tensile stress of 50 % of the minimum tensile yield for new joints and 60 % for used joints.

PROPERTIES OF DRILL PIPES AND TOOL JOINTS (continued)
(API RP 7G)

			4 1/2 - 16,60 #				4 1/2 - 20,00 #			
Nominal size		1								
Grade		2	105		135		E			
Type (1)		3	IU	EU	IU	EU	IEU	IEU	IEU	EU
Tool joint data	Size and style	4	4 1/2 H90	4 1/2IF NC50	4 1/2XH NC46	4 1/2IF NC50	4 1/2XH NC46	4 1/2 FH	4 1/2 H90	4 1/2IF NC50
	Outside diameter in	5	6	6 3/8	6 1/4	6 3/8	6 1/4	6	6	6 3/8
	mm	6	152,4	161,9	158,8	161,9	158,8	152,4	152,4	161,9
	Inside diameter in	7	3 1/4	3 3/4	2 3/4	3 1/2	3	3	3	3 5/8
	mm	8	82,6	95,2	69,9	88,9	76,2	76,2	76,2	92,1
	Bevel diameter in	9	5 3/4	5 59/64	5 23/32	5 59/64	5 23/32	5 23/32	5 3/4	5 59/64
	mm	10	146,0	150,4	145,3	150,4	145,3	145,3	146,1	150,4
Weight pipe + tool joint	lb/ft	11	18,09	18,10	18,54	18,32	22,18	21,73	21,73	22,50
	kg/m	12	26,92	26,94	27,59	27,26	33,01	32,34	32,34	33,48
Capacity	gal/ft	13	0,584	0,599	0,584	0,599	0,531	0,531	0,531	0,544
	l/m	14	7,25	7,44	7,25	7,44	6,60	6,60	6,60	6,76
open end displacement	gal/ft	15	0,276	0,276	0,283	0,279	0,339	0,332	0,332	0,343
	l/m	16	3,43	3,43	3,51	3,47	4,21	4,12	4,12	4,26
closed end displacement	gal/ft	17	0,860	0,875	0,867	0,878	0,870	0,863	0,863	0,887
	l/m	18	10,68	10,87	10,76	10,91	10,81	10,72	10,72	11,02
Tensile yield Pipe	10^3lb	19	462,8	462,8	595,0	595,0	412,4	412,4	412,4	412,4
	10^3daN	20	205,9	105,9	264,7	264,7	183,4	183,4	183,4	183,4
Tool joint	10^3lb	21	938,2	944,0	1183,9	1110,2	1048,4	976,2	976,2	1030,9
	10^3daN	22	417,3	419,9	526,6	493,8	466,4	434,2	434,2	458,6
Torsional yield Pipe (2)	ft-lb	23	43130	43130	55450	55450	36900	36900	36900	36900
	m-daN	24	5850	5850	7520	7520	5000	5000	5000	5000
Tool joint (3)	ft-lb	25	38900	37700	44900	44700	39600	34800	45200	41200
	m-daN	26	5270	5100	6090	6060	5370	4720	6130	5590
Class I Outside diam. tool joint	in	27	6	6 3/8	6 1/4	6 3/8	6 1/4	6	6	6 3/8
	mm	28	152,4	161,9	158,8	161,9	158,8	152,4	152,4	161,9
Make-up torque	ft-lb	29	19500	18900	22500	22400	19800	17400	22600	20600
	m-daN	30	2640	2560	3050	3040	2690	2360	3060	2790
Premium Outside diam. tool joint	in	31	5 11/32	5 29/32	5 25/32	6 1/16	5 1/2	5 15/32	5 7/16	5 13/16
	mm	32	135,7	150,0	146,8	154,0	139,7	138,9	138,1	147,6
Make-up torque	ft-lb	33	11900	16600	21200	21000	14300	14200	14300	14100
	m-daN	34	1610	2250	2870	2850	1940	1920	1940	1910
Class II Outside diam. tool joint	in	35	5 1/4	5 25/32	5 5/8	5 15/16	5 3/8	5 11/32	5 11/32	5 23/32
	mm	36	133,3	146,8	142,9	150,8	136,5	135,7	135,7	145,3
Make-up torque	ft-lb	37	9500	13200	17300	17500	11400	11400	11900	11600
	m-daN	38	1290	1790	2350	2370	1550	1550	1610	1570
Class III Outside diam. tool joint	in	39	5 3/16	5 23/32	5 1/2	5 27/32	5 5/16	5 1/4	5 1/4	5 21/32
	mm	40	131,8	145,3	139,7	148,6	134,9	133,3	133,3	143,7
Make-up torque	ft-lb	41	8000	11600	14300	14900	10000	9400	9500	9900
	m-daN	42	1080	1570	1940	2020	1360	1270	1290	1340

(1) EU = external upset. IU = internal upset. IEU = internal external upset.

(2) The torsional yield strength is based on a shear strength of 57,7 % of the minimum yield strength.

(3) The torsional yield strengths for tool-joints are calculated on the pin and box. The smallest of the two torsional strengths thus obtained is the theoretical torsional strength of the tool joint.

PROPERTIES OF DRILL PIPES AND TOOL JOINTS (continued)
(API RP 7G)

1	4 1/2 - 20,00 #							5 - 19,50 #		
2	95				105		135	E	95	
3	IEU	IEU	IEU	EU	IEU	EU	EU	IEU	IEU	IEU
4	4 1/2 FH	4 1/2XH NC46	4 1/2 H90	4 1/2 IF NC50	4 1/2XH NC46	4 1/2IF NC50	4 1/2IF NC50	5XH NC50	5XH NC50	5 H90
5 / 6	6 / 152,4	6 1/4 / 158,8	6 / 152,4	6 3/8 / 161,9	6 1/4 / 158,8	6 3/8 / 161,9	6 5/8 / 168,3	6 3/8 / 161,9	6 3/8 / 161,9	6 1/2 / 165,1
7 / 8	2 1/2 / 63,5	2 3/4 / 69,9	3 / 76,2	3 1/2 / 88,9	2 1/2 / 63,5	3 1/2 / 88,9	2 7/8 / 73,0	3 3/4 / 95,2	3 1/2 / 88,9	3 1/4 / 82,6
9 / 10	5 23/32 / 145,3	5 23/32 / 145,3	5 3/4 / 146,0	5 59/64 / 150,4	5 23/32 / 145,3	5 59/64 / 150,4	6 3/8 / 161,9	5 59/64 / 150,4	5 59/64 / 150,4	6 1/8 / 155,6
11 / 12	21,69 / 32,28	22,33 / 33,23	21,73 / 32,34	23,11 / 34,39	22,45 / 33,41	23,11 / 34,39	23,49 / 34,96	20,99 / 31,24	21,09 / 31,39	21,57 / 32,10
13 / 14	0,531 / 6,60	0,531 / 6,60	0,544 / 6,76	0,544 / 6,76	0,531 / 6,60	0,544 / 6,76	0,544 / 6,76	0,738 / 9,16	0,738 / 9,16	0,738 / 9,16
15 / 16	0,331 / 4,11	0,341 / 4,23	0,332 / 4,12	0,353 / 4,38	0,343 / 4,26	0,353 / 4,38	0,358 / 4,45	0,320 / 3,98	0,322 / 4,00	0,329 / 4,09
17 / 18	0,862 / 10,71	0,872 / 10,83	0,876 / 10,88	0,897 / 11,14	0,874 / 10,86	0,897 / 11,14	0,902 / 11,21	1,058 / 13,14	1,060 / 13,16	1,067 / 13,25
19 / 20	522,3 / 232,3	522,3 / 232,3	522,3 / 232,3	522,3 / 232,3	577,3 / 256,8	577,3 / 256,8	742,2 / 330,1	395,6 / 176,0	501,1 / 222,9	501,1 / 222,9
21 / 22	1235,4 / 549,5	1183,9 / 526,6	976,2 / 434,2	1110,2 / 493,8	1307,6 / 581,6	1110,2 / 493,8	1506,2 / 670	944,0 / 419,9	1110,2 / 493,8	1269,2 / 564,6
23 / 24	46740 / 6340	46740 / 6340	46740 / 6340	46660 / 6330	51660 / 7000	46660 / 6330	66300 / 8990	41170 / 5580	52140 / 7070	52140 / 7070
25 / 26	44300 / 6000	44900 / 6090	45200 / 6130	44700 / 6060	49600 / 6720	44700 / 6060	61000 / 8270	37700 / 5100	44700 / 6060	51400 / 6970
27 / 28	6 / 152,4	6 1/4 / 158,8	6 / 152,4	6 3/8 / 161,9	6 1/4 / 158,8	6 3/8 / 161,9	6 5/8 / 168,3	6 3/8 / 161,9	6 3/8 / 161,9	6 1/2 / 165,1
29 / 30	22200 / 3010	22500 / 3050	22600 / 3060	22400 / 3040	24800 / 3360	22400 / 3040	30400 / 4120	18900 / 2560	22400 / 3040	26000 / 3525
31 / 32	5 5/8 / 142,9	5 21/32 / 143,7	5 9/16 / 141,3	5 15/16 / 150,8	5 23/32 / 145,3	6 / 152,4	6 7/32 / 158,0	5 7/8 / 149,2	6 1/32 / 153,2	5 25/32 / 146,8
33 / 34	17800 / 2410	18100 / 2450	17600 / 2390	17500 / 2370	19600 / 2660	19200 / 2600	25500 / 3460	15800 / 2140	20200 / 2740	19900 / 2700
35 / 36	5 15/32 / 138,9	5 1/2 / 139,7	5 7/16 / 138,1	5 27/32 / 148,8	5 9/16 / 141,3	5 7/8 / 149,2	6 1/16 / 154,0	5 25/32 / 146,8	5 29/32 / 150,0	5 23/32 / 145,3
37 / 38	14200 / 1925	14300 / 1940	14300 / 1940	14900 / 2020	15800 / 2140	15800 / 2140	21000 / 2850	13300 / 1800	16700 / 2260	16300 / 2210
39 / 40	5 3/8 / 136,5	5 13/32 / 137,3	5 3/8 / 136,5	5 3/4 / 146,0	5 15/32 / 138,9	5 25/32 / 146,8	5 15/16 / 150,8	5 11/16 / 144,5	5 13/16 / 147,6	5 5/8 / 142,9
41 / 42	12100 / 1640	12100 / 1640	12700 / 1720	12400 / 1680	13500 / 1830	13200 / 1790	17500 / 2370	10800 / 1460	14100 / 1910	13600 / 1840

(4) Calculations for recommended tool joint make-up torque are based on a tensile stress of 50 % of the minimum tensile yield for new joints and 60 % for used joints.

PROPERTIES OF DRILL PIPES AND TOOL JOINTS (continued)
(API RP 7G)

				5 · 19,50 #			5 · 25,60 #						
Nominal size			1	105	135		E		95		105		135
Grade			2	105	135		E		95		105		
Type (1)			3	IEU	IEU	IEU	IEU	IEU	IEU	IEU	IEU	IEU	
	Size and style		4	5XH NC50	5XH NC50	5 1/2 FH	5XH NC50	5 1/2 FH	5XH NC50	5 1/2 FH	5XH NC50	5 1/2 FH	5 1/4 FH
Tool joint data	Outside diameter	in	5	6 1/2	6 5/8	7 1/4	6 3/8	7	6 1/2	7	6 5/8	7 1/4	7 1/
		mm	6	165,1	168,3	184,2	161,9	177,8	165,1	177,8	168,3	184,2	184
	Inside diameter	in	7	3 1/4	2 3/4	3 1/2	3 1/2	3 1/2	3	3 1/2	2 3/4	3 1/2	3 1/
		mm	8	82,6	69,8	88,9	88,9	88,9	76,2	88,9	69,8	88,9	82,
	Bevel diameter	in	9	5 59/64	5 59/64	6 23/32	5 59/64	6 23/32	5 59/64	6 23/32	5 59/64	6 23/32	6 23/
		mm	10	150,4	150,4	170,7	150,4	170,7	150,4	170,7	150,4	170,7	170
Weight pipe + tool joint		lb/ft	11	21,50	22,09	23,24	27,01	28,50	28,30	28,62	28,11	28,99	29,2
		kg/m	12	31,99	32,87	34,58	40,20	42,41	42,11	42,59	41,83	43,14	43,5
Capacity		gal/ft	13	0,738	0,738	0,738	0,653	0,653	0,652	0,653	0,651	0,653	0,65
		l/m	14	9,16	9,16	9,16	8,11	8,11	8,10	8,11	8,09	8,11	8,11
open end displacement		gal/ft	15	0,329	0,337	0,355	0,412	0,435	0,432	0,437	0,429	0,443	0,44
		l/m	16	4,08	4,19	4,41	5,12	5,40	5,36	5,43	5,33	5,50	5,54
closed end displacement		gal/ft	17	1,067	1,075	1,093	1,065	1,088	1,084	1,090	1,080	1,096	1,09
		l/m	18	13,24	13,35	13,57	13,23	13,51	13,46	13,54	13,42	13,61	13,6
Tensile yield	Pipe	10³lb	19	553,8	712,1	712,1	530,2	530,2	671,5	671,5	742,2	742,2	954
		10³daN	20	246,3	316,8	316,8	235,8	235,8	298,7	298,7	330,1	330,1	424
	Tool joint	10³lb	21	1269,2	1552,0	1619,2	1110,2	1619,3	1416,5	1619,3	1552,0	1619,3	1778
		10³daN	22	564,6	690,4	720,3	493,8	720,3	630,1	720,3	690,4	720,3	791
Torsional yield	Pipe (2)	ft-lb	23	57600	74100	74100	52260	52260	66190	66190	73160	73160	9406
		m-daN	24	7810	10050	10050	7090	7090	8970	8970	9920	9920	1275
	Tool joint (3)	ft-lb	25	51400	63400	72500	44900	62200	57000	62200	63400	72500	7770
		m-daN	26	6970	8600	9830	6090	8430	7730	8430	8600	9830	1053
Class I	Outside diam. tool joint	in	27	6 1/2	6 5/8	7 1/4	6 3/8	7	6 1/2	7	6 5/8	7 1/4	7 1/
		mm	28	165,1	168,3	184,2	161,9	177,8	165,1	177,8	168,3	184,2	184
	Make-up torque	ft-lb	29	25800	31800	36300	22400	31500	28600	31500	31800	36300	3890
		m-daN	30	3500	4310	4920	3040	4270	3880	4270	4310	4920	527
Premium	Outside diam. tool joint	in	31	6 3/32	6 5/16	6 3/4	6 1/32	6 1/2	6 3/16	6 21/32	6 9/32	6 23/32	6 15/
		mm	32	154,8	160,3	171,4	153,2	165,1	157,2	169,1	159,5	170,7	176
	Make-up torque	ft-lb	33	21900	28400	28800	20200	20200	24700	25500	27500	27700	355
		m-daN	34	2970	3850	3900	2740	2740	3350	3460	3730	3760	48
Class II	Outside diam. tool joint	in	35	5 31/32	6 1/8	6 19/32	5 29/32	6 3/8	6 1/32	6 1/2	6 1/8	6 9/16	6 3/
		mm	36	151,6	155,6	167,5	150,0	161,9	153,2	165,1	155,6	166,7	171
	Make-up torque	ft-lb	37	18400	22900	23400	16700	16100	22000	20200	22900	22300	2880
		m-daN	38	2500	3100	3170	2270	2180	2980	2740	3100	3020	390
Class III	Outside diam. tool joint	in	39	5 25/32	6	6 15/32	5 25/32	6 5/16	5 15/16	6 13/32	6	6 15/32	6 3/
		mm	40	146,8	152,4	164,3	146,8	160,3	150,8	162,7	152,4	164,3	168
	Make-up torque	ft-lb	41	15000	19300	19200	13300	14100	17500	17200	19300	19200	2450
		m-daN	42	2030	2620	2600	1800	1910	2370	2330	2620	2600	332

(1) EU = external upset. IU = internal upset. IEU = internal external upset.

(2) The torsional yield strength is based on a shear strength of 57,7 % of the minimum yield strength.

(3) The torsional yield strengths for tool-joints are calculated on the pin and box. The smallest of the two torsional strengths thus obtained is the theoretical torsional strength of the tool joint.

PROPERTIES OF DRILL PIPES AND TOOL JOINTS (continued)
(API RP 7G)

1	5 1/2 - 21,90 #					5 1/2 - 24,70 #				6 5/8 - 25,20 #
2	E	95		105	135	E	95	105	135	E
3	IEU	IEU	IEU	IEU	IEU	IEU	IEU	IEU	IEU	IEU
4	5 1/2 FH	5 1/2 FH	5 1/2 H90	5 1/2 FH	5 1/2 FH	5 1/2 FH	5 1/2 FH	5 1/2 FH	5 1/2 FH	6 5/8 FH
7 / 6	7 / 177,8	7 / 177,8	7 / 177,8	7 1/4 / 184,1	7 1/2 / 190,5	7 / 177,8	7 1/4 / 184,1	7 1/4 / 184,1	7 1/2 / 190,5	8 / 203,2
7 / 8	4 / 101,6	3 3/4 / 95,2	3 1/2 / 88,9	3 1/2 / 88,9	3 / 76,2	4 / 101,6	3 1/2 / 88,9	3 1/2 / 88,9	3 / 76,2	5 / 127,0
9 / 10	6 23/32 / 170,7	6 23/32 / 170,7	6 3/4 / 171,5	6 23/32 / 170,7	7 3/32 / 183,4	6 23/32 / 170,7	6 23/32 / 170,7	6 23/32 / 170,7	7 3/32 / 183,4	7 3/4 / 196,5
11 / 12	23,94 / 35,63	24,24 / 36,07	25,00 / 37,21	24,98 / 37,17	25,86 / 38,48	26,86 / 39,97	27,89 / 41,50	27,89 / 41,50	28,74 / 42,78	27,14 / 40,39
13 / 14	0,914 / 11,35	0,914 / 11,35	0,914 / 11,35	0,914 / 11,35	0,914 / 11,35	0,872 / 10,83	0,872 / 10,83	0,872 / 10,83	0,872 / 10,83	1,425 / 17,70
15 / 16	0,366 / 4,54	0,370 / 4,59	0,382 / 4,74	0,382 / 4,74	0,395 / 4,90	0,410 / 5,09	0,426 / 5,29	0,426 / 5,29	0,439 / 5,45	0,415 / 5,15
17 / 18	1,280 / 15,89	1,284 / 15,94	1,296 / 16,09	1,296 / 16,09	1,309 / 16,25	1,282 / 15,92	1,298 / 16,12	1,298 / 16,12	1,311 / 16,28	1,840 / 22,95
19 / 20	437,1 / 194,4	553,7 / 246,3	553,7 / 246,3	612,0 / 272,2	786,8 / 350	497,2 / 221,2	629,8 / 280,1	696,1 / 310,0	895,0 / 398,1	489,5 / 217,7
21 / 22	1265,8 / 563,1	1448,4 / 644,3	1265,7 / 563,0	1619,3 / 720,3	1925,5 / 856,5	1265,8 / 563,1	1619,3 / 720,3	1619,3 / 720,3	1925,5 / 856,5	1448,8 / 644,5
23 / 24	50710 / 6875	64230 / 8710	64230 / 8710	70990 / 9625	91280 / 12380	56570 / 7670	71660 / 9720	79200 / 10740	101830 / 13810	70580 / 9570
25 / 26	56300 / 7630	62200 / 8430	59000 / 8000	72500 / 9830	86800 / 11770	56300 / 7630	72500 / 9830	72500 / 9830	86800 / 11770	74200 / 10060
27 / 28	7 / 177,8	7 / 177,8	7 / 177,8	7 1/4 / 184,1	7 1/2 / 190,5	7 / 177,8	7 1/4 / 184,1	7 1/4 / 184,1	7 1/2 / 190,5	8 / 203,2
29 / 30	28000 / 3800	31500 / 4270	30300 / 4110	36300 / 4920	43700 / 5920	28000 / 3800	36300 / 4920	36300 / 4920	43700 / 5920	38000 / 5150
31 / 32	6 15/32 / 164,3	6 5/8 / 168,3	6 3/16 / 157,2	6 23/32 / 170,7	6 15/16 / 176,2	6 17/32 / 165,9	6 23/32 / 170,7	6 25/32 / 172,2	7 1/32 / 178,6	
33 / 34	19200 / 2600	24500 / 3320	24500 / 3320	27700 / 3760	35500 / 4810	21300 / 2890	27700 / 3760	29900 / 4050	39000 / 5290	
35 / 36	6 3/8 / 161,9	6 1/2 / 165,1	6 1/16 / 154,0	6 9/16 / 166,7	6 3/4 / 171,4	6 7/16 / 163,5	6 9/16 / 166,7	6 5/8 / 168,3	6 27/32 / 173,8	
37 / 38	16200 / 2200	20300 / 2750	20400 / 2770	22400 / 3040	28800 / 3900	18200 / 2470	22400 / 3040	24500 / 2320	32100 / 4350	
39 / 40	6 9/32 / 159,5	6 13/32 / 162,7	5 31/32 / 151,6	6 15/32 / 164,3	6 5/8 / 168,3	6 11/32 / 161,1	6 15/32 / 164,3	6 17/32 / 165,9	6 11/32 / 161,1	
41 / 42	13200 / 1790	17200 / 2330	17400 / 2360	19200 / 2600	24500 / 3320	15200 / 2060	19200 / 2600	21300 / 2890	26600 / 3610	

(4) Calculations for recommended tool joint make-up torque are based on a tensile stress of 50 % of the minimum tensile yield for new joints and 60 % for used joints.

PROPERTIES OF DRILL PIPES AND TOOL JOINTS (continued)
(API RP 7G)

Nominal size		1												
Grade		2												
Type (1)		3												
Tool joint data	Size and style	4												
	Outside diameter	in	5												
		mm	6												
	Inside diameter	in	7												
		mm	8												
	Bevel diameter	in	9												
		mm	10												
Weight pipe + tool joint	lb/ft	11												
	kg/m	12												
Capacity	gal/ft	13												
	l/m	14												
open end displacement	gal/ft	15												
	l/m	16												
closed end displacement	gal/ft	17												
	l/m	18												
Tensile yield	Pipe	10^3lb	19												
		10^3da	20												
	Tool joint	10^3lb	21												
		10^3daN	22												
Torsional yield	Pipe (2)	ft-lb	23												
		m-daN	24												
	Tool joint (3)	ft-lb	25												
		m-daN	26												
Class I	Outside diam. tool joint	in	27												
		mm	28												
	Make-up torque	ft-lb	29												
		m-daN	30												
Premium	Outside diam. tool joint	in	31												
		mm	32												
	Make-up torque	ft-lb	33												
		m-daN	34												
Class II	Outside diam. tool joint	in	35												
		mm	36												
	Make-up torque	ft-lb	37												
		m-daN	38												
Class III	Outside diam. tool joint	in	39												
		mm	40												
	Make-up torque	ft-lb	41												
		m-daN	42												

(1) EU = external upset. IU = internal upset. IEU = internal external upset.

(2) The torsional yield strength is based on a shear strength of 57,7 % of the minimum yield strength.

(3) The torsional yield strengths for tool-joints are calculated on the pin and box. The smallest of the two torsional strengths thus obtained is the theoretical torsional strength of the tool joint.

DIMENSIONAL DATA OF NEW AND USED DRILL PIPES (API RP 7G)

Nominal outside diameter (in)	Nominal weight (lb/ft)	CLASS I (new)				PREMIUM CLASS			CLASS II			CLASS III		
		Inside diameter (mm)	Outside diameter (mm)	Wall thickness (mm)	Cross section area (mm²)	Outside diameter (1)	Wall thickness (mm)	Cross section area (4) (mm²)	Outside diameter (1) (mm)	Wall thickness (mm)	Cross section area (4) (mm²)	Outside diameter (mm)	Wall thickness (mm¹)	Cross section area (mm²)
2 3/8	4.85	50.7	60.3	4.83	841	58.4	3.86	662	58.4	3.15	662	56.7	2.67	508
2 3/8	6.65	46.1	60.3	7.11	1189	57.5	5.69	926	57.5	4.62	926	55.0	3.91	706
2 7/8	6.85	62.0	73.0	5.51	1169	70.8	4.42	920	70.8	3.58	920	68.9	3.02	708
2 7/8	10.40	54.6	73.0	9.19	1844	69.3	7.67	1433	69.3	5.97	1433	66.1	5.05	1090
3 1/2	9.50	76.0	88.9	6.45	1671	86.3	5.16	1316	86.3	4.19	1316	84.1	3.56	1014
3 1/2	13.30	70.1	88.9	9.35	2336	85.2	7.47	1825	85.2	6.07	1825	81.9	5.13	1395
3 1/2	15.50	66.1	88.9	11.40	2777	84.3	9.12	2156	84.3	7.42	2156	80.3	6.27	1639
4	11.85	88.3	101.6	6.65	1985	98.9	5.33	1566	98.9	4.32	1566	96.6	3.66	1208
4	14.00	84.8	101.6	8.38	2455	98.3	6.10	1944	98.3	4.95	1944	95.3	4.19	1483
4 1/2	13.75	100.5	114.3	6.88	2323	111.5	5.51	1835	111.5	4.47	1835	109.1	3.78	1417
4 1/2	16.60	97.2	114.3	8.56	2844	110.9	6.86	2237	110.9	5.56	2237	107.9	4.70	1723
4 1/2	20.00	92.5	114.3	10.92	3547	109.9	8.74	2778	109.9	7.11	2778	106.1	6.02	2129
5	16.25	112.0	127.0	7.52	2812	124.0	6.02	2230	124.0	4.88	2230	121.1	4.14	1723
5	19.50	108.6	127.0	9.19	3403	123.3	7.37	2679	123.3	5.97	2679	120.1	5.05	2065
5	25.60	101.6	127.0	12.70	4560	121.9	10.16	3567	121.9	8.26	3567	117.5	6.99	2732
5 1/2	21.90	121.4	139.7	9.17	3760	136.0	7.34	2967	136.0	5.97	2967	132.8	5.05	2289
5 1/2	24.70	118.6	139.7	10.54	4277	135.5	8.43	3366	135.5	6.86	3366	131.8	5.79	2592

(1) Outside diameters for Premium Class and Class II are the result of a remaining wall thickness of 80 % of nominal wall thickness due to uniform wear.
(2) Minimum wall thickness of 65 % of nominal wall thickness due to eccentric wear.
(3) Minimum wall thickness of 55 % of nominal wall thickness due to eccentric wear.
(4) Cross section areas for Premium and Class II are calculated with a wall 80 % of nominal wall thickness.

HEAVY-WALL DRILL PIPES (Drilco hevi-wate)
(1977 Drilco Drilling Assembly handbook)

	Nominal size		3 1/2" -26 #	4" - 28 #	4 1/2" -42 #	5" - 50 #	4 1/2" - 42 #	5" - 50 #
	Range		II	II	II	II	III	III
Drill pipe body	Inside diameter	mm in	52,4 2 1/16	65,1 2 9/16	69,8 2 3/4	76,2 3	69,8 2 3/4	76,2 3
	Wall thickness	mm in	18,2 0,719	18,2 0,719	22,2 0,875	25,4 1,000	22,2 0,875	25,4 1,000
	Cross-sect. area	mm² in²	4051 6,280	4779 7,409	6427 9,965	8106 12,566	6427 9,965	8106 12,566
	End upset	mm in	92,1 3 5/8	104,8 4 1/8	117,5 4 5/8	130,2 5 1/8	117,5 4 5/8	130,2 5 1/8
	Central upset	mm in	101,6 4	114,3 4 1/2	127,0 5	139,7 5 1/2	127,0 5	139,7 5 1/2
Tool joint	Style		NC38 (3 1/2IF)	NC40 (4FH)	NC46 (4IF)	NC50 (4 1/2IF)	NC46 (4IF)	NC50 (4 1/2IF)
	Outside diameter	mm in	120,6 4 3/4	133,3 5 1/4	158,7 6 1/4	165,1 6 1/2	158,7 6 1/4	165,1 6 1/2
	inside diameter	mm in	55,6 2 3/16	68,3 2 11/16	73,0 2 7/8	77,8 3 1/16	73,0 2 7/8	77,8 3 1/16
Weight : pipe + tool joint		kg/m lb/ft	37,7 25,3	44,2 29,7	61,0 41,0	73,5 49,3	59,4 39,9	72,2 48,5
Capacity		l/m gal/ft	2,19 0,177	3,37 0,271	3,87 0,312	4,61 0,371	3,87 0,312	4,61 0,371
Open end displacement		l/m gal/ft	4,81 0,387	5,64 0,454	7,79 0,627	9,36 0,754	7,57 0,610	9,20 0,741
Closed end displacement		l/m gal/ft	7,00 0,564	9,01 0,725	11,66 0,939	13,97 1,125	11,44 0,922	13,81 1,112
Tensile yield	Pipe	10³daN 10³lb	153 345	181 408	244 548	307 691	244 548	307 691
	Tool joint	10³daN 10³lb	333 749	316 711	456 1025	563 1266	456 1025	563 1266
Torsional yield	Pipe	m-daN ft-lb	2654 19575	3747 27635	5520 40715	7660 56495	5520 40715	7660 56495
	Tool joint	m-daN ft-lb	2383 17575	3190 23525	5260 38800	6966 51375	5260 38800	6966 51375
Make-up torque		m-daN ft-lb	1342 9900	1797 13250	2956 21800	3985 29400	2956 21800	3985 29400

Note : Hevi-wate drill pipes have extra length tool joints (box 21'', 53,3 cm - pin 27'', 68,6 cm) and a central wear pad of 24'', 61 cm in length for range II drill pipes and 2 central wear pads of 34'', 86,4 cm in length each for range III drill pipes.

CYLINDRICAL DRILL COLLARS (1)
(API Spec 7 and Drilco)
Dimensions. Connections. Weight. Make-up torque
(30 or 31 ft lengths)

Drill collar number	Diameter			Optional connections		Weight (kg/m)	Make-up torque (3)	
	Outside diameter (in)	Inside diameter (in)	Bevel diameter (mm)	Type	Bevel diameter (mm)		minimum (m-daN)	recommended (m-daN)
NC 23-31 *	3 1/8	1 1/4	76,2	—	—	32,7	445	490
NC 26-35	3 1/2	1 1/2	82,9	2 3/8 IF	82,9	45,0	625	690
NC 31-41	4 1/8	1 1/2	100,4	2 7/8 IF	100,4	51,6	900	990
NC 35-47	4 3/4 (2)	2 1/4	114,7	NC 38-(3 1/2 XH)	116,3	69,7	1250	1350
NC 38-50	5	2 1/4	121,0	—	—	79,3	1750	1900
NC 40-55 +	5 1/2	2 1/4	132,2	—	—	100,2	2300	2550
NC 40-57 +	5 3/4	2 1/4	136,9	4 H 90-(4 FH)	139,7	113,3	2300	2550
NC 44-57 +	5 3/4	2 1/4	139,7	4 1/2 FH	140,5	113,3	2800	3100
NC 44-60	6	2 1/4	144,5	4 H 90-(4 1/2 FH)	139,7	122,9	3150	3500
NC 44-60	6	2 13/16	144,5	NC 46-(4 IF)	145,3	111,6	2450	2700
NC 44-62	6 1/4	2 1/4	149,2	4 1/2 H 90-(NC 46)	152,4	134,7	3150	3500
NC 46-62	6 1/4	2 13/16	150,0	4 1/2 H 90	152,4	123,6	3000	3300
NC 46-65	6 1/2	2 1/4	154,8	5 H 90	155,6	148,1	3800	4200
NC 46-65	6 1/2	2 13/16	154,8	5 H 90	155,6	136,4	3000	3300
5 H 90	6 5/8	2 13/16	155,6	—	—	142,9	4000	4400
NC 46-67	6 3/4 (2)	2 1/4	159,5	5 H 90	161,9	161,5	3800	4200
NC 50-67 +	6 3/4 (2)	2 13/16	159,9	5 H 90-(5 1/2 REG)	161,9	149,4	4350	4800
NC 50-70	7	2 1/4	164,7	5 1/2 H 90-(5 1/2 REG)	168,3	174,9	5150	5650
NC 50-70	7	2 13/16	164,7	5 1/2 H 90-(5 1/2 REG)	168,3	163,2	4350	4800
NC 50-72	7 1/4	2 13/16	169,5	5 1/2 H 90-(5 1/2 REG)	168,3	177,2	4350	4800
NC 56-75 +	7 1/2	2 13/16	180,6	—	—	191,9	6500	7150
NC 56-75 +	7 1/2	3	180,6	—	—	187,5	6100	6700
NC 56-77	7 3/4 (2)	2 13/16	185,3	6 5/8 REG-(5 1/2 FH)	186,1	207,0	6500	7150
NC 56-77 +	7 3/4	3	185,3	6 5/8 REG-(5 1/2 FH)	186,1	202,4	6100	6700
NC 56-80	8	2 13/16	190,1	6 5/8 REG-(6 5/8 H 90)	190,9	222,7	6500	7150
NC 56-80 +	8	3	190,1	6 5/8 REG-(6 5/8 H 90)	190,9	218,8	6100	6700
6 5/8 REG	8 1/4	2 13/16	195,7	6 5/8 H 90	190,5	238,7	7200	7900
NC 61-82 +	8 1/4	3	198,4	6 5/8 H 90	190,5	235,1	8700	9550
NC 61-85 +	8 1/2	2 13/16	203,2	—	—	255,5	9200	10100
NC 61-90	9	2 13/16	212,7	7 5/8 REG	214,3	290,2	9200	10100
NC 61-90 +	9	3	212,7	7 5/8 REG	214,3	285,7	8800	9700
7 5/8 REG	9 1/2 (2)	3	223,8	7 5/8 H 90	234,9	322,9	11950	13150
NC 70-97	9 3/4	3	232,6	7 5/8 H 90	322,9	342,3	14250	15650
NC 70-100	10	3	237,3	7 5/8 H 90-(7 5/8 REG)	244,5	361,6	14250	15650
NC 77-110	11	3	260,7	—	—	445,0	19400	21350
—	11 1/4	3	—	8 5/8 H 90-(8 5/8 REG)	273,0	467,3	17350 (5)	19100
— (4)	14	3	—	8 5/8 H 90-(8 5/8 REG)	273,0	742,4	17350 (5)	19100

+ Non API.

* The drill collar number consists of two parts separated by a hyphen. The first part is the connection number in the NC style. The second part, consisting of 2 or 3 digits, indicates the drill collar outside diameter in units and tenths of inch.

(1) Basis of calculations for recommended tool joint make-up torque assumed the use of a thread compound containing 40-60 % by weight of finely powdered metallic zinc or 60 % by weight of finely powdered metallic lead.
(2) Diameters recommended by the French Technical Committee.
(3) Make-up torques are for connections given in the first column.
(4) 14'' Drill collars have a restricted diameter of 11 1/4'' at each end.
(5) Make-up torques for low-torque connections.

SQUARE DRILL COLLARS

Hole size (in)	Square size A (in)	Rib width C (in)	Across corners D (in)	Fishing neck and tong space E (in)	Standard bore B (in)	Connection		Make-up torque (m-daN)		Total weight for 30 ft (kg)
						recommended	optional	recommended	maximum	
4 1/4	3 3/4	5/8	4 11/16	3 1/2	1 1/2	2 3/8 IF	NC 26	625	690	526
5 5/8	4 1/2	3/4	5 9/16	4 1/2	1 3/4	NC 35	2 7/8 IF	1200	1300	772
6 1/8	5	1 1/8	6 1/16	5	2	3 1/2 FH	3 1/2 XH	1600	1750	915
6 1/4	5	1	6 3/16	5	2	3 1/2 FH	3 1/2 XH	1600	1750	935
6 1/2	5 1/2	1 1/2	6 7/16	5 1/2	2	4 FH	NC 40	2500	2750	1090
6 5/8	5 1/2	1 3/8	6 9/16	5 1/2	2	4 FH	NC 40	2500	2750	1148
6 3/4	5 1/2	1 1/4	6 11/16	5 1/2	2	4 FH	NC 40	2500	2750	1172
7 7/8	6 1/2	1 1/2	7 11/16	6 1/2	2 1/4	4 1/2 IF	NC 50	3800	4200	1685
8 1/2	7	1 1/2	8 7/16	7	2 1/4	4 1/2 IF	NC 50	3800	4200	1960
8 5/8	7	1 1/2	8 9/16	7	2 1/4	4 1/2 IF	NC 50	3800	4200	1980
8 3/4	7	1 1/2	8 11/16	7	2 1/4	4 1/2 IF	NC 50	3800	4200	2020
9	7	1 3/8	8 15/16	7	2 1/4	4 1/2 IF	NC 50	3800	4200	2040
9 5/8	8	1 7/8	9 9/16	8	2 13/16	6 5/8 REG	6 5/8 H 90	7200	7900	2570
9 7/8	8	1 5/8	9 13/16	8	2 13/16	6 5/8 REG	6 5/8 H 90	7200	7900	2605
10 5/8	9	2 3/8	10 9/16	9	2 13/16	7 5/8 REG	6 5/8 FH	11250	12350	3140
11	9	2	10 15/16	9	2 13/16	7 5/8 REG	6 5/8 FH	11250	12350	3180
12 1/4	10	2 1/4	12 3/16	9 3/4	2 13/16	7 5/8 H 90	NC 70	12350 (1)	13600	3865
13 3/4	11	2 1/4	13 11/16	9 3/4	2 13/16	7 5/8 H 90	NC 70	12350 (1)	13600	5340
15	12	2 3/16	14 15/16	9 3/4	2 13/16	7 5/8 H 90	NC 70	12350 (1)	13600	6260
17 1/2	14	2 1/2	17 7/16	9 3/4	2 13/16	7 5/8 H 90	NC 70	12350 (1)	13600	7420

(1) Low-torque connections.

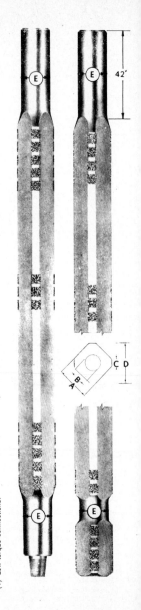

SPIRAL DRILL COLLARS (No-Wall-Stick)
(Hycalog patent - Drilco 1976)

Cross section	Size OD (in)	Depth of cut e (mm)	Number of spirals	Direction	Pitch (mm)	Length of cylindrical end — pin end (mm) mini	pin end maxi	box end (mm) mini	box end maxi
Cross section for drill collars 3 7/8 to 7'' OD	3 7/8''	1.98 ± 0.79	3	to the right	914.4 ± 25.4				
	4'' to 4 3/8''	4.76 ± 0.79	3	to the right	914.4 ± 25.4				
	4 1/2'' to 5 1/8''	5.56 ± 0.79	3	to the right	965.2 ± 25.4	304.8	457.2	457.2	609.6
	5 1/4'' to 5 3/4''	6.35 ± 0.79	3	to the right	1066.8 ± 25.4				
	5 7/8'' to 6 3/8''	7.14 ± 1.59	3	to the right	1066.8 ± 25.4				
	6 1/2'' to 7''	7.94 ± 1.59	3	to the right	1168.4 ± 25.4				
	7''	7.94 ± 1.59	3	to the right	1625.6 ± 25.4				
Cross section for drill collars 7 1/8 to 12'' OD	7 1/8 to 7 7/8''	8.73 ± 1.59	3	to the right	1625.6 ± 25.4				
	8'' to 8 7/8''	9.53 ± 1.59	3	to the right	1727.2 ± 25.4				
	9'' to 9 7/8''	10.32 ± 2.38	3	to the right	1828.8 ± 25.4	304.8	457.2	457.2	609.6
	10'' to 10 7/8''	11.11 ± 2.38	3	to the right	1930.4 ± 25.4				
	11'' to 12''	11.91 ± 2.38	3	to the right	2032.0 ± 25.4				

Note : The weight of a round drill-collar will be reduced approximately 4 % by No-Wall-Stick Conversion.

STRESS-RELIEF GROOVE
FOR DRILL COLLAR CONNECTIONS
(API Spec 7)

Number or Size and Style of connection	Length, shoulder face to groove of box member L_X (1)		Diameter of pin member at groove D_{RG} (2)	
	(mm)	(in)	(mm)	(in)
NC 35	85,7	3 3/8	82,15	3 15/64
NC 38 - 3 1/2 IF	92,1	3 5/8	89,3	3 33/64
NC 40 - 4 FH	104,8	4 1/8	96,0	3 25/32
NC 44	104,8	4 1/8	106,4	4 3/16
NC 46 - 4 IF	104,8	4 1/8	109,9	4 21/64
NC 50 - 4 1/2 IF	104,8	4 1/8	120,6	4 3/4
NC 56	117,5	4 5/8	134,5	5 19/64
NC 61	130,2	5 1/8	148,8	5 55/64
NC 70	142,9	5 5/8	171,1	6 47/64
NC 77	155,6	6 1/8	188,5	7 27/64
4 1/2 FH	92,1	3 5/8	106,8	4 13/64
5 1/2 REG	111,2	4 3/8	123,4	4 55/64
6 5/8 REG	117,5	4 5/8	137,5	5 27/64
7 5/8 REG	123,8	4 7/8	162,7	6 13/32

(1) Tol (+ 0, − 1/8 in) (+ 0, − 3,2 mm),
(2) Tol (+ 0, − 1/32 in) (+ 0, − 0,8 mm).

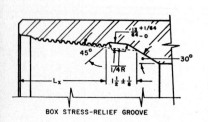

BOX STRESS-RELIEF GROOVE

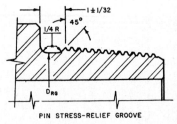

PIN STRESS-RELIEF GROOVE

SHOULDER MODIFICATIONS
FOR LOW-TORQUE CONNECTIONS

7 5/8 REG Section   7 5/8 H 90 Section

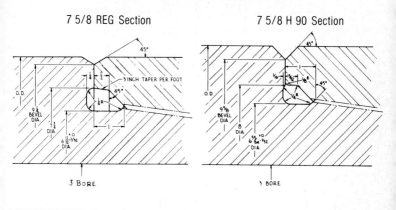

RECOMMENDED MAKE UP TORQUE : 62,000 FT. LBS.   RECOMMENDED MAKE UP TORQUE : 73,000 FT. LBS.

DIMENSIONS OF LOW-TORQUE SHOULDERS

Connection size and style	Outside diameter (in)	Bevel diameter L		Inside diameter of shoulder E		Width of flat H N : normal R : modified		
		(in)	(mm)	(in)	(mm)	(in)	(mm)	N or R
7 H 90	8 1/4	8	203,2	6 9/16	166,7	23/32	18,25	N
	8 1/2	8 1/4	209,5	6 9/16	166,7	27/32	21,40	N
	8 3/4	8 1/2	215,9	7 1/8	181,0	11/16	17,45	R
	9	8 5/8	219,1	7 1/8	181,0	3/4	19,05	R
7 5 8 H 90	9 1/2	9 1/4	234,9	7 29/64	189,6	57/64	22,60	N
	9 3/4	9 1/4	234,9	8	203,2	5/8	15,90	R
	10	9 5/8	244,5	8	203,2	13/16	20,60	R
	10 1/4	9 5/8	244,5	8	203,2	13/16	20,60	R
7 5 8 REG	9 1/2	8 7/8	225,4	7 3/32	180,2	57/64	22,60	N
	9 3/4	9 1/4	234,9	7 3/4	196,8	3/4	19,05	R
	10	9 1/4	234,9	7 3/4	196,8	3/4	19,05	R
8 5 8 H 90	10 1/2	10 3/8	263,5	8 11/32	211,9	1 1/64	25,8	N
	10 3/4	10 1/2	266,7	9 3/8	238,1	9/16	14,3	R
	11	10 1/2	266,7	9 3/8	238,1	9/16	14,3	R
	11 1/4	10 3/4	273,0	9 3/8	238,1	11/16	17,5	R
8 5 8 REG	10 1/2	9 3/4	247,7	8 1/16	204,8	27/32	21,4	N
	10 3/4	10 1/2	266,7	9	228,6	3/4	19,05	R
	11	10 1/2	266,7	9	228,6	3/4	19,05	R

DRILL COLLAR GROOVE AND ELEVATOR BORE DIMENSIONS (API RP 7G)

	Groove dimensions based on drill collar OD								Elevator bore based on drill collar OD				
	Elevator groove					Slip groove				Top bore +0 / −1/32" +0 / −0.8 mm		Bottom bore +1/16" / −0 +1.6 mm / −0	
Drill collar OD range (in)	Depth A (1)		R		C (2)	Depth B (1)		D (2)		OD minus		OD plus	
	mm	(in)	(mm)	(in)	(°)	(mm)	(in)	(°)		(mm)	(in)	(mm)	(in)
4 to 4 5/8	5.6	7/32	3.2	1/8	4	4.8	3/16	3 1/2		7.9	5/16	3.2	1/8
4 3/4 to 5 5/8	6.4	1/4	3.2	1/8	5	4.8	3/16	3 1/2		9.5	3/8	3.2	1/8
5 3/4 to 6 5/8	7.9	5/16	3.2	1/8	6	6.4	1/4	5		12.7	1/2	3.2	1/8
6 3/4 to 8 5/8	9.5	3/8	4.8	3/16	7 1/2	6.4	1/4	5		14.3	9/16	3.2	1/8
8 3/4 and up	11.1	7/16	6.4	1/4	9	6.4	1/4	5		15.9	5/8	3.2	1/8

(1) A and B dimensions are from nominal OD of new drill collar.
(2) Angle C and D dimensions are reference and approximate.

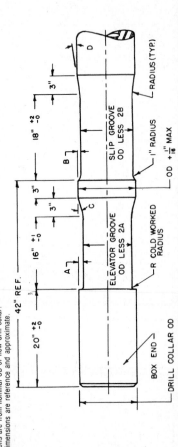

DRILL COLLAR WEIGHT (kilogrammes per metre)

Outside diameter (in)	Inside diameter (in)														
	1 1/4	1 1/2	1 3/4	2	2 1/4	2 1/2	2 3/4	2 13/16	2 7/8	3	3 1/4	3 1/2	3 3/4	4	4 1/4
3 1/2	42.5	45.0													
3 3/4	49.7	47.06	43.8												
4 1/8	61.3	58.6	55.5	51.6											
4 1/4	65.5	62.8	59.5	55.8											
4 1/2	74.3	71.6	68.3	64.6	60.4	55.7									
4 5/8	78.7	76.0	72.8	69.1	64.9	60.1									
4 3/4	83.5	80.8	77.5	73.8	69.7	64.9									
5	93.2	90.5	87.2	83.5	79.3	74.6	69.4								
5 1/4	103.3	100.6	97.3	93.6	89.4	84.7	79.5	78.0	76.4	73.8					
5 1/2	114.0	111.3	108.0	104.3	100.2	95.4	90.2	88.7	87.4	84.5					
5 3/4	125.2	122.5	119.2	115.5	111.3	106.6	101.0	99.8	98.5	95.7	89.4				
6	136.8	134.1	130.8	127.1	122.9	118.2	113.0	111.6	110.1	107.3	101.1	94.4	87.06		
6 1/4	148.5	145.8	142.6	138.9	134.7	129.9	124.7	123.6	121.9	119.5	112.8	106.1	98.8		
6 3/8	155.1	152.4	149.2	145.4	141.1	136.5	131.3	129.9	128.5	125.6	119.4	112.7	105.5		
6 1/2	161.9	159.2	156.0	152.2	148.1	143.3	138.1	136.4	135.3	132.5	126.2	119.5	112.2		
6 5/8	168.1	165.4	162.2	158.4	154.2	149.5	144.3	142.9	141.5	138.6	132.4	125.7	118.5		
6 3/4	175.3	172.6	169.4	165.6	161.5	156.7	151.5	149.4	148.7	145.8	139.6	132.9	125.6	118.0	
7	188.7	186.0	182.8	179.0	174.9	170.1	164.9	163.2	162.1	159.2	153.0	146.3	139.0	131.4	
7 1/4	202.1	199.4	196.1	192.4	188.3	183.5	178.3	177.2	175.5	172.6	166.4	159.7	152.4	144.8	
7 1/2	217.0	214.3	211.0	207.3	203.1	198.4	193.2	191.9	190.3	187.5	181.3	174.6	167.3	159.7	
7 3/4	231.9	229.2	225.9	222.2	218.0	213.3	208.0	207.0	205.2	202.4	196.1	189.4	182.2	174.6	
8	248.2	245.5	242.3	238.6	234.4	229.6	224.4	222.7	221.6	218.8	212.9	205.8	198.5	190.9	182.8
8 1/4	264.6	261.9	258.6	254.9	250.8	246.0	240.8	238.7	238.0	235.1	228.9	222.2	214.9	207.3	199.1
8 1/2		278.3	275.0	271.3	267.1	262.4	257.2	255.5	254.3	251.5	245.3	238.6	231.3	223.7	215.5
8 3/4		296.1	292.9	289.2	285.0	280.2	275.0	272.9	272.0	269.4	263.1	256.4	249.1	241.5	233.3
9			309.2	305.5	301.4	296.6	291.4	290.2	288.6	285.7	279.5	272.8	265.5	257.9	249.6
9 1/4			328.6	324.9	320.7	316.0	310.7	308.2	307.9	305.1	298.8	292.1	284.8	277.2	269.7
9 1/2			346.4	342.7	338.6	333.8	328.6	326.9	325.8	322.9	316.7	310.0	302.7	295.1	286.9
9 3/4			365.8	367.1	357.9	353.1	347.9	345.9	345.0	342.3	336.0	329.3	322.0	314.5	306.3
10					377.3	372.5	367.3	365.6	364.5	361.6	355.4	348.7	341.6	333.8	325.6
10 1/2					417.4	412.7	407.5	408.0	404.6	401.8	395.6	388.9	381.6	374.0	365.8
10 3/4					439.8	435.0	429.8	427.3	426.9	424.1	417.9	411.2	403.9	396.3	388.1
11						455.8	450.6	449.0	447.8	445.0	438.7	432.0	424.7	417.1	409.0
11 1/4						478.2	472.9	470.9	470.1	467.3	461.0	454.3	447.0	439.5	431.3
12						548.1	542.9	540.3	540.1	537.2	531.0	524.3	517.0	509.4	501.2
14							748.1	746.8	745.3	742.4	736.2	729.5	722.4	714.6	706.5

DRILL COLLAR CONNECTION SELECTION CHART
API numbered connections
(Tool Pusher's Manual)

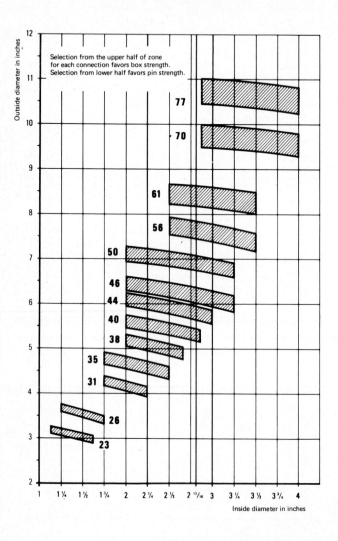

Selection from the upper half of zone
for each connection favors box strength.
Selection from lower half favors pin strength.

Outside diameter in inches

Inside diameter in inches

DRILL COLLAR CONNECTION SELECTION CHART
Hughes H 90

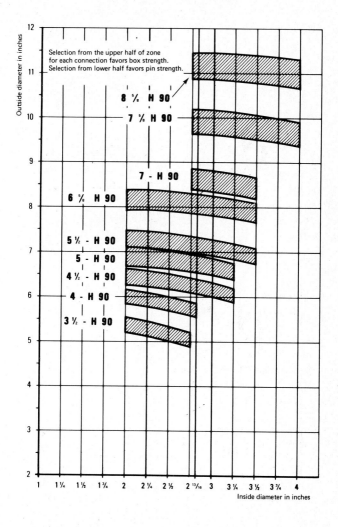

Outside diameter in inches

Selection from the upper half of zone
for each connection favors box strength.
Selection from lower half favors pin strength.

8 ⅝ H 90

7 ⅝ H 90

7 - H 90

6 ⅝ H 90

5 ½ - H 90

5 - H 90

4 ½ - H 90

4 - H 90

3 ½ - H 90

Inside diameter in inches

DRILL COLLAR CONNECTION SELECTION CHART
Regular. Full Hole

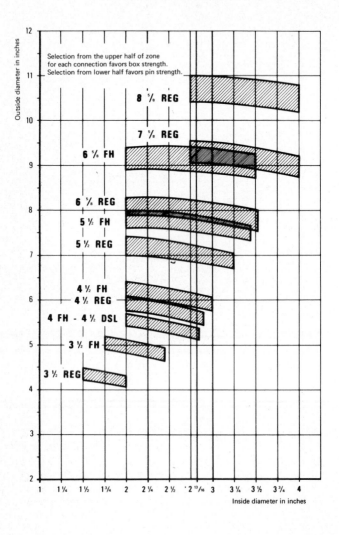

Selection from the upper half of zone
for each connection favors box strength.
Selection from lower half favors pin strength.

Outside diameter in inches

8 ⅝ REG

7 ⅝ REG

6 ⅝ FH

6 ⅝ REG

5 ½ FH

5 ½ REG

4 ½ FH
4 ½ REG

4 FH - 4 ½ DSL

3 ½ FH

3 ½ REG

Inside diameter in inches

DRILL COLLAR CONNECTION SELECTION CHART
Internal Flush. Xtra Hole. Slim-Hole.
Double Streamline

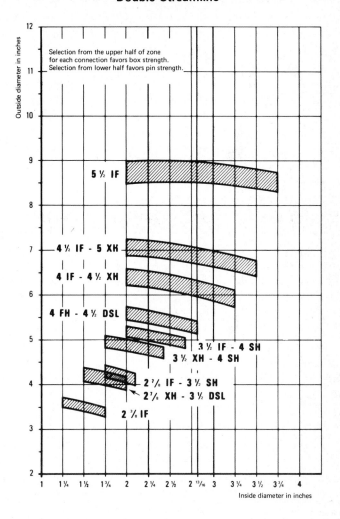

Selection from the upper half of zone
for each connection favors box strength.
Selection from lower half favors pin strength.

Outside diameter in inches

5 ½ IF

4 ½ IF - 5 XH

4 IF - 4 ½ XH

4 FH - 4 ½ DSL

3 ½ IF - 4 SH

3 ½ XH - 4 SH

2 ⅞ IF - 3 ½ SH

2 ⅞ XH - 3 ½ DSL

2 ⅞ IF

Inside diameter in inches

DRILL COLLARS • RECOMMENDED MAKE-UP TORQUE
(API RP 7G)

Outside diameter (in)	BORE OF DRILL COLLAR (in) MINIMUM AND RECOMMENDED MAKE-UP TORQUE (m-daN) (1)						Weaker member (2)
	API NC 23						
	1	1 1/4	1 1/2	1 3/4	2	2 1/4	F
3 1/8	445-490	445-490	350-385				M
3 1/4	540-	460-505	350-385				M
3 3/8	540-	460-505	350-385				M
	API NC 26 - 2 3/8 IF - 2 7/8 SH						
	1	1 1/4	1 1/2	1 3/4	2	2 1/4	
3 1/2		625-690	625-690	500-550			M
3 3/4		750-825	640-825	500-550			M
	API NC 31 - 2 7/8 IF - 3 1/2 SH						
	1 1/4	1 1/2	1 3/4	2	2 1/4	2 1/2	
3 7/8	625-690	625-690	625-690	625-690			F
4 1/8	990-1090	990-1090	990-1090	900-990			M
4 1/4	1200-1320	1200-1320	1100-1200	900-990			M
4 1/2	1350-1480	1250-1370	1100-1200	900-990			M
	2 7/8 XH - 2 7/8 EH - 3 1/2 DSL - 2 7/8 MO						
	1 1/4	1 1/2	1 3/4	2	2 1/4	2 1/2	
3 3/4	550-600	550-600	550-600				F
3 7/8	720-790	720-790	720-790				F
4 1/8	1080-1200	1080-1200	1000-1100				M
	API NC 35						
	1 1/4	1 1/2	1 3/4	2	2 1/4	2 1/2	F
4 1/2			1200-1300	1200-1300	1200-1300	1000-1100	M
4 3/4			1650-1800	1450-1600	1250-1350	1000-1100	M
	3 1/2 XH - 3 1/2 EH - 4 SH - 3 1/2 MO						
	1 1/4	1 1/2	1 3/4	2	2 1/4	2 1/2	
4 1/4			690-750	690-750	690-750	690-750	F
4 1/2			1150-1250	1150-1250	1150-1250	1100-1200	M
4 3/4			1600-1750	1600-1750	1350-1500	1100-1200	M
5			1800-2000	1600-1750	1350-1500	1100-1200	M
5 1/4			1800-2000	1600-1750	1350-1500	1100-1200	M
	API NC 38 - 3 1/2 IF - 3 1/2 WO - 4 1/2 SH						
	1 3/4	2	2 1/4	2 1/2	2 13/16	3	
4 3/4	1350-1500	1350-1500	1350-1500	1350-1500	1125-1250		M
5	1850-2050	1850-2050	1750-1900	1500-1650	1125-1250		M
5 1/4	2150-2350	2000-2200	1750-1900	1500-1650	1125-1250		M
5 1/2	2150-2350	2000-2200	1750-1900	1500-1650	1125-1250		M

(1) First digit minimum, second digit recommended make-up torque.
(2) Notation in this column indicates cross section area (3/4'' from base on pin (M) or 3/8'' from shoulder on box (F) is smaller in the member indicated.

DRILL COLLARS • RECOMMENDED MAKE-UP TORQUE (continued) (API RP 7G)

3 1/2 H 90

Outside diameter (in)	1 3/4	2	2 1/4	2 1/2	2 13/16	3	Weaker member (2)
4 3/4	1200-1300	1200-1300	1200-1300	1200-1300	1200-1300		F
5	1700-1850	1700-1850	1700-1850	1700-1850	1400-1550		M
5 1/4	2300-2550	2200-2400	2050-2250	1750-1900	1400-1550		M
5 1/2	2500-2750	2200-2400	2050-2250	1750-1900	1400-1550		M

API NC 40 - 4 FH - 4 1/2 DSL - 4 MO

Outside diameter (in)	1 3/4	2	2 1/4	2 1/2	2 13/16	3	Weaker member (2)
5	1450-1600	1450-1600	1450-1600	1450-1600	1450-1600		F
5 1/4	2050-2250	2050-2250	2050-2250	2000-2200	1650-1800		M
5 1/2	2650-2900	2500-2750	2300-2550	2000-2200	1650-1800		M
5 3/4	2750-3000	2500-2750	2300-2550	2000-2200	1650-1800		M
6	2750-3000	2500-2750	2300-2550	2000-2200	1650-1800		M

4 H 90

Outside diameter (in)	1 3/4	2	2 1/4	2 1/2	2 13/16	3	Weaker member (2)
5 1/4		1700-1850	1700-1850	1700-1850	1700-1850		F
5 1/2		2350-2600	2350-2600	2350-2600	2250-2500		M
5 3/4		3000-3300	2900-3200	2650-2900	2250-2500		M
6		3200-3500	2900-3200	2650-2900	2250-2500		M
6 1/4		3200-3500	2900-3200	2650-2900	2250-2500		M

4 1/2 REG

Outside diameter (in)	1 3/4	2	2 1/4	2 1/2	2 13/16	3	Weaker member (2)
5 1/2		2100-2300	2100-2300	2100-2300	2100-2300		F
5 3/4		2750-3000	2750-3000	2650-2900	2200-2400		M
6		3150-3500	2950-3250	2650-2900	2200-2400		M
6 1 4		3150-3500	2950-3250	2650-2900	2200-2400		M

API NC 44

Outside diameter (in)	1 3/4	2	2 1/4	2 1/2	2 13/16	3	Weaker member (2)
5 3/4		2800-3100	2800-3100	2800-3100	2450-2700		M
6		3400-3750	3150-3500	2850-3150	2450-2700		M
6 1/4		3400-3750	3150-3500	2850-3150	2450-2700		M
6 1/2		3400-3750	3150-3500	2850-3150	2450-2700		M

4 1/2 FH

Outside diameter (in)	1 3/4	2	2 1/4	2 1/2	2 13/16	3	Weaker member (2)
5 1/2		1750-1900	1750-1900	1750-1900	1750-1900	1750-1900	F
5 3/4		2450-2700	2450-2700	2450-2700	2450-2700	2400-2650	M
6		3150-3500	3150-3500	3100-3450	2700-2950	2400-2650	M
6 1/4		3650-4000	3400-3750	3100-3450	2700-2950	2400-2650	M
6 1/2		3650-4000	3400-3750	3100-3450	2700-2950	2400-2650	M

For notes(1) and (2), see page B 40.

DRILL COLLARS • RECOMMENDED MAKE-UP TORQUE (continued)
(API RP 7G)

API NC 46-4 1/2 XH - 4 1/2 EH - 4 IF - 4 1/2 SIF - 4″ WO - 4 1/2 MO - 5 DSL

Outside diameter (in) / BORE OF DRILL COLLAR (in) MINIMUM AND RECOMMENDED MAKE-UP TORQUE (m-daN) (1) / Weaker member (2)

Outside diameter (in)	1 3/4	2	2 1/4	2 1/2	2 13/16	3	Weaker member (2)
5 3/4			2400-2650	2400-2650	2400-2650	2400-2650	F
6			3150-3500	3150-3500	3000-3300	2750-3000	M
6 1/4			3800-4200	3450-3800	3000-3300	2750-3000	M
6 1/2			3800-4200	3450-3800	3000-3300	2750-3000	M
6 3/4			3800-4200	3450-3800	3000-3300	2750-3000	M

4 1/2 H 90

Outside diameter (in)	1 3/4	2	2 1/4	2 1/2	2 13/16	3	Weaker member (2)
5 3/4			2400-2650	2400-2650	2400-2650	2400-2650	F
6			3150-3500	3150-3500	3100-3400	2850-3150	M
6 1/4			3850-4250	3550-3900	3100-3400	2850-3150	M
6 1/2			3850-4250	3550-3900	3100-3400	2850-3150	M
6 3/4			3850-4250	3550-3900	3100-3400	2850-3150	M

5 H 90

Outside diameter (in)	1 3/4	2	2 1/4	2 1/2	2 13/16	3	Weaker member (2)
6 1/4			3400-3750	3400-3750	3400-3750	3400-3750	F
6 1/2			4250-4700	4250-4700	4000-4400	3650-4000	M
6 3/4			4750-5200	4450-4900	4000-4400	3650-4000	M
7			4750-5200	4450-4900	4000-4400	3650-4000	M
7 1/4			4750-5200	4450-4900	4000-4400	3650-4000	M

5 1/2 REG

Outside diameter (in)	1 3/4	2	2 1/4	2 1/2	2 13/16	3	Weaker member (2)
6 3/4			4250-4700	4250-4700	4250-4700	4250-4700	F
7			5300-5850	5300-5850	4900-5400	4550-5000	M
7 1/4			5700-6250	5350-5900	4900-5400	4550-5000	M
7 1/2			5700-6250	5350-5900	4900-5400	4550-5000	M

5 1/2 H90

Outside diameter (in)	1 3/4	2	2 1/4	2 1/2	2 13/16	3	Weaker member (2)
6 3/4			4600-5050	4600-5050	4600-5050	4600-5050	M
7			5650-6200	5400-5950	4950-5450	4600-5050	M
7 1/4			5750-6300	5400-5950	4950-5450	4600-5050	M
7 1/2			5750-6300	5400-5950	4950-5450	4600-5050	M

API NC 50 - 4 1/2 IF - 5 XH - 5 EH - 5 SIF - 5 1/2 DSL - 4 1/2 WO - 5 MO

Outside diameter (in)	2	2 1/4	2 1/2	2 13/16	3	3 1/4	Weaker member (2)
6 1/4		3100-3400	3100-3400	3100-3400	3100-3400	3100-3400	F
6 1/2		4000-4400	4000-4400	4000-4400	4000-4400	3600-3950	M
6 3/4		4900-5400	4800-5300	4350-4800	4050-4450	3600-3950	M
7		5150-5650	4800-5300	4350-4800	4050-4450	3600-3950	M
7 1/4		5150-5650	4800-5300	4350-4800	4050-4450	3600-3950	M

For notes (1) and (2), see page B 40.

DRILL COLLARS • RECOMMENDED MAKE-UP TORQUE (continued)
(API RP 7G)

Outside diameter (in)	BORE OF DRILL COLLAR (in) MINIMUM AND RECOMMENDED MAKE-UP TORQUE (m-daN) (1)						Weaker member (2)
	5 1/2 FH						
	2 1/2	2 13/16	3	3 1/4	3 1/2	3 3/4	
7	4400-4850	4400-4850	4400-4850	4400-4850			F
7 1/4	5500-6050	5500-6050	5500-6050	5500-6050			F
7 1/2	6650-7300	6350-7000	6100-6700	5650-6200			M
7 3/4	6900-7600	6350-7000	6100-6700	5650-6200			M
	API NC 56						
	2 1/2	2 13/16	3	3 1/4	3 1/2	3 3/4	
7 1/4	5400-5950	5400-5950	5400-5950	5400-5950			F
7 1/2	6550-7200	6500-7150	6100-6700	5700-6250			M
7 3/4	6900-7600	6500-7150	6100-6700	5700-6250			M
8	6900-7600	6500-7150	6100-6700	5700-6250			M
	6 5/8 REG						
	2 1/2	2 13/16	3	3 1/4	3 1/2	3 3/4	
7 1/2	6250-6900	6250-6900	6250-6900	6250-6900			F
7 3/4	7450-8200	7200-7900	6800-7500	6350-7000			M
8	7750-8500	7200-7900	6800-7500	6350-7000			M
8 1/4	7750-8500	7200-7900	6800-7500	6350-7000			M
	6 5/8 H 90						
	2 1/2	2 13/16	3	3 1/4	3 1/2	3 3/4	
7 1/2	6250-6900	6250-6900	6250-6900	6250-6900			F
7 3/4	7450-8200	7450-8200	7200-7900	6700-7250			M
8	8000-8800	7600-8350	7200-7900	6700-7250			M
8 1/4	8000-8800	7600-8350	7200-7900	6700-7250			M
	API NC 61						
	2 1/2	2 13/16	3	3 1/4	3 1/2	3 3/4	
8	7300-8050	7300-8050	7300-8050	7300-8050			F
8 1/4	8700-9550	8700-9550	8700-9550	8300-9150			M
8 1/2	9750-10700	9200-10100	8800-9700	8300-9150			M
8 3/4	9750-10700	9200-10100	8800-9700	8300-9150			M
9	9750-10700	9200-10100	8800-9700	8300-9150			M
	5 1/2 IF						
	2 1/2	2 13/16	3	3 1/4	3 1/2	3 3/4	
8	7600-8350	7600-8350	7600-8350	7600-8350	7600-8350		F
8 1/4	8950-9850	8950-9850	8950-9850	8550-9400	8000-8800		M
8 1/2	10050-11050	9500-10450	9100-10000	8550-9400	8000-8800		M
8 3/4	10050-11050	9500-10450	9100-10000	8550-9400	8000-8800		M
9	10050-11050	9500-10450	9100-10000	8550-9400	8000-8800		M
9 1/4	10050-11050	9500-10450	9100-10000	8550-9400	8000-8800		M

For notes (1) and (2), see page B 40.

DRILL COLLARS • RECOMMENDED MAKE-UP TORQUE (continued)
(API RP 7G)

Outside diameter (in)	BORE OF DRILL COLLAR (in) MINIMUM AND RECOMMENDED MAKE-UP TORQUE (m-daN) (1)						Weaker member (2)
	6 5/8 FH						
	2 13/16	3	3 1/4	3 1/2	3 3/4	4	
8 1/2	9100-10000	9100-10000	9100-10000	9100-10000	9000-9900		M
8 3/4	10600-11650	10600-11650	10300-11350	9750-10700	9000-9900		M
9	11250-12350	10850-11950	10300-11350	9750-10700	9000-9900		M
9 1/4	11250-12350	10850-11950	10300-11350	9570-10700	9000-9900		M
9 1/2	11250-12350	10850-11950	10300-11350	9750-10700	9000-9900		M
	API NC 70						
	2 13/16	3	3 1/4	3 1/2	3 3/4	4	
9 1/4	11950-12050	11950-12050	11950-12050	11950-12050	11950-12050		F
9 1/2	13700-15100	13700-15100	13550-14900	12900-14200	12200-13400		M
9 3/4	14500-16000	14250-15650	13550-14900	12900-14200	12200-13400		M
10	14500-16000	14250-15650	13550-14900	12900-14200	12200-13400		M
	API NC 77						
	2 13/16	3	3 1/4	3 1/2	3 3/4	4	
10		14500-15950	14500-15950	14500-15950	14500-15950		F
10 1/4		16550-18200	16550-18200	16550-18200	16550-18200		F
10 1/2		18700-20600	18700-20600	18050-19850	17350-19100		M
10 3/4		19400-21350	18700-20600	18050-19850	17350-19100		M
11		19400-21350	18700-20600	18050-19850	17350-19100		M
11 1/4		19400-21350	18700-20600	18050-19850	17350-19100		M
	STANDARD SHOULDERS						
	7 H 90						
	2 13/16	3	3 1/4	3 1/2	3 3/4	4	
8	7200-7900	7200-7900	7200-7900	7200-7900			F
8 1/4	8550-9400	8550-9400	8550-9400	8200-9000			M
8 1/2 (4)	9700-10650	9300-10250	8800-9700	8200-9000			M
	7 5/8 REG						
	2 13/16	3	3 1/4	3 1/2	3 3/4	4	
8 1/2	8150-8950	8150-8950	8150-8950	8150-8950	8150-8950		F
8 3/4	9650-10600	9650-10600	9650-10600	9650-10600	9650-10600		F
9	11250-12350	11250-12350	11250-12350	10700-11900	10050-11050		M
9 1/4		11950-13150	11250-12350	10700-11900	10050-11050		M
9 1/2 (4)		11950-13150	11250-12350	10700-11900	10050-11050		M
	7 5/8 H 90						
	2 13/16	3	3 1/4	3 1/2	3 3/4	4	
9	9750-10700	9750-10700	9750-10700	9750-10700	9750-10700		F
9 1/4	11600-12750	11600-12750	11600-12750	11600-12750	11600-12750		F
9 1/2 (4)	13300-14650	13300-14650	13300-14650	13300-14650	12900-14200		M

For notes (1) and (2), see page B 40.

DRILL COLLARS • RECOMMENDED MAKE-UP TORQUE (continued)
(API RP 7G)

Outside diameter (in)	BORE OF DRILL COLLAR (in) MINIMUM AND RECOMMENDED MAKE-UP TORQUE (m-daN) (1)						Wearker member (2)
				8 5/8 REG			
	2 13/16	3	3 1/4	3 1/2	3 3/4	4	
10		14650-16100	14650-16100	14650-16100	14650-16100		F
10 1/4		16700-18350	16700-18350	16700-18350	16700-18350		M
10 1/2 (4)		18850-20750	18150-20750	17500-20750	16700-18350		M
				8 5/8 H 90			
	2 13/16	3	3 1/4	3 1/2	3 3/4	4	
10 1/4		15250-16750	15250-16750	15250-16750	15250-16750		F
10 1/2 (4)		17400-19150	17400-19150	17400-19150	17400-19150		F
			LOW TORQUE SHOULDERS				
				7 H 90			
	2 13/16	3	3 1/4	3 1/2	3 3/4	4	
8 3/4	9150-10050	9150-10050	9050-9950	8400-9250			M
9	10050-11050	9650-10600	9050-9950	8400-9250			M
				7 5/8 REG			
	2 13/16	3	3 1/4	3 1/2	3 3/4	4	
9 1/4		9750-10700	9750-10700	9750-10700	9750-10700		F
9 1/2		11500-12650	11500-12650	11100-12200	10450-11500		M
9 3/4		12350-13600	11800-13000	11100-12200	10450-11500		M
10		12350-13600	11800-13000	11100-12200	10450-11500		M
				7 5/8 H 90			
	2 13/16	3	3 1/4	3 1/2	3 3/4	4	
9 3/4		12350-13600	12350-13600	12350-13600	12350-13600		F
10		14250-15650	14250-15650	14000-15500	13300-14650		M
10 1/4		15250-16750	14650-16100	14000-15500	13300-14650		M
10 1/2		15250-16750	14650-16100	14000-15500	13300-14650		M
				8 5/8 REG			
	2 13/16	3	3 1/4	3 1/2	3 3/4	4	
10 3/4		15200-16700	15200-16700	15200-16700	15200-16700		F
11		17500-19250	17500-19250	17500-19250	17500-19250		F
				8 5/8 H 90			
	2 13/16	3	3 1/4	3 1/2	3 3/4	4	
10 3/4		12550-14800	12550-14800	12550-14800	12550-14800		F
11		14900-16400	14900-16400	14900-16400	14900-16400		F
11 1/4		17350-19100	17350-19100	17350-19100	17350-19100		F
14 (6)		19100-21000	19100-21000	19100-21000	19100-21000		F

For notes (1) and (2), see page B 40.

KELLYS (API Spec 7)

SQUARE KELLYS — LH BOX CONNECTION (UPPER UPSET), LOWER UPSET, R H PIN CONNECTION; CORNER CONFIGURATION MANUFACTURER'S OPTION

HEXAGONAL KELLYS — LH BOX CONNECTION (UPPER UPSET), LOWER UPSET, R H PIN CONNECTION; CORNER CONFIGURATION MANUFACTURER'S OPTION

SQUARE KELLYS

Kelly size (1) G (in)	Inside diameter E (in)	LENGTH Overall (A) Std	Overall (A) Opt	Drive section (B) Std	Drive section (B) Opt	UPPER BOX Size and style Std	Size and style Opt	Outside dia. (C) Std	(C) Opt	LOWER PIN Size and style Std	Size and style Opt	Outside dia. (D) Std	(D) Opt	across flats G (in)	across corners F (in)	TOTAL WEIGHT Std	Opt
2 1/2	1 1/4	12.19	—	11.28	—	6 5/8 REG	4 1/2 REG	7 3/4	5 3/4	NC 26	2 3/8 IF	3 3/8	—	2 1/2	3 9/32	404	358
3 1/2	1 3/4	12.19	—	11.28	—	6 5/8 REG	4 1/2 REG	7 3/4	5 3/4	NC 31	2 7/8 IF	4 1/8	—	3	3 15/16	490	445
4 1/4	2 1/4	12.19	—	11.28	—	6 5/8 REG	4 1/2 REG	7 3/4	5 3/4	NC 38	3 1/2 IF	4 3/4	—	3 1/2	4 17/32	600	552
4 1/4	2 13/16	12.19	16.46	11.28	15.54	6 5/8 REG	—	7 3/4	5 3/4	NC 46	4 IF	6 1/8	—	4 1/4	5 9/16	840	788
5 1/4	3 1/4	12.19	16.46	11.28	15.54	6 5/8 REG	—	7 3/4	—	NC 56	5 1/2 FH	6 3/8	—	5 1/4	6 29/32	1260	—

HEXAGONAL KELLYS

Kelly size (1) G (in)	Inside diameter E (in)	LENGTH Overall (A) Std	Overall (A) Opt	Drive section (B) Std	Drive section (B) Opt	UPPER BOX Size and style Std	Size and style Opt	Outside dia. (C) Std	(C) Opt	LOWER PIN Size and style Std	Size and style Opt	Outside dia. (D) Std	(D) Opt	across flats G (in)	across corners F (in)	TOTAL WEIGHT Std	Opt
3	1 1/2	12.19	—	11.28	—	6 5/8 REG	4 1/2 REG	7 3/4	5 3/4	NC 26	2 3/8 IF	3 3/8	—	3	3 3/8	440	395
3 1/2	1 3/4	12.19	—	11.28	—	6 5/8 REG	4 1/2 REG	7 3/4	5 3/4	NC 31	2 7/8 IF	4 1/8	—	3 1/2	3 15/16	567	532
4 1/4	2 1/4	12.19	16.46	11.28	15.54	6 5/8 REG	—	7 3/4	5 3/4	NC 38	3 1/2 IF	4 3/4	—	4 1/4	4 25/32	886	840
5 1/4	3 1/4	12.19	16.46	11.28	15.54	6 5/8 REG	—	7 3/4	—	NC 46	4 IF	6	—	5 1/4	5 29/32	990	—
5 1/4	3 1/4	12.19	16.46	11.28	15.54	6 5/8 REG	—	7 3/4	—	NC 50	4 1/2 IF	6 1/8	—	5 1/4	5 29/32	965	—
5 1/4	2 13/16	12.19	16.46	11.28	15.54	6 5/8 REG	—	7 3/4	—	NC 46	4 IF	6 1/8	—	5 1/4	5 29/32	1007	—
5 1/4	2 13/16	12.19	16.46	11.28	15.54	6 5/8 REG	—	7 3/4	—	NC 50	4 1/2 IF	6 3/8	—	5 1/4	5 29/32	1007	—
6	3 1/2	12.19	16.46	11.28	15.54	6 5/8 REG	—	7 3/4	—	NC 56	5 1/2 FH	7	—	6	6 13/16	1095	—

(1) Size of square or hexagonal kellys is the same as G (across flats).

STRENGTH OF KELLYS (1) (API RP 7G)

Kelly size and style (in)	Inside diameter (in)	Lower pin connection Size and style	Outside diameter (in)	Minimum recommended casing OD (2) (in)	Tensile yield lower pin connection (3) (10³ daN)	drive section (10³ daN)	Torsional yield lower pin connection (m-daN)	drive section (m-daN)	Yield in bending through corners drive section (m-daN)	through faces drive section (m-daN)
SQUARE										
2 1/2	1 1/4	NC 26 (2 3/8 IF)	3 3/8	4 1/2	185	242	1310	2060	2045	3000
3	1 3/4	NC 31 (2 7/8 IF)	4 1/8	5 1/2	238	317	1960	3268	3010	4935
3 1/2	2 1/4	NC 38 (3 1/2 IF)	4 3/4	6 5/8	322	394	3080	4800	4895	7500
4 1/4	2 13/16	NC 46 (4 IF)	6 1/4	8 5/8	468	582	5335	8350	8540	13190
4 1/4	2 13/16	NC 50 (4 1/2 IF)	6 3/8	8 5/8	632	570	7760	8530	8730	13370
5 1/4	3 1/4	5 1/2 FH	7	9 5/8	715	925	9900	16750	17040	25780
HEXAGONAL										
3	1 1/2	NC 26 (2 3/8 IF)	3 3/8	4 1/2	158	293	1125	3390	3620	3080
3 1/2	1 7/8	NC 31 (2 7/8 IF)	4 1/8	5 1/2	220	385	1815	5220	5600	4745
4 1/4	2 1/4	NC 38 (3 1/2 IF)	4 3/4	6 5/8	322	570	3080	9430	10070	8530
5 1/4	3	NC 46 (4 IF)	6 1/4	8 5/8	426	820	4805	16950	18290	15410
5 1/4	3 1/4	NC 50 (4 1/2 IF)	6 3/8	8 5/8	512	758	6335	15930	17390	14520
6	3 1/2	5 1/2 FH	7	9 5/8	650	1052	8990	24950	26990	22690

(1) All values have no safety factor and are based on 110 000 psi minimum tensile yield strength and a shear strength of 57.7 % of the minimum tensile yield strength.
(2) Clearance between protector rubber on kelly saver sub and casing inside diameter should also be checked.
(3) Tensile area calculated at root of thread 3/4 inch from pin shoulder.

ROTARY DRILLING HOSE (API Spec 7)

DIMENSIONS - TEST PRESSURES - WORKING PRESSURES

Inside diameter (in)	Standard length (1) (m)	(ft)	Line pipe thread size (in)	Test Grade A bar	psi	Grade B bar	psi	Grade C bar	psi	Grade D bar	psi	Working Grade A bar	psi	Grade B bar	psi	Grade C bar	psi	Grade D bar	psi
2	10.7	35	2 1/2	207	3000	275	4000					103	1500	138	2000				
	12.2	40	2 1/2	207	3000	275	4000					103	1500	138	2000				
2 1/2	15.3	50	3	207	3000	275	4000	515	7500			103	1500	138	2000	275	4000		
	16.8	55	3	207	3000	275	4000	515	7500			103	1500	138	2000	275	4000		
3	16.8	55	4					515	7500	690	10000					275	4000	345	5000
	18.3	60	4					515	7500	690	10000					275	4000	345	5000
	21.4	70	4					515	7500	690	10000					275	4000	345	5000
	22.9	75	4					515	7500	690	10000					275	4000	345	5000
3 1/2	16.8	55	4					515	7500	690	10000					275	4000	345	5000
	18.3	60	4					515	7500	690	10000					275	4000	345	5000
	21.4	70	4					515	7500	690	10000					275	4000	345	5000
	22.9	75	4					515	7500	690	10000					275	4000	345	5000

(1) Non standard lengths in 5 ft (1.50 m) increments may be marked with API monogram provided the hose meets all other requirements of this specification.

HOSE LENGTH :

$$L = \frac{L_t}{2} + \pi R + S$$

STAND PIPE HEIGHT :

$$H_s = \frac{L_t}{2} + Z$$

with :

L : length of hose in feet or metres,
L_t : length of hose travel in feet or metres,

R : minimum bending radius of hose in feet or metres,
R = 0.9 m (3 ft) for 2" hose,
R = 1.2 m (4 ft) for 2 1/2" cm and 3" hose,
R = 1.4 m (4 1/2 ft) for 3 1/2" hose,
S : allowance for contraction in L due to maximum recommended working pressure in feet or metres, which is 0.3 m (1 ft) for all sizes of hose.

H_s : vertical height of stand pipe in feet or metres,
L_t : length of hose travel in feet or metres,
Z : height, in ft or m, from the top of the derrick floor to the end of hose at the swivel when the swivel is in its lowest drilling position.

ROTARY TABLE OPENING And SQUARE DRIVE MASTER BUSHING (API Spec 7)

Nominal table size (in)	ROTARY-TABLE OPENING								SQUARE DRIVE MASTER BUSHING									
	A		B		C		D max.		A_1		B_1		C_1		D_1		Concentricity TIR	
	cm	in	cm	in	cm	in	cm	in	cm	in	cm	in	cm	in	cm	in	mm	in
17 1/2	44.45	17 1/2	46.20	18 3/16	13.33	5 1/4	4.445	1 3/4	44.29	17 7/16	46.04	18 1/8	13.33	5 1/4	4.445	1 3/4	0.794	1/32
20 1/2	52.07	20 1/2	53.82	21 3/16	13.33	5 1/4	4.445	1 3/4	51.91	20 7/16	53.66	21 1/8	13.33	5 1/4	4.445	1 3/4	0.794	1/32
27 1/2	69.85	27 1/2	71.60	28 3/16	13.33	5 1/4	4.445	1 3/4	69.69	27 7/16	71.28	28 1/16	13.33	5 1/4	4.445	1 3/4	0.794	1/32
37 1/2	95.25	37 1/2	—	—	—	—	—	—	95.08	37 7/16	—	—	—	—	—	—	—	—
49 1/2	124.73	49 1/2	—	—	—	—	—	—	—	—	—	—	—	—	—	—	—	—

Note : Pipe slips and master bushing must have a taper of 4 in per foot on diameter (33,33 %), that is an angle of 9°27'45''

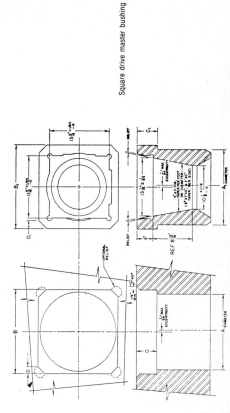

Rotary-table opening

Square drive master bushing

FOUR-PIN DRIVE KELLY BUSHING AND MASTER BUSHING
(API Spec 7)

Nominal table size (in)	F ± 1/16 in (± 1,6 mm)		G ± 0,005 in (± 0,13 mm)		H		I ± 0,005 in (± 0,13 mm)		J +1/16 -0 [+ 1,6 - 0]		K + 1/16 -0 [+ 1,6 - 0]	
	mm	in	mm	in	mm	in	mm	in	mm	in	mm	in
17 1/2	482,6	19	65,2	2,565	107,9	4 1/4	62,8	2,472	365,1	14 3/8	257,2	10 1/8
20 1/2	584,2	23	65,2	2,565	107,9	4 1/4	62,8	2,472	412,7	16 1/4	311,1	12 1/4
27 1/2	654,2	25 3/4	86,2	3,395	107,9	4 1/4	82,9	3,265	482,6	19	381,0	15
37 1/2	654,1	25 3/4	86,2	3,395	107,9	4 1/4	82,9	3,265	—	—	—	—
49 1/2	—	—	—	—	—	—	—	—	—	—	—	—

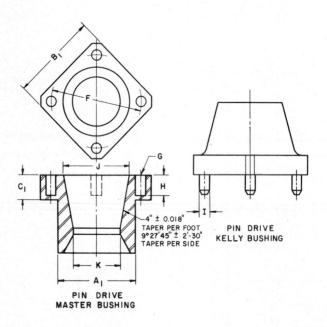

PIN DRIVE
MASTER BUSHING

PIN DRIVE
KELLY BUSHING

C PHYSICAL PROPERTIES OF CASING, TUBING, LINE PIPE

Summary

TENSILE REQUIREMENTS
I. Casing and tubing (API Stds 5A, 5AC, 5AX)

TENSILE PROPERTIES	UNITS	Grades										
		H40	J55	K55 (2)	C75 (1)	N80	C95 (1)(2)	O95 (1)(4)	P105 (3)	P110	Q125 (4)	V150 (4)
Color band indentification (5)		1 black	1 green	2 green	1 blue	1 red	1 brown		1 white	1 white		
Minimum yield strength	hbar	27,5	37,9	37,9	51,7	55,1	65,5	65,5	72,4	75,8	86,1	103,4
	psi	40 000	55 000	55 000	75 000	80 000	95 000	95 000	105 000	110 000	125 000	150 000
Maximum yield strength	hbar		55,1	55,1	62,0	75,8	75,8	86,1	93,0	96,5	106,8	
	psi		80 000	80 000	90 000	110 000	110 000	125 000	135 000	140 000	155 000	
Minimum tensile strength	hbar	41,3	51,7	65,5	65,5	68,9	72,4	75,8	82,7	86,1	93,0	110,3
	psi	60 000	75 000	95 000	95 000	100 000	105 000	110 000	120 000	125 000	135 000	160 000
Average yield strength	hbar	34,4	44,8	51,7	58,6	62,0	69,0	72,4	82,7	86,1	93,0	110,3
	psi	50 000	65 000	75 000	85 000	90 000	100 000	105 000	120 000	125 000	135 000	160 000
Elongation min per cent (API)		29,5	24	19,5	19,5	18,5	18	17	16	15	13	11

(1) Special corrosion   (2) For casing only   (3) For tubing only   (4) Non API grades.

(5) Special clearance couplings shall be painted the color indicative of the steel grade from which the couplings are manufactured and shall also be painted with a black band around the centre.

II. Line pipe (API Stds 5L, 5LX)

PROPERTIES	GRADES	A	B	X42	X46	X52	X56*	X60*	X65*	X70
Minimum yield strength	hbar	20,7	24,1	28,9	31,7	35,9	38,5	41,4	44,8	48,3
	psi	30 000	35 000	42 000	46 000	52 000	56 000	60 000	65 000	70 000
Minimum tensile strength	hbar	33,1	41,4	41,4	43,5	45,5	49,0	51,7	53,1	56,5
	psi	48 000	60 000	60 000	63 000	66 000	71 000	75 000	77 000	82 000
Elongation (%)		35/21	30/18	25/17,5	23/13	22/10				

* For grades X 52, X 56, X 60, X 65 pipes 20 in outside diameter and larger with wall thickness 9,52 mm (0,375 in) and less the minimum tensile strength is respectively 49,6 hbars (72 000 psi), 51,7 (75 000 psi), 53,8 hbars (78 000 psi) and 55,1 hbars (80 000 psi).

API CASINGS LIST
(API Stds 5A, 5AC, 5AX)

Nominal outside diam. (in)	Nominal weight (lb/ft)	Wall thickness (mm)	H	J	K	75	N	95	P	short	long	buttress	X-line
4 1/2	9,50	5,21	●	●	●					x			
	10,50	5,69		●	●					x			
	11,60	6,35		●	●					x		x	
	11,60	6,35				●	●	●	●		x	x	
	13,50	7,37					●	●	●		x	x	
	15,10	8,56							●			x	
5	11,50	5,59		●	●					x			
	13,00	6,43		●	●					x		x	
	15,00	7,52		●	●					x		x	x
	15,00	7,52				●	●	●	●		x	x	x
	18,00	9,19				●	●	●	●		x	x	x
5 1/2	14,00	6,20	●	●	●					x			
	15,50	6,98		●	●					x		x	x
	17,00	7,72		●	●					x		x	x
	17,00	7,72				●	●	●	●			x	x
	20,00	9,17				●	●	●	●			x	x
	23,00	10,54				●	●	●	●			x	x
6 5/8	20,00	7,32	●							x			
	20,00	7,32		●	●					x			
	24,00	8,94		●	●					x		x	x
	24,00	8,94				●	●	●	●			x	x
	28,00	10,59				●	●	●	●			x	x
	32,00	12,06				●	●	●	●			x	x
7	17,00	5,87	●							x			
	20,00	6,91	●	●	●					x			
	23,00	8,05		●	●					x		x	x
	23,00	8,05				●	●	●				x	x
	26,00	9,19		●	●					x		x	x
	26,00	9,19				●	●	●	●			x	x
	29,00	10,36				●	●	●	●			x	x
	32,00	11,51				●	●	●	●			x	x
	35,00	12,65				●	●	●	●			x	x
	38,00	13,72				●	●	●	●			x	x
7 5/8	24,00	7,62	●							x			
	26,40	8,33		●	●					x			
	26,40	8,33				●	●	●				x	x
	29,70	9,52				●	●	●	●			x	x
	33,70	10,92				●	●	●	●			x	x
	39,00	12,70				●	●	●	●			x	x

Nominal outside diam. (in)	Nominal weight (lb/ft)	Wall thickness (mm)	H	J	K	75	N	95	P	short	long	buttress	X-line
8 5/8	24,00	6,71		●	●					x			
	28,00	7,72		●	●					x			
	32,00	8,94	●							x			
	32,00	18,94		●	●					x		x	x
	36,00	10,16		●	●					x		x	x
	36,00	10,16				●	●	●	●			x	x
	40,00	11,43				●	●	●	●			x	x
	44,00	12,70				●	●	●	●			x	x
	49,00	14,15					●	●	●			x	x
9 5/8	32,30	7,92	●							x			
	36,00	8,94	●							x			
	36,00	8,94		●	●					x		x	x
	40,00	10,03		●	●					x		x	x
	40,00	10,03				●	●	●	●			x	x
	43,50	11,05				●	●	●	●			x	x
	47,00	11,99				●	●	●	●			x	x
	53,50	13,84				●	●	●	●			x	x
10 3/4	32,75	7,09	●							x			
	40,50	8,89	●							x			
	40,50	8,89		●	●					x		x	
	45,50	10,16		●	●					x		x	x
	51,00	11,43				●	●	●	●		x	x	x
	55,50	12,57						●	●		x	x	x
	60,70	13,84							●		x	x	x
	65,70	15,11							●		x	x	
11 3/4	42,00	8,46	●							x			
	47,00	9,52		●	●					x		x	
	54,00	11,05		●	●					x		x	
	60,00	12,42				●	●	●	●		x		x	
13 3/8	48,00	8,38	●							x			
	54,40	9,65		●	●					x			
	61,00	10,92		●	●					x		x	
	68,00	12,19		●	●					x		x	
	72,00	13,06				●	●	●	●		x		x	
16	65,00	9,52	●							x			
	75,00	11,13		●	●					x		x	
	84,00	12,57		●	●					x		x	
18 5/8	87,50	11,05	●	●	●					x			
	87,50	11,05	●	●	●					x			
20	94,00	11,13	●	●	●					x	x		x
	94,00	11,13		●	●							x	
	106,50	12,70		●	●					x	x	x	
	133,00	16,13		●	●					x	x	x	

	4 1/2 - (114,3)												5 - (127,0)					
2	**16,90***				17,70*				18,80*				11,50		13,00			
3	9.65				10.20				10.92				5.59		6.43			
4	95,00				93,90				92,46				115,8		114,1			
5	3203				3335				3572				2132		2434			
6	7.10				6.95				6.73				10.53		10.22			
7	10.30				10.30				10.30				12.69		12.69			
8	24.92				26.20				27.86				16.73		19.11			
9	C95*	P110*	Q125*	V150*	C95*	P110*	Q125*	V150*	C95*	P110*	Q125*	V150*	K55	C75*	K55	C75*	N80*	C95*	P110*
10	1013	1173	1333	1578	1065	1235	1402	1682	1132	1311	1490	1788	211	240	285	344	354	382	402
11	968	1121	1274	1528	1021	1182	1343	1612	1095	1268	1441	1729	292	398	336	458	489	580	672
12	208	241	274	328	219	253	288	345	232	270	306	367	81	111	92	125	134	158	184
13													141,3		141,3				
14													112,65		111,0				
15													16,98		19,33				
16													—		19,37				
17													65,5	78	82	100	106	118	140
18													74	88	89	107	112	118	140
19													200		270				
20													141,3		141,3				
21													136,5		136,5				
22													112,65		111,0				
23															19,41				
24															19,22				
25													120	125	137	144	152	163	193
26													120	125	137	—	152	—	193
27													120	125	137	144	152	163	193
28													120	125	137	—	152	—	193
29													164	164	143	143	143	143	143
30													238	207	197	—	197	—	197
31													115	115	101	101	101	101	101
32													167	146	138	—	138	—	138
34	129,7				129,7				129,7						141,3				
35	—				—				—						136,0				
36	91,8				90,7				89,3						111,0				
37	25,15				26,42				28,06						19,43				
38	—				—				—						19,22				
39	212	252	274	326	224	265	289	343	238	282	307	364			137	144	152	163	193
40	—	252	274	—	—	265	289	—	—	282	307	—			137	—	152	—	193
41															137	144	152	163	193
42															137	—	152	—	193
43	112	112	112	112	107	107	107	107	100	100	100	100			146	146	146	146	146
44					146	129									210	—	210	—	210
45															99	99	99	99	99
46															144	—	144	—	144
47	880	880	—	—	—	—	—	—	1080	1080	—	—			540	690	690	690	690

#	K55	C75	N80	C95	P110	Q125*	V150*	C75	N80	C95	P110	Q125*	V150*	C75*	N80*	C95*	P110*	Q125*	V150*
1	-5 - (127,0)																		
2	15.00							18.00						20.80*					
3	7.52							9.19						10.72					
4	112.0							108.6						105.6					
5	2822							3403						3915					
6	9.85							9.26						8.75					
7	12.69							12.69						12.69					
8	22.13							26.68						30.70					
9	K55	C75	N80	C95	P110	Q125*	V150*	C75	N80	C95	P110	Q125*	V150*	C75*	N80*	C95*	P110*	Q125*	V150*
10	383	480	500	558	609	654	707	689	723	828	927	1023	1165	800	851	1012	1171	1330	1575
11	393	536	572	678	787	893	1071	655	699	830	961	1092	1310	764	815	967	1120	1273	1528
12	107	146	156	185	214	243	292	175	188	222	258	293	351	203	216	257	297	335	405
13	141.3							141.3						141.3					
14	108.8							105.4						102.4					
15	22.34							—											
16	22.39							26.90											
17	101	121	129	145	173	188	228	155	165	185	220	238	291	184	195	220	260	283	346
18	109	131	138	145	173	188	228	167	176	185	220	238	291	198	210	220	262	283	346
19	300	400	425	480	565			510	540	615	720								
20	141.3							141.3						141.3					
21	136.5							136.5						136.5					
22	108.8							105.4						102.4					
23	22.41							26.91						30.92					
24	22.22							26.72						30.71					
25	160	166	176	189	224	244	290	205	216	232	274	300	356	220	233	245	291	314	373
26	160	—	176	—	224	244	—	—	216	—	274	300	—	—	245	—	310	338	—
27	150	150	157	165	197	214	252	150	158	165	197	212	252	149	157	165	197	212	252
28	158	—	176	—	223	244	—	—	197	—	252	252	—	—	197	—	252	252	—
29	124	124	124	124	124	124	124	102	102	102	102	102	102	89	89	89	89	89	89
30	180	—	170	—	168	148	—	—	141	—	140	120	—	—	—	—	121	107	—
31	87	87	87	87	87	87	87	72	72	72	72	72	72	63	63	63	63	63	63
32	126	—	119	—	118	104	—	—	99	—	98	86	—	—	—	—	—	75	—
33																			
34	141.3							141.3						141.3					
35	136.0							136.0						136.0					
36	108.8							105.4						102.4					
37	22.44							26.94						30.92					
38	22.23							26.73						30.71					
39	160	166	176	189	223	244	290	205	216	232	274	300	356	220	233	245	291	314	373
40	160	—	176	—	223	244	—	—	216	—	274	300	—	—	245	—	310	338	—
41	150	150	157	165	197	214	252	150	158	165	197	212	252	149	157	165	197	212	252
42	158	—	176	—	223	244	—	—	197	—	252	252	—	—	197	—	252	252	—
43	126	126	126	126	126	126	126	105	105	105	105	105	105	91	91	91	91	91	91
44	183	—	173	—	172	151	—	—	144	—	143	126	—	—	125	—	124	110	—
45	85	85	85	85	85	85	85	71	71	71	71	71	71	61	61	61	61	61	61
46	124	—	117	—	116	102	—	—	98	—	97	85	—	—	84	—	83	73	—
47	730	880	880	880	880	—		1080	1080	1080	1080			1180	1180	1180	1180		
48	136.1							136.1											
49	—							—											
50	105.2							105.4											
51	22.36							26.77											
52	—																		
53	185	185	194	204	243	262	311	198	208	218	260	282	334						
54	—	—	—	—	—	—	—	—	—										
55	107	107	107	107	107	107	107	89	89	89	89	89	89						
56	—							—											
57																			

1	5 1/2 - (139,7)																		
	14.00		15.50					17.00						20.00					
2	14.00		15.50					17.00						20.00					
3	6.20		6.98					7.72						9.17					
4	127.3		125.7					124.3						121.4					
5	2599		2912					3201						3760					
6	12.73		12.41					12.13						11.58					
7	15.33		15.36					15.36						15.36					
8	20.41		22.86					25.13						29.51					
9	K55	C75*	K55	C75*	N80*	C95*	P110*	K55	C75	N80	C95	P110	Q125*	K55*	C75	N80	C95	P110	Q125*
10	215	245	279	335	344	370	387	339	419	433	478	514	545	456	582	609	690	764	834
11	294	401	332	452	483	573	664	367	500	534	634	734	834	435	594	634	752	871	989
12	99	134	·110	151	160	191	221	121	165	176	210	242	275	143	194	208	246	285	324
13	153.7		153.7					153.7						153.7					
14	124.1		122.6					121.1						118.2					
15	20.66		23.08					25.33						—					
16	—		23.12					25.38						29.71					
17	845	102	99	119	126	144	168	111.5	135	144	163	191	213	138	166	177	202	235	263
18	91	110	107	128	136	146	174	119.5	145	155	166	198	213	149	179	190	205	244	263
19	255		300					365	440	470	535	625		545	580	660	770		
20	153.7		153.7					153.7						153.7					
21			149.2					149.2						149.2					
22	124.1		122.6					121.1						118.2					
23			23.15					25.39						29.72					
24			22.96					25.20						29.52					
25	145	153	162	170	180	194	229	179	188	199	214	252	275	210	220	233	250	296	323
26	145	—	162	—	180	—	229	179	—	199	—	252	275	210	—	233	—	296	323
27	145	153	162	170	179	188	224	170	170	179	188	224	242	170	170	179	188	224	287
28	145	—	162	—	180	—	229	179	—	199	—	252	275	179	—	224	—	296	287
29	145	145	129	129	129	129	129	118	118	118	118	118	118	100	100	100	100	100	100
30	211	—	188	—	150	—	176	171	—	162	—	160	141	146	—	138	—	136	120
31	104	104	93	93	93	93	93	85	85	85	85	85	85	72	72	72	72	72	72
32	152	—	135	—	108	—	127	123	—	116	—	115	102	105	—	99	—	98	86
33																			
34			153.7					153.7						153.7					
35			149.0					149.0						149.0					
36			122.6					121.1						118.2					
37			23.18					25.43						29.75					
38			22.98					25.22						29.55					
39			163	170	180	194	229	179	188	199	213	252	275	210	220	233	250	297	323
40			163	—	180	—	229	179	—	199	—	252	275	210	—	233	—	297	323
41			163	170	179	188	224	170	170	179	188	224	242	170	170	179	188	224	242
42			163	—	180	—	229	179	—	199	—	252	275	179	—	224	—	287	287
43			132	132	132	132	132	120	120	120	120	120	120	103	103	103	103	103	103
44			192	—	181	—	180	175	—	165	—	164	144	149	—	140	—	140	124
45			94	94	94	94	94	85	85	85	85	85	85	73	73	73	73	73	73
46			137	—	129	—	128	124	—	117	—	116	102	106	—	100	—	100	88
47			690	790	790	790	790	790	880	880	880	880	—	940	1080	1080	1080	1080	—
48			148.8					148.8						148.8					
49			146.8					146.8						146.8					
50			119.9					119.0						118.2					
51			23.12					25.35						29.57					
52			23.05					25.28						29.50					
53			191	191	201	210	250	209	209	220	231	275	297	221	221	232	245	291	313
54			191	191	201	210	250	209	209	220	231	275	297	212	212	225	236	280	302
55			110	110	110	110	110	105	105	105	105	105	105	89	89	89	89	89	89
56			110	110	110	110		100	100	100	100	100	100	86	86	86	86	86	86
57																			

	5 1 2 - (139,7)							6 5 8 - (168,3)											
1	K55*	C75	N80	C95	P110	Q125*	V150*	K55	C75	N80*	C95*	P110*	K55	C75	N80	C95	P110	Q125*	V150*
2	23.00							20.00					24.00						
3	10.54							7.32					8.94						
4	118.6							153.6					150.4						
5	4277							3699					4475						
6	11.05							18.55					17.77						
7	15.36							22.34					22.34						
8	33.54							29.03					35.09						
9	K55*	C75	N80	C95	P110	Q125*	V150*	K55	C75	N80*	C95*	P110*	K55	C75	N80	C95	P110	Q125*	V150*
10	529	721	770	891	1001	1108	1268	205	231	240	262	278	314	384	397	434	463	485	507
11	501	683	729	865	1002	1138	1366	288	393	420	498	577	352	481	513	608	705	803	962
12	163	221	236	279	324	368	441	140	191	204	242	280	169	231	246	293	339	385	462
13	153.7							187.7					187.7						
14	115.4							150.5					147.2						
15	—							29.53					35.56						
16	33.72							29.65					35.67						
17	162	196	207	237	276	308	372	118	145	154	175	206	152	185	197	225	263	294	355
18	174	210	223	240	286	308	372	129	157	166	189	222	165	202	213	243	285	312	381
19		640	680	770	900			390					500	610	650	745	865		
20	153.7							187.7					187.7						
21	149.2							177.8					177.8						
22	115.4							150.5					147.2						
23	33.72							29.68					35.69						
24	33.54							29.14					35.15						
25	239	240	252	264	315	341	403	202	214	226	245	289	244	259	273	295	349	382	454
26	239	—	265	—	337	367	—	202	—	226	—	289	244	—	273	—	349	382	—
27	170	170	179	188	224	242	287	202	213	223	235	280	213	213	224	235	280	302	358
28	179	—	224	—	287	287	—	202	—	226	—	289	224	—	273	—	349	358	—
29	88	88	88	88	88	88	88	167	167	167	167	167	138	138	138	138	138	138	138
30	128	—	121	—	120	106	—	243	—	229	—	227	201	—	190	—	188	—	—
31	63	63	63	63	63	63	63	90	90	90	90	90	74	74	74	74	74	74	74
32	92	—	87	—	86	76	—	121	—	124	—	123	108	—	102	—	101	—	—
33																			
34	153.7							187.7					187.7						
35	149.0							177.8					177.8						
36	115.4							150.5					147.2						
37	33.74							29.74					35.73						
38	33.54							29.19					35.18						
39	239	240	252	264	315	341	403	202	214	226	245	289	244	259	273	295	349	382	454
40	239	—	265	—	337	367	—	202	—	226	—	289	244	—	273	—	349	382	—
41	170	170	179	188	224	242	287	202	213	223	235	280	213	213	224	235	280	302	358
42	179	—	224	—	287	287	—	202	—	226	—	289	224	—	273	—	349	358	—
43	90	90	90	90	90	90	90	169	169	169	169	169	140	140	140	140	140	140	140
44	131	—	124	—	104	123	—	246	—	232	—	230	204	—	193	—	191	168	—
45	64	64	64	64	64	64	64	93	93	93	93	93	77	77	77	77	77	77	77
46	93	—	88	—	74	87	—	135	—	128	—	127	112	—	106	—	105	92	—
47	1080	1220	1220	1220	1220	—	—	1080	1080	1080	1080	1080	1080	1080	1080	1080	1080	—	—
48	148.8												177.8						
49	146.8												176.0						
50	115.4												145.5						
51	33.57												35.27						
52	33.50												35.20						
53	244	244	256	269	321	346	410						269	269	283	296	353	382	453
54	212	213	224	236	280	302	385						269	269	283	296	353	382	453
55	87	87	87	87	87	87	87						92	92	92	92	92	92	92
56	76	76	76	76	76	76	76						92	92	92	92	92	92	92
57																			

	6 5/8 - (168,3)														7 - (177,8)				
1																			
2	28.00							32.00							17.00		20.00		
3	10.59							12.06							5.87		6.91		
4	147.1							144.2							166.1		164.0		
5	5247							5921							3170		3709		
6	16.99							16.33							21.66		21.12		
7	22.34							22.34							24.88		24.88		
8	41.18							46.47							24.87		29.10		
9	K55*	C75	N80	C95	P110	Q125*	V150*	K55*	C75	N80	C95	P110	Q125*	V150*	H40	K55*	H40	K55*	C75*
10	425	540	563	634	699	758	837	505	678	712	814	910	1003	1139	100	112	136	157	183
11	418	570	607	721	836	949	1139	476	649	692	822	952	1081	1298	159	219	188	258	352
12	199	271	290	343	397	452	542	224	305	326	388	448	510	612	87	120	102	140	192
13	187.7							187.7							194.5		194.5		
14	143.9							141.0							162.9		160.8		
15	—							—							25.32		29.53		
16	41.64							46.87							—		—		
17	186	226	240	275	320	358	433	224	261	278	317	371	415	500	53	78	78	113	138
18	202	245	260	295	347	380	464	233	284	300	342	401	440	538	68	100	86	125	152
19		750	790	900	1060				860	915	1050	1060			165		240	340	
20	187.7							187.7							194.5		194.5		
21	177.8							177.8							187.3		187.3		
22	143.9							141.0							162.9		160.8		
23	41.66							46.88											
24	41.12							46.33											
25	286	303	320	347	410	448	533	322	342	361	391	462	505	601	112	171	130	201	218
26	286	—	320	—	410	448	—	322	—	361	—	462	505	—	112	171	130	201	—
27	213	213	224	235	280	302	358	213	213	224	235	280	302	358	112	171	130	201	218
28	224	—	280	—	358	358	—	224	—	280	—	358	358	—	112	171	130	201	—
29	118	118	118	118	118	118	118	104	104	104	104	104	104	104	181	181	155	155	155
30	172	—	162	—	160	142	—	151	—	143	—	142	125	—	249	263	211	225	—
31	63	63	63	63	63	63	63	56	56	56	56	56	56	56	113	113	97	97	97
32	92	—	87	—	86	76	—	81	—	77	—	77	67	—	156	154	132	141	—
33																			
34	187.7							187.7											
35	177.8							177.8											
36	143.9							141.0											
37	41.70							46.83											
38	41.15							46.32											
39	286	303	320	347	410	448	533	322	342	361	391	462	505	601					
40	286	—	320	—	410	448	—	322	—	361	—	462	505	—					
41	213	213	224	235	280	302	358	213	213	224	235	280	302	358					
42	224	—	280	—	358	358	—	224	—	280	—	358	358	—					
43	120	120	120	120	120	120	120	106	106	106	106	106	106	106					
44	175	—	165	—	164	144	—	154	—	146	—		127	—					
45	65	65	65	65	65	65	65	58	58	58	58	58	58	58					
46	95	—	89	—	89	78	—	84	—	80	—	79	70	—					
47	1180	1180	1180	1180	1180	—	—	1270	1270	1270	1270	1270	—	—					
48	177.8							177.8											
49	176.0							176.0											
50	143.9							141.0											
51	41.19							46.41											
52	41.12							46.33											
53	288	288	303	318	379	409	485	318	319	336	352	420	453	537					
54	286	286	301	316	377	407	483	286	286	301	316	377	407	483					
55	83	83	83	83	83	83	83	82	82	82	82	82	82	82					
56	83	83	83	83	83	83	83	74	74	74	74	74	74	74					
57																			

#	K55	C75	N80	C95	P110*	K55	C75	N80	C95	P110	Q125*	V150*	K55*	C75	N80	C95	P110	Q125*	V150*
1	7 - (177,8)																		
2	23.00					26.00							29.00						
3	8.05					9.19							10.36						
4	161.7					159.4							157.1						
5	4293					4870							5451						
6	20.53					19.95							19.38						
7	24.88					24.88							24.88						
8	33.68					38.19							42.74						
9	K55	C75	N80	C95	P110*	K55	C75	N80	C95	P110	Q125*	V150*	K55*	C75	N80	C95	P110	Q125*	V150*
10	225	260	264	286	307	298	362	373	405	429	445	475	373	466	484	539	587	629	676
11	301	410	437	519	601	343	468	499	593	687	778	936	387	527	563	668	774	879	1056
12	162	222	237	281	325	184	251	268	319	369	420	503	206	282	300	357	413	469	563
13	194.5					194.5							194.5						
14	158.5					156.2							153.9						
15	34.07					38.55							—						
16	34.18					38.65							43.14						
17	137	168	179	205	239	161	198	209	241	280	314	379	185	227	238	276	323	361	436
18	152	185	197	224	262	178	217	231	264	308	342	416	205	249	265	303	353	393	479
19	460	560	600	685		540	660	700	800	940					760	805	925	1075	
20	194.5					194.5							194.5						
21	187.3					187.3							187.3						
22	158.5					156.2							153.9						
23	34.21					38.66							43.15						
24	33.78					38.23							42.73						
25	232	247	261	283	334	262	280	295	321	379	415	493	308	329	347	376	444	487	579
26	232	—	261	—	334	262	—	295	—	379	—		308	—	347	—	444	487	—
27	230	230	242	253	302	230	230	242	253	302	326	387	230	230	242	253	302	326	387
28	232	—	261	—	334	242	—	295	—	379	387		242	—	302	—	387	387	
29	133	133	133	133	133	118	118	118	118	118	118	118	105	105	105	105	105	105	105
30	194	—	187	—	182	171	—	162	—	160	142	—	153	—	145	—	143	126	—
31	84	84	84	84	84	74	74	74	74	74	74	74	66	66	66	66	66	66	66
32	122	—	115	—	114	107	—	101	—	100	88	—	96	—	90	—	90	79	—
33																			
34	194.5					194.5							194.5						
35	187.3					187.3							187.3						
36	158.5					156.2							153.9						
37	34.25					38.71							43.18						
38	33.82					38.28							42.75						
39	232	247	261	283	334	262	280	295	321	379	415	493	308	329	347	376	444	487	579
40	232	—	261	—	334	262	—	295	—	379	415	—	308	—	347	—	444	487	—
41	230	230	242	253	302	230	230	242	253	302	326	387	230	230	242	253	302	326	387
42	232	—	261	—	334	242	—	295	—	379	387		242	—	302	—	387	387	
43	136	136	136	136	136	120	120	120	120	120	120	120	107	107	107	107	107	107	107
44	198	—	187	—	185	175	—	165	—	164	144	—	156	—	147	—	146	128	—
45	86	86	86	86	86	76	76	76	76	76	76	76	68	68	68	68	68	68	68
46	125	—	118	—	117	111	—	105	—	104	91	—	99	—	93	—	93	82	—
47	1080	1080	1080	1080	1080	1180	1180	1180	1180	1180	—	—	1270	1270	1270	1270	1270	—	—
48	187.7					187.7							187.7						
49	185.7					185.7							185.7						
50	158.5					156.2							153.9						
51	33.98					38.35							42.81						
52	33.90					38.27							42.73						
53	281	281	295	310	370	285	285	299	318	375	405	480	304	304	320	337	401	433	513
54	281	281	295	310	370	285	285	299	318	375	405	480	298	299	315	321	393	425	504
55	100	100	100	100	100	89	89	89	89	89	89	89	85	85	85	85	85	85	85
56	100	100	100	100	100	89	89	89	89	89	89	89	84	84	84	84	84	84	84
57																			

7 - (177,8)

#	K55*	C75	N80	C95	P110	Q125*	V150*	C75	N80	C95	P110	Q125*	V150*	C75	N80	C95	P110	Q125*	V150*
2	32.00							35.00						36.00					
3	11.51							12.65						13.72					
4	154.8							152.5						150.4					
5	6011							6563						7070					
6	18.82							18.27						17.77					
7	24.88							24.88						24.88					
8	47.19							51.51						55.50					
9	K55*	C75	N80	C95	P110	Q125*	V150*	C75	N80	C95	P110	Q125*	V150*	C75	N80	C95	P110	Q125*	V150*
10	446	568	593	671	742	808	898	669	702	802	897	988	1119	736	785	925	1042	1155	1326
11	429	585	625	742	859	976	1171	644	687	816	944	1072	1287	698	745	884	1024	1163	1395
12	227	310	331	393	455	517	621	339	362	430	497	565	679	365	390	463	536	608	731
13	194.5							194.5						194.5					
14	151.6							149.3						147.2					
15	—							—						—					
16	47.49							51.76						55.70					
17	209	256	271	311	363	404	490	284	301	345	403	451	544	309	329	377	440	492	594
18	231	281	298	342	398	442	539	312	332	379	442	491	598	336	362	414	483	537	652
19		855	910	1035	1210			950	1010	1155	1345			1035	1100	1250	1470		
20	194.5							194.5						194.5					
21	187.3							187.3						187.3					
22	151.6							149.3						147.2					
23	47.50							51.76						55.69					
24	47.07							51.34						55.26					
25	325	346	366	396	468	512	609	364	383	402	479	517	613	364	383	402	479	517	613
26	325	—	366	—	468	512	—	—	399	—	511	560	—	—	430	—	550	603	—
27	230	230	242	253	302	326	387	230	242	253	302	326	387	230	242	253	302	326	387
28	242	—	302	—	387	387	—	—	302	—	387	387	—	—	302	—	387	387	—
29	95	95	95	95	95	95	95	87	87	87	87	87	87	81	81	81	81	81	81
30	138	—	131	—	130	—	—	—	120	—	119	104	—	—	111	—	110	97	—
31	60	60	60	60	60	60	60	55	55	55	55	55	55	51	51	51	51	51	51
32	87	—	82	—	81	—	—	—	75	—	76	66	—	—	70	—	69	62	—
33	1200	1200	1200	1200	1200	1200	1200	1400	1400	1400	1400	1400	1400	1500	1500	1500	1500	1500	1500
34	194.5							194.5						194.5					
35	187.3							187.3						187.3					
36	151.6							149.3						147.2					
37	47.50							51.76						55.66					
38	47.07							51.30						55.22					
39	325	346	366	396	468	512	609	364	383	402	479	517	613	364	383	402	479	517	613
40	325	—	366	—	468	512	—	—	399	—	511	560	—	—	430	—	550	603	—
41	230	230	242	253	302	326	387	230	242	253	302	326	387	230	242	253	302	326	387
42	242	—	302	—	387	387	—	—	302	—	387	387	—	—	302	—	387	387	—
43	97	97	97	97	97	97	97	89	89	89	89	89	89	83	83	83	83	83	83
44	141	—	133	—	132	116	—	—	122	—	121	107	—	—	114	—	113	100	—
45	61	61	61	61	61	61	61	56	56	56	56	56	56	52	52	52	52	52	52
46	89	—	84	—	83	73	—	—	77	—	76	67	—	—	72	—	71	62	—
47	1270	1270	1270	1270	1270	—	—	1370	1370	1370	1370	—	—	1370	1370	1370	1370		
48	187.7							191.3						191.3					
49	185.5							187.7						187.7					
50	151.6							149.3						147.2					
51	47.16							51.56						55.49					
52	47.08							51.43						55.36					
53	337	337	356	374	445	481	570	378	397		497	537	637	407	429	450	536	580	686
54	298	299	315	331	393	445	504	338	356	374	445	481	570	338	356	374	445	481	570
55	86	86	86	86	86	86	86	88	88	88	88	88	88	88	88	88	88	88	88
56	76	76	76	76	76	76	76	78	78	78	78	78	78	73	73	73	73	73	73
57																			

1	7 - (177,8)								
2	41.00*				44.00*				
3	14.98				16.25				
4	147.8				145.3				
5	7718				8243				
6	17.21				16.64				
7	24.88				24.88				
8	60.07				64.66				
9	C95*	P110*	Q125*	V150*	C95*	P110*	Q125*	V150*	
10	1011	1171	1331	1571	1088	1260	1432	1718	
11	966	1118	1271	1527	1048	1214	1379	1655	
12	502	581	660	792	541	626	710	853	
13									
14									
15									
16									
17									
18									
19									
20									
21									
22									
23									
24									
25									
26									
27									
28									
29									
30									
31									
32									
33									
34	194.5				194.5				
35	187.3				187.3				
36	144.7				142.1				
37	60.19				64.70				
38	59.76				64.25				
39	402	479	517	613	402	479	517	613	
40	—	596	613	—	—	613	613	—	
41	253	302	326	387	253	302	326	387	
42	—	387	387	--	—	387	387	—	
43	76	76	76	76	71	71	71	71	
44	—			—	—			—	
45	48	48	48	48	45	45	45	45	
46	—			—	—			—	
47	1460	1460	—	—	1670	1670	—	—	
48									
49									
50									
51									
52									
53									
54									
55									
56									
57									

7 5/8 - (193,7)

#	H40	K55*	K55	C75	N80	C95	P110*	C75	N80	C95	P110	Q125*	V150*	C75	N80	C95	P110	Q125*	V150*
	24.00		**26.40**					**29.70**						**33.70**					
2	24.00		26.40					29.70						33.70					
3	7.62		8.33					9.52						10.92					
4	178.5		177.0					174.7						171.9					
5	4454		4851					5510						6271					
6	25.02		24.61					23.97						23.21					
7	29.58		29.58					29.58						29.58					
8	34.96		38.07					43.26						49.21					
9	H40	K55*	K55	C75	N80	C95	P110*	C75	N80	C95	P110	Q125*	V150*	C75	N80	C95	P110	Q125*	V150*
10	141	160	199	226	234	256	271	322	330	353	368	391	418	436	452	501	541	576	611
11	190	261	285	390	415	493	571	445	475	564	653	741	890	510	545	647	749	851	1020
12	122	168	184	249	267	317	367	284	302	359	417	474	570	319	345	410	475	540	647
13	215.9		215.9					215.9						215.9					
14	175.3		173.8					171.5						168.7					
15	35.68		38.76					—						—					
16	—		38.93					44.05						49.93					
17	94	136	152	187	198	230	264	219	233	270	311	349	421	257	273	317	368	409	495
18	137	150	167	205	217	248	290	241	255	293	342	383	462	282	299	342	400	448	541
19	285			510	625	660	760	730	775	890	1040			860	900	1045	1220		
20			215.9					215.9						215.9					
21			206.4					206.4						206.4					
22			173.8					171.5						168.7					
23			39.01					44.11						49.98					
24			38.35					43.46						49.33					
25			257	276	293	318	376	314	333	361	427	467	556	358	379	411	486	532	633
26			257	—	293	—	376	—	333	—	427	467	—	—	379	—	486	532	—
27			257	276	293	318	376	314	333	353	421	455	540	320	337	353	421	455	540
28			257	—	293	—	376	—	333	—	427	467	—	—	379	—	486	532	—
29			168	168	168	168	168	148	148	148	148	148	148	130	130	130	130	130	130
30			245	—	232	—	230	—	204	—	202	178	—	—	179	—	178	156	—
31			103	103	103	103	103	91	91	91	91	91	91	82	82	82	82	82	82
32			150	—	142	—	141	—	125	—	124	109	—	—	110	93	109	96	—
33																			
34			216.0					216.0						216.0					
35			206.4					206.4						206.4					
36			173.8					171.9						168.7					
37			39.06					44.17						50.00					
38			38.40					43.50						49.34					
39			257	276	293	318	376	314	333	361	427	467	556	358	379	411	486	532	633
40			257	—	293	—	376	—	333	—	427	467	—	—	379	—	486	532	—
41			257	276	293	318	376	314	333	353	421	455	540	320	337	353	421	455	540
42			257	—	293	—	376	—	333	—	427	467	—	—	379	—	486	532	—
43			172	172	172	172	172	151	151	151	151	151	151	133	133	133	133	133	133
44			250	—	237	—	235	—	208	—	206	181	—	—	183	—	181	160	—
45			106	106	106	106	106	94	94	94	94	94	94	82	82	82	82	82	82
46			154	—	146	—	145	—	129	—	128	113	—	—	129	—	112	98	—
47			1080	1080	1080	1080	1080	1180	1180	1180	1180	—	—	1270	1270	1270	1270	—	—
48			203.5					203.5						203.5					
49			201.2					201.2						201.2					
50			173.8					171.5						168.7					
51			38.36					43.38						49.21					
52			38.25					43.27						49.10					
53			311	311	326	343	410	311	326	343	410	442	525	341	358	376	448	484	574
54			311	311	326	343	410	311	326	343	410	442	525	331	348	366	434	470	557
55			98	98	98	98	98	86	86	86	86	86	86	83	83	83	83	83	83
56			98	98	98	98	98	86	86	86	86	86	86	80	80	80	80	80	80
57																			

#	C75	N80	C95	P110	Q125*	V150*	K55	C75*	H40	K55	C75*	N80*	C95*	P110*	K55	C75*	N80*	C95*	P110*
1	7 5/8 - (193,7)								8 5/8 - (219,1)										
2	39.00						24.00		28.00						32.00				
3	12.70						6.71		7.72						8.94				
4	168.3						205.7		203.6						201.2				
5	7221						4474		5127						5903				
6	22.25						33.23		32.59						31.79				
7	29.58						37.87		37.87						37.87				
8	56.63						35.11		40.25						46.32				
9	C75	N80	C95	P110	Q125*	V150*	K55	C75*	H40	K55	C75*	N80*	C95*	P110*	K55	C75*	N80*	C95*	P110*
10	581	608	688	763	832	927	94	99	113	129	146	149	152	152	174	203	210	226	236
11	594	633	753	870	989	1187	203	277	170	234	319	340	404	468	271	369	394	468	542
12	373	397	472	546	622	746	169	231	141	194	265	283	336	389	223	304	325	386	447
13	215.9						244.5		244.5						244.5				
14	165.1						202.5		200.5						198.0				
15	—						36.18		41.21						47.26				
16	57.30						—		—						47.57				
17	304	323	371	433	485	585	117	145	103	150	184	196	225	262	179	221	235	270	314
18	334	354	406	474	531	640	128	171	116	167	206	219	251	294	201	247	262	301	351
19	1015	1080	1240	1440			355		315						610				
20	215.9								244.5						244.5				
21	206.4								231.8						231.8				
22	165.1								200.5						198.0				
23	57.34								41.65						47.60				
24	56.69								40.64						46.59				
25	412	436	473	559	613	729			174	266	290	305	334	391	306	333	352	384	453
26	—	436	—	559	613	—			174	266	—	305	—	391	306	—	352	—	453
27	320	337	353	421	455	540			174	266	290	305	334	391	306	333	352	384	453
28	—	421	—	540	540	—			174	266	—	305	—	391	306	—	352	—	453
29	113	113	113	113	113	113			216	216	216	216	216	216	177	177	177	177	177
30	—	156	—	154	136	—			297	314	—	297	—	295	258	—	244	—	242
31	69	69	69	69	69	69			112	112	112	112	112	112	97	97	97	97	97
32	—	96	—	95	83	—			154	163	—	154	—	152	141	—	133	—	132
33																			
34	216.0								244.5						244.5				
35	206.4								231.8						231.8				
36	165.1								200.5						198.0				
37	57.33								41.65						47.70				
38	56.66								40.64						46.69				
39	412	436	473	559	613	729			174	266	290	305	334	391	306	333	352	384	453
40	—	436	—	559	613	—			174	266	—	305	—	391	306	—	352	—	453
41	320	337	353	421	455	540			174	266	290	305	334	391	306	333	352	384	453
42	—	421	—	540	540	—			174	266	—	305	—	391	306	—	352	—	453
43	115	115	115	115	115	115			207	207	207	207	207	207	180	180	180	180	180
44	—	158	—	157	138	—			285	299	—	285	—	282	260	—	248	—	245
45	71	71	71	71	71	71			115	115	115	115	115	115	100	100	100	100	100
46	—	98	—	97	85	—			158	166	—	158	—	157	144	—	138	—	136
47	1370	1370	1370	1370			—	—	—	1080	1080	1080	1080	1080	1180	1180	1180	1180	1180
48	203.5														231.6				
49	201.2														229.4				
50	165.1														195.8				
51	56.58														46.92				
52	56.47														46.77				
53	378	398	418	497	538	637									386	386	406	425	508
54	331	348	366	435	470	557									386	386	406	425	508
55	80	80	80	80	80	80									100	100	100	100	100
56	70	70	70	70	70	70									100	100	100	100	100
57																			

#	8 5/8 - (219,1)																		
1																			
2	36.00						40.00							44.00					
3	10.16						11.43							12.70					
4	198.8						196.2							193.7					
5	666˙						7456							8234					
6	31.04						30.23							29.47					
7	37.87						37.87							37.87					
8	52.34						58.52							64.63					
9	K55	C75	N80	C95	P110*	Q125*	K55*	C75	N80	C95	P110	Q125*	V150*	C75	N80	C95	P110	Q125*	V150*
10	238	277	283	301	324	340	303	369	381	414	440	458	485	461	479	533	579	620	665
11	308	420	448	532	616	699	346	472	504	598	692	787	944	525	560	665	769	874	1049
12	252	344	367	436	505	575	283	388	411	489	565	642	771	425	454	539	624	708	850
13	244.5						244.5							244.5					
14	195.6						193.0							190.5					
15	53.22						—							—					
16	53.50						59.60							65.62					
17	207	257	273	314	366	410	234	294	313	359	419	470	565	331	352	404	472	529	637
18	234	288	305	350	409	459	268	330	350	401	469	526	634	371	394	452	527	591	713
19	710	875	930	1055					1000	1055	1225	1430		1130	1200	1375	1600		
20	244.5						244.5							244.5					
21	231.8						231.8							231.8					
22	195.6						193.0							190.5					
23	53.53						59.53							65.64					
24	52.52						58.61							64.63					
25	346	376	397	434	512	562	388	421	444	486	573	629	748	465	491	536	633	694	827
26	346	—	397	—	512	562	388	—	444	—	573	629	—	—	491	—	633	694	—
27	346	365	385	403	481	519	365	365	384	403	481	519	615	365	385	403	481	519	615
28	346	—	397	—	512	562	384	—	444	—	573	615	—	—	481	—	615	615	—
29	157	157	157	157	157	157	140	140	140	140	140	140	140	127	127	127	127	127	127
30	229	—	216	—	214	188	205	—	193	—	191	169	—	—	175	—	173	153	—
31	86	86	86	86	86	86	77	77	77	77	77	77	77	70	70	70	70	70	70
32	125	—	118	—	117	103	172	—	106	—	105	92	—	—	96	—	95	83	—
33																			
34	244.5						244.5							244.5					
35	231.8						231.8							231.8					
36	195.6						193.0							190.5					
37	53.61						59.61							65.60					
38	52.60						58.60							64.59					
39	346	376	397	434	512	562	388	421	444	486	573	629	748	465	491	536	633	694	827
40	346	—	397	—	512	562	388	—	444	—	573	629	—	—	491	—	633	694	—
41	346	365	385	403	481	519	365	365	384	403	481	519	615	365	385	403	481	519	615
42	346	—	397	—	512	562	384	—	444	—	573	615	—	—	481	—	615	615	—
43	159	159	159	159	159	159	143	143	143	143	143	143	143	129	129	129	129	129	129
44	230	—	219	—	217	191	207	—	197	—	195	172	—	—	177	—	176	155	—
45	88	88	88	88	88	88	79	79	79	79	79	79	79	71	71	71	71	71	71
46	127	—	121	—	120	106	114	—	109	—	108	95	—	—	98	—	97	85	—
47	1270	1270	1270	1270	1270	—	1470	1470	1470	1470	1470	1470	—	1570	1570	1570	1570	—	—
48	231.6						231.6							231.6					
49	229.4						229.4							229.4					
50	195.6						193.0							190.5					
51	52.69						58.70							64.70					
52	52.53						58.55							64.62					
53	387	387	407	428	509	550	419	419	441	463	551	595	705	447	471	495	589	637	754
54	387	387	407	428	509	550	393	393	414	435	518	559	663	393	414	435	518	559	663
55	89	89	89	89	89	89	86	86	86	86	86	86	86	83	83	83	83	83	83
56	89	89	89	89	89	89	80	80	80	80	80	80	80	73	73	73	73	73	73
57																			

1	8 5/8 - (219,1)						9 5/8 - (244,5)												
2	49,00						32,30		36,00					40,00					
3	14,15						7,92		8,94					10,03					
4	190,8						228,7		226,6					224,4					
5	9108						5889		6615					7390					
6	28,59						41,08		40,33					39,55					
7	37,87						47,10		47,10					47,10					
8	71,43						46,22		51,92					58,00					
9	C75	N80	C95	P110	Q125*	V150*	H40	K55*	K55	C75*	N80*	C95*	P110*	K55	C75	N80	C95	P110*	Q125*
10	565	591	668	739	804	893	96,5	108	139	160	163	169	171	177	206	213	229	240	243
11	585	623	740	857	974	1168	156	215	243	331	353	419	485	272	372	396	471	545	619
12	471	502	596	689	784	941	162	223	250	342	364	433	501	280	381	406	483	560	637
13	244,5						269,9		269,9					269,9					
14	187,6						224,7		222,6					220,5					
15	—						47,33		52,96					58,97					
16	72,39						—		53,37					59,36					
17	372	396	454	531	594	718	113	162	188	234	248	286	334	216	268	285	328	383	430
18	417	443	508	593	665	802	130	188	217	266	285	328	383	249	308	328	377	439	492
19	1270	1350	1550	1800			340		660					760	940	995	1135		
20	244,5								269,9					269,9					
21	231,8								257,2					257,2					
22	187,6								222,6					220,5					
23	72,40								53,33					59,32					
24	71,39								52,21					58,20					
25	514	543	593	700	769	915			336	368	390	427	503	375	411	436	477	562	619
26	—	543	—	700	769	—			336	—	390	—	503	375	—	436	—	562	619
27	365	385	403	481	519	615			336	368	390	427	503	375	401	423	444	529	571
28	—	481	—	615	615	—			336	—	390	427	503	375	—	436	—	562	619
29	115	115	115	115	115	115			176	176	176	176	176	157	157	157	157	157	157
30	—	158	—	157	138	—			256	—	242	—	240	229	—	217	—	215	190
31	63	63	63	63	63	63			96	96	96	96	96	87	87	87	87	87	87
32	—	86	—	86	75	—			140	—	133	—	131	126	—	119	—	118	104
33																			
34	244,5								269,9					269,9					
35	231,8								257,0					257,0					
36	187,6								222,6					220,5					
37	72,34								53,40					59,36					
38	71,32								52,26					58,23					
39	514	543	593	700	769	915			336	368	390	427	503	375	411	436	477	562	619
40	—	543	—	700	769	—			336	—	390	—	503	375	—	436	—	562	619
41	365	385	403	481	519	615			336	368	390	427	503	375	401	423	444	529	571
42	—	481	—	615	615	—			336	—	390	427	503	375	—	436	—	562	619
43	117	117	117	117	117	117			178	178	178	178	178	160	160	160	160	160	160
44	—	161	—	160	140	—			257	—	245	—	243	231	—	220	—	218	192
45	64	64	64	64	64	64			98	98	98	98	98	88	88	88	88	88	88
46	—	88	—	87	77	—			143	—	135	—	134	128	—	121	—	120	105
47	1670	1670	1670	1670	—	—			1180	1180	1180	1180	1180	1270	1270	1270	1270	1270	—
48														256,5					
49														254,5					
50														218,4					
51														58,48					
52														58,33					
53														434	434	456	479	570	616
54														434	434	456	479	570	616
55														89	89	89	89	89	89
56														89	89	89	89	89	89
57																			

	9 5/8 - (244,5)																	
1																		
2	43.50						47.00						53.50					
3	11.05						11.99						13.84					
4	222.4						220.5						216.8					
5	8103						8756						10030					
6	38.95						38.19						36.92					
7	47.10						47.10						47.10					
8	63.60						68.73						78.72					
9	C75	N80	C95	P110	Q125*	V150*	C75	N80	C95	P110	Q125*	V150*	C75	N80	C95	P110	Q125*	V150*
10	259	263	285	305	319	328	320	328	350	366	389	415	440	456	505	547	582	619
11	409	436	519	600	682	818	444	474	562	651	740	888	512	547	650	752	854	1025
12	418	445	531	613	697	837	451	482	572	659	755	915	517	552	656	759	864	1037
13	269.9						269.9						269.9					
14	218.4						216.5						212.8					
15	—						..						—					
16	64.88						69.94						79.78					
17	—	—	—	—	—	—	—	—	—	—	—	—	—	—	—	—	—	—
18	344	367	421	431	551	579	379	402	462	539	605	636	444	472	542	633	709	745
19	1050	1115	1280	1495			1150	1225	1410	1640			1355	1440	1650*	1925		
20	269.9						269.9						269.9					
21	257.2						257.2						257.2					
22	218.4						216.5						212.8					
23	64.84						69.89						79.74					
24	63.72						68.78						78.62					
25	451	477	523	617	679	808	488	516	566	667	734	874	559	590	648	764	840	1001
26	—	477	—	617	679	—	—	516	—	667	734	—	—	590	—	764	840	—
27	411	423	444	529	571	677	401	523	444	529	571	677	401	423	444	529	571	677
28	—	477	—	617	677	—	—	516	—	667	677	—	—	590	—	677	677	—
29	144	144	144	144	144	144	133	133	133	133	133	133	116	116	116	116	116	116
30	—	197	—	196	173	—	—	183	—	181	160	—	—	160	—	158	139	—
31	79	79	79	79	79	79	73	73	73	73	73	73	64	64	64	64	64	64
32	—	108	—	107	95	—	—	100	—	99	87	—	—	87	—	87	76	—
33																		
34	269.9						269.9						269.9					
35	257.0						257.0						257.0					
36	218.4						216.5						212.8					
37	64.85						69.88						79.67					
38	63.71						68.74						78.54					
39	451	477	523	617	679	808	488	516	566	667	734	874	559	590	648	764	840	1001
40	..	477	—	617	679	—	—	516	—	667	734	—	—	590	—	764	840	—
41	411	423	444	529	571	677	401	423	444	529	571	677	401	423	444	529	571	677
42	—	477	—	617	677	—	—	516	—	667	677	—	—	529	—	677	677	—
43	146	146	146	146	146	146	135	135	135	135	135	135	118	118	118	118	118	118
44	—	201	—	199	175	—	—	186	—	184	162	—	—	162	—	161	142	—
45	80	80	80	80	80	80	74	74	74	74	74	74	64	64	64	64	64	64
46	—	110	—	109	96	—	—	102	—	101	89	—	—	88	—	87	77	—
47	1370	1370	1370	1370	—	—	1570	1570	1570	1570	—	—	1760	1760	1760	1760	—	—
48	256.5						256.5						256.5					
49	254.5						254.5						254.5					
50	218.4						216.5						212.8					
51	63.85						68.83						78.67					
52	63.69						68.68						78.51					
53	434	456	479	570	616	731	459	483	507	604	652	773	522	549	577	686	742	898
54	434	456	479	570	616	731	459	483	507	604	652	773	468	492	518	616	666	789
55	82	82	82	82	82	82	80	80	80	80	80	80	79	79	79	79	79	79
56	81	81	81	81	81	81	80	80	80	80	80	80	71	71	71	71	71	71
57																		

#	|	| 9 5/8 - (244,5)						|				|		
1		**9 5/8 - (244,5)**													
2		58.40*						61.10*				71.80*			
3		15.11						15.87				19.05			
4		214.3						212.7				206.4			
5		10890						11401				13491			
6		36.14						35.53				33.46			
7		47.10						47.10				47.10			
8		85.40						89.40				105.90			
9	K55*	C75*	N80*	C95*	P110*	Q125*	V150*	C95*	P110*	Q125*	V150*	P110*	Q125*	V150*	
10	413	522	544	612	672	727	798	675	747	814	905	1060	1175	1350	
11	410	559	597	709	820	932	1120	744	862	979	1175	1034	1175	1410	
12	412	562	600	713	826	938	1127	746	865	980	1178	1023	1164	1395	
13				269.9				269.9				269.9			
14				210.3				208.8				202.4			
15				—				—				—			
16				—				—				—			
17	343	425	450	519	574	680	819	547	621	716	863	772	865	1043	
18	394	488	518	596	695	780	941	629	733	822	988	885	993	1198	
19															
20				269.9				269.9				269.9			
21				257.0				257.0				257.0			
22				210.3				208.8				202.4			
23				86.30				90.23				106.43			
24				85.16				89.09				105.29			
25	552	607	641	703	830	913	1086	779	918	1010	1202	966	1043	1236	
26	552	—	641	—	830	913	—	—	918	1010	—	1028	1131	—	
27	401	401	423	444	529	571	677	444	529	571	677	529	571	677	
28	423	—	529	—	677	677	—	—	677	677	—	677	677	—	
29	107	107	107	107	107	107	107	102	102	102	102	86	86	86	
30	156	—	147	—	146	128	—	—	139	122	—	118	103	—	
31	59	59	59	59	59	59	59	56	56	56	56	47	47	47	
32	86	—	81	—	80	77	—	—	76	67	—	64	57	—	
33															
34				269.9				269.9				269.9			
35				257.0				257.0				257.0			
36				210.3				208.8				202.4			
37				86.30				90.23				106.43			
38				85.16				89.09				105.29			
39	552	607	641	703	830	913	1086	779	918	1010	1202	966	1043	1236	
40	552	—	641	—	830	913	—	—	918	1010	—	1028	1131	—	
41	401	401	423	444	529	571	677	444	529	571	677	529	571	677	
42	423	—	529	—	677	677	—	—	677	677	—	677	677	—	
43	104	104	104	104	104	104	104	98	98	98	98	87	87	87	
44	150	—	143	—	142	125	—	—	134	118	—	119	104	—	
45	57	57	57	57	57	57	57	52	52	52	52	43	43	43	
46	83	—	78	—	78	69	—	—	71	63	—	59	52	—	
47	1860	1860	1860	1860	1860	—	—	—	—	—	—	—	—	—	
48															
49															
50															
51															
52															
53															
54															
55															
56															
57															

10 3/4 - (273)

#	H40	K55*	H40	K55	C75*	N80*	C95*	K55	C75*	N80*	C95*	P110*	Q125*	K55	C75	N80	C95	P110	Q125*
2	32.75		40.50					45.50						51.00					
3	7.09		8.89					10.16						11.43					
4	258.9		255.3					252.7						250.2					
5	5921		7377					8391						9394					
6	52.60		51.15					50.15						49.13					
7	58.74		58.74					58.74						58.74					
8	46.47		57.91					65.87						73.73					
9	H40	K55*	H40	K55	C75*	N80*	C95*	K55	C75*	N80*	C95*	P110*	Q125*	K55	C75	N80	C95	P110	Q125*
10	61	61	98	109	118	119	119	144	166	171	176	180	180	186	214	222	240	253	259
11	126	172	157	216	295	314	373	247	337	359	427	494	561	278	379	404	480	556	632
12	163	224	200	280	381	406	483	318	434	463	549	637	723	356	486	518	614	712	809
13	298.4		298.4					298.4						298.4					
14	254.9		251.3					248.8						246.2					
15	47.79		59.11					66.98						74.76					
16	—		—					—						—					
17	91	150	138	200	214	228	306	235	291	304	359	421	470	269	336	357	412	480	540
18	—	—																	
19	275		425	610				710						820	1020	1085	1240	1465	
20	298.4		298.4					298.4						298.4					
21	285.8		285.8					285.8						285.8					
22	254.9		251.3					248.8						246.2					
23			59.47					67.31						75.07					
24			58.24					66.08						73.83					
25	193	292	241	364	405	429	473	414	461	488	538	633	698	463	516	545	602	708	782
26	193	292	241	364	—	429	—	414	—	488	—	633	698	463	—	545	—	708	782
27	193	292	241	364	405	429	473	414	454	478	502	597	645	454	454	478	502	597	645
28	193	292	241	364	—	429	—	414	—	488	—	633	698	463	—	545	—	708	765
29	219	219	175	175	175	175	175	154	154	154	154	154	154	138	138	138	138	138	138
30	301	319	241	255	—	241	—	224	—	212	—	210	185	200	—	189	—	188	166
31	120	120	96	96	96	96	96	85	85	85	85	85	85	76	76	76	76	76	76
32	165	175	133	140	—	133	—	123	—	117	—	115	102	110	—	104	—	103	91
33																			
34			298.4					298.4						298.4					
35			285.8					285.8						285.8					
36			251.3					248.8						246.2					
37			59.59					67.41						75.11					
38			58.36					66.18						73.88					
39			241	364	405	429	473	414	461	488	538	633	698	463	516	545	602	708	782
40			241	364	—	429	—	414	—	488	—	633	698	463	—	545	—	708	782
41			241	364	405	429	473	414	454	478	502	597	645	454	454	478	502	597	645
42			241	364	—	429	—	414	—	488	—	633	698	463	—	545	—	708	765
43			177	177	177	177	177	156	156	156	156	156	156	139	139	139	139	139	139
44			243	257	—	243	—	227	—	215	—	213	187	202	—	191	—	190	167
45			99	99	99	99	99	87	87	87	87	87	87	78	78	78	78	78	78
46			136	144	—	136	—	127	—	120	—	119	114	113	—	107	—	106	94
47			—	1180	1180	1180	1180	1370	1370	1370	1370	1370	—	1570	1570	1570	1570	1570	—
48								291.1						291.1					
49								—						—					
50								248.8						246.2					
51								66.33						74.57					
52								—						—					
53								549	549	577	607	723	771	615	615	647	680	809	875
54								—	—	—	—	—	—	—	—	—	—	—	—
55								100	100	100	100	100	100	100	100	100	100	100	100
56								—						—					
57																			

10 3/4 - (273)

#	K55*	C75	N80*	C95	P110	Q125*	C75*	N80*	C95*	P110	Q125*	V150*	K55*	C75*	N80*	C95*	P110	Q125*	V150*
2	55.50						60.70						65.70						
3	12.57						13.84						15.11						
4	247.91						245.4						242.8						
5	10288						11273						12246						
6	48.27						47.26						46.30						
7	58.74						58.74						58.74						
8	80.75						88.40						96.03						
10	234	272	277	296	319	334	346	356	384	404	419	452	339	419	435	479	515	547	575
11	306	416	445	528	611	694	459	489	581	673	765	918	367	501	534	634	734	835	1002
12	390	532	567	673	780	885	583	622	738	855	971	1167	464	633	676	801	931	1056	1267
13	298.4						298.4						298.4						
14	243.9						241.4						238.9						
15	81.69						89.33						96.88						
16	—						—						' —						
17	299	374	397	459	535	599	478	507	510	594	668	805	366	525	558	561	654	735	885
18	—	—	—	—	—	—	—	—	—	—	—	—	—	—	—	—	—	—	—
19		1140	1210	1395	1630				1810								1995		
20	298.4						298.4						298.4						
21	285.8						285.8						285.8						
22	243.9						241.4						238.9						
23	81.98						89.59						97.12						
24	80.74						88.35						95.88						
25	507	545	597	659	776	856	619	655	722	850	938	1128	604	673	711	787	924	1028	1215
26	507	—	597	—	776	856	—	655	—	850	938		604	—	711	—	924	1028	
27	454	454	478	502	597	645	454	478	502	597	645	765	454	454	478	502	597	645	765
28	478	—	597	—	765	765	—	597	—	765	765	—	478	—	597	—	765	765	—
29	126	126	126	126	126	126	115	115	115	115	115	115	106	106	106	106	106	106	106
30	183	—	173	—	171	150	—	158	—	156	138	—	154	—	145	—	144	127	—
31	69	69	69	69	69	69	63	63	63	63	63	63	58	58	58	58	58	58	58
32	100	—	95	—	94	83	—	87	—	86	76	—	84	—	80	—	79	70	—
34	298.4						298.4						298.4						
35	285.8						285.8						285.8						
36	243.9						241.4						238.9						
37	82.00						89.51						97.00						
38	80.77						88.28						95.77						
39	507	545	597	659	776	856	619	655	722	850	938	1128	604	673	711	787	924	1028	1215
40	507	—	597	—	776	856	—	655	—	850	938		604	—	711	—	924	1028	
41	454	454	478	502	597	645	454	478	502	597	645	765	454	454	478	502	597	645	765
42	478	—	597	—	765	765	—	597	—	765	765	—	478	—	597	—	765	765	—
43	127	127	127	127	127	127	116	116	116	116	116	116	107	107	107	107	107	107	107
44	185	—	175	—	173	152	—	160	—	158	139	—	156	—	147	—	146	128	—
45	71	71	71	71	71	71	65	65	65	65	65	65	60	60	60	60	60	60	60
46	103	—	98	—	97	85	—	89	—	89	78	—	87	—	83	—	82	72	—
47	1670	1670	1670	1670	1670	—	1820	1820	1820	1820	—	—	1910	1910	1910	1910	1910	—	—
48	291.1						291.1												
49	—						—												
50	243.9						241.4												
51	81.47						89.07												
53	674	674	709	744	887	957	682	718	753	890	970	1150							
54	—	—	—	—	—	—	—	—	—	—	—	—							
55	100	100	100	100	100	100	91	91	91	91	91	91							
56	—	—	—	—	—	—	—	—	—	—	—	—							

	11 3/4 - (298,4)										13 3/8 - (339,7)						
1																	
2	42.00		47.00		54.00		60.00				48.00		54.50				
3	8.46		9.52		11.05		12.42				8.38		9.65				
4	281.5		279.4		276.3		273.6				323.0		320.4				
5	7706		8646		9976		11161				8725		10009				
6	62.24		61.31		59.96		58.79				81.89		80.63				
7	70.16		70.16		70.16		70.16				90.80		90.80				
8	60.47		67.86		78.30		87.60				68.49		78.56				
9	H40	K55*	K55	C75*	K55	C75*	K55	C75	N80	C95	H40	K55*	K55	C75*	N80*	C95*	P110*
10	74	77	104	112	143	166	183	212	219	237	53	53	78	78	78	78	78
11	137	188	212	289	245	335	276	376	402	480	119	164	188	257	274	326	377
12	212	292	328	446	378	516	423	577	615	730	241	330	379	517	551	655	725
13	323.8		323.8		323.8		323.8				365.1		365.1				
14	277.6		275.4		272.4		269.7				319.0		316.5				
15	61.81		69.11		79.44		88.63				70.0		78.96				
16	—		—		—		—				—						
17	136	196	226	284	269	338	308	387	410	474	143	205	243	306	326	377	440
18	—	—	—	—	—	—	—	—	—	—	—	—					
19	415		690		820		940	1175	1250	1440	435		740				
20			323.8		323.8		323.8						365.1				
21			—		—		—						—				
22			275.4		272.4		269.7						316.5				
23			69.49		79.77		88.94						80.39				
24			—		—		—						—				
25			415	469	480	540	537	605	640	710			460	531	562	629	737
26			415	—	480	—	537	—	640	—			460	—	562	—	737
27			—		—		—	—	—	—			—	—	—	—	—
28			—		—		—	—	—	—			—	—	—	—	—
29			163	163	141	141	126	126	126	126			160	160	160	160	160
30			237	—	205	—	183	—	174	—			232	—	220	—	218
31			—		—		—						—				
32			—		—												
33																	
34													365.1				
35													—				
36													316.5				
37													80.54				
38													—				
39													460	530	561	628	736
40													460	—	561	—	736
41													—	—	—	—	—
42													—	—	—	—	—
43													162	162	162	162	162
44													236	—	223	—	221
45													—	—	—	—	—
46													—	—	—	—	—
47													1470	1470	1470	1470	1470	—
48																	
49																	
50																	
51																	
52																	
53																	
54																	
55																	
56																	
57																	

#	K55	C75*	N80*	C95*	P110*	K55	C75*	N80*	C95*	P110*	K55	C75	N80	C95	P110*
1	13 3/8 - (339,7)														
2	61.00					68.00					72.00				
3	10.92					12.19					13.06				
4	317.9					315.3					313.6				
5	11282					12545					13392				
6	79.37					78.08					77.24				
7	90.80					90.80					90.80				
8	88.55					98.47					105.16				
9	K55	C75*	N80*	C95*	P110*	K55	C75*	N80*	C95*	P110*	K55	C75	N80	C95	P110*
10	106	114	115	115	115	134	153	156	161	161	153	179	184	194	199
11	214	291	310	369	427	238	325	346	411	476	255	348	371	441	510
12	428	584	622	738	816	475	648	691	821	908	507	692	738	878	970
13	365.1					365.1					365.1				
14	313.9					311.4					309.7				
15	89.84					99.61					106.30				
16	—					—					—				
17	281	354	377	437	508	319	402	428	495	577	344	435	462	536	624
18	—	—	—	—	—	—	—	—	—	—					
19	855					970							1320	1410	1630
20	365.1					365.1					365.1				
21	—					—					—				
22	313.9					311.4					309.7				
23	90.22					100.00					106.60				
24	—					—					—				
25	519	597	633	708	832	578	665	704	787	925	618	710	752	841	987
26	519	—	633	—	832	578	—	704	—	925	618	—	752	—	987
27	—	—	—	—	—	—	—	—	—	—	—	—	—	—	—
28	—	—	—	—	—	—	—	—	—	—	—	—	—	—	—
29	141	141	141	141	141	127	127	127	127	127	119	119	119	119	119
30	206	—	195	—	194	185	—	175	—	173	173	—	164	—	162
31	—	—	—	—	—	—	—	—	—	—	—	—	—	—	—
32	—	—	—	—	—	—	—	—	—	—	—	—	—	—	—
33															
34	365.1					365.1					365.1				
35	—					—					—				
36	313.9					311.4					309.7				
37	90.34					100.08					106.65				
38	—					—					—				
39	519	597	633	708	832	578	665	704	787	925	618	710	752	841	987
40	519	—	633	--	832	578	—	704	—	925	618	—	752	—	987
41	—	—	—	—	—	—	—	—	—	—	—	—	—	—	—
42	—	—	—	—	—	—	—	—	—	—	—	—	—	—	—
43	144	144	144	144	144	129	129	129	129	129	121	121	121	121	121
44	209	—	198		196		—	177		176	176	—	166	—	165
45	—	—	—	—	—	—	—	—	—	—	—	—	—	—	—
46	—	—	—	—	—	—	—	—	..	—	—	—	—	—	—
47	1570	1570	1570	1570	1570	1670	1670	1670	1670	1670	1760	1760	1760	1760	1760
48															
49															
50															
51															
52															
53															
54															
55															
56															
57															

1	16 - (406,4)			18 5/8 - (473,1)	20 - (508,0)		
2	65.00	75.00	84.00	87.50	94.00	106.5	133.0
3	9.52	11.13	12.57	11.05	11.13	12.70	16.13
4	387.4	384.1	381.3	451.0	485.7	482.6	475.7
5	11876	13815	15556	16038	17366	19762	24923
6	117.87	115.87	114.19	159.74	185.28	182.92	177.76
7	129.94	129.94	129.94	176.20	202.96	202.96	202.96
8	93.21	108.32	122.09	125.88	136.16	155.15	195.62
9	H40 K55*	K55 C75*	K55 C75*	K55 C75* N80*	K55 C75* N80*	K55 C75*	K55 C75*
10	46 47	70 70	97 102	43 43 43	36 36 36	53 53	103 110
11	113 156	181 248	205 280	155 211 225	145 198 211	166 226	211 287
12	327 449	524 714	589 804	608 829 884	658 898 958	749 1022	945 1289
13	431.8	431.8	431.8	508.0	533.4	533.4	533.4
14	382.6	379.4	376.5	446.2	480.9	477.8	470.9
15	95.14	110.05	123.64	129.22	138.39	157.09	196.98
16	—	—	—	— — —	139.03	157.60	197.46
17	195 276	334 424	385 488	352 450 480	366 469 498	427 546	557 713
18	—	—	—	— — —	425 541 577	494 632	644 825
19	595	1020	1170	1075	1290	1505	1970
20		431.8	431.8				
21		—	—				
22		379.4	376.5				
23		110.39	123.89				
24		—	—				
25		591 706	667 795				
26		591 —	667 —				
27		— —	— —				
28		— —	— —				
29		138 138	122 122				
30		200 —	178 —				
31		— —	— —				
32		— —	— —				
33							
34							
35							
36							
37							
38							
39							
40							
41							
42							
43							
44							
45							
46							
47							
48							
49							
50							
51							
52							
53							
54							
55							
56							
57							

CASING JOINTS

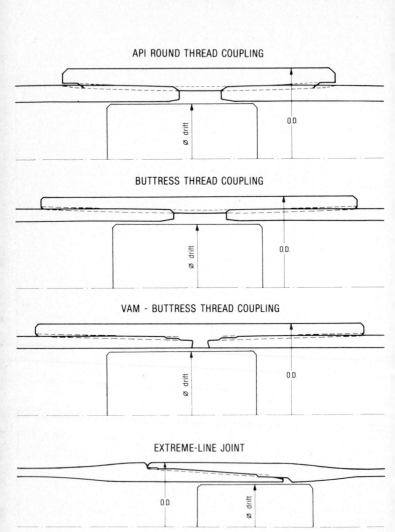

API ROUND THREAD COUPLING

BUTTRESS THREAD COUPLING

VAM - BUTTRESS THREAD COUPLING

EXTREME-LINE JOINT

API CASING THREAD FORMS
Round thread form

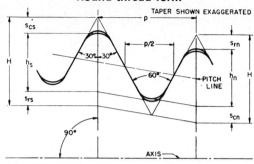

Taper on diameter 6,25 % (0,750 in per ft) number of threads per inch : 8, pitch p = 3,175 mm

$$H = 0,866\ p \qquad = 2,750\ mm \qquad t_b = 0,120\ p + 0,051 = 0,432\ mm$$
$$h = 0,626\ p - 0,178 = 1,810\ mm \qquad t_s = 0,120\ p + 0,127 = 0,508\ mm$$

Buttress and VAM thread form

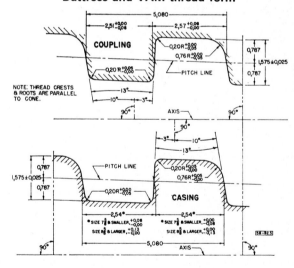

Taper on diameter : Size 13 3/8'' and smaller : 6,25 %, Size 16'' and larger : 8,33 %, number of threads per inch : 5.

Thread crests and roots are parallel to cone for casing sizes from 4 1/2'' to 13 3/8''.

Thread crests and roots are parallel to axis for casing sizes 16'' and larger.

API CASING THREAD FORMS
(continued)

Extreme-line form - Casing sizes from 5″ to 7 5/8″

Taper 12,5 % - 6 threads per inch

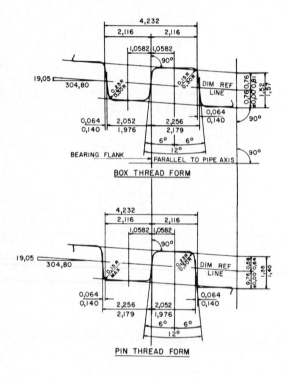

BOX THREAD FORM

PIN THREAD FORM

API CASING THREAD FORMS
(continued)

Extreme-line form - Casing sizes from 8 5/8" to 10 3/4"

Taper 10,42 % - 5 threads per inch

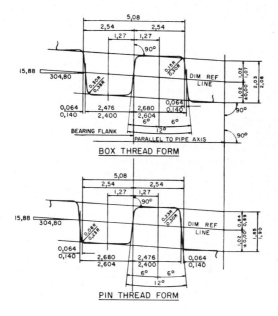

BOX THREAD FORM

PIN THREAD FORM

ELLIPSE OF BIAXIAL YIELD STRESS
EFFECT OF TENSILE LOAD ON COLLAPSE RESISTANCE

$$\frac{P_{CA}}{P_{CO}} = \sqrt{1 - 0{,}75 \left[\frac{T_A}{R_{tm}}\right]^2} - 0{,}5 \ \frac{T_A}{R_{tm}}$$

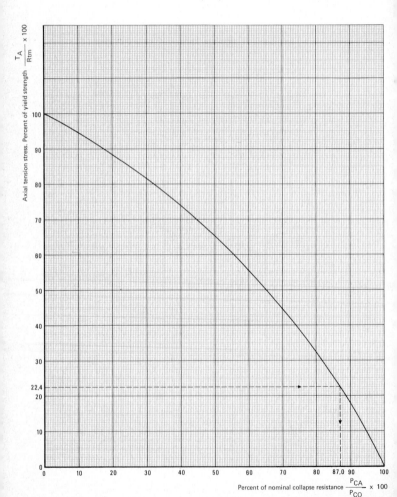

EFFECT OF TENSILE LOAD ON COLLAPSE RESISTANCE
(API Bul 5C3)

FORMULAS

$$P_{CA} = \left(\sqrt{1 - 0.75 \left[\frac{T_A}{R_{tn}}\right]^2} - 0.5 \frac{T_A}{R_{tn}} \right) P_{CO} \quad \text{metric system}$$

$$P_{CA} = \left(\sqrt{1 - 0.75 \left[\frac{S_A}{Y_p}\right]^2} - 0.5 \frac{S_A}{Y_p} \right) P_{CO} \quad \text{english system}$$

with :

English system	Metric system	
P_{CA}	P_{CA}	Minimum collapse pressure under axial tension stress, psi or bar.
P_{CO}	P_{CO}	Minimum collapse pressure without axial tension stress, psi or bar.
S_A	T_A	Axial tension stress, psi or bar.
Y_p	R_{tn}	Minimum yield strength of the pipe, psi or bar.

This formula is based on the Hencky-von Mises maximum strain energy of distorsion theory of yielding. It is applicable where the collapse pressure is directly proportional to yield strength. This condition is met by the API collapse formulas for yield collapse and approximately by the API formula for plastic collapse. It is not applicable to the elastic collapse formula.

The part of the ellipse for collapse under tension, after Holmquist and Nadai, is given on the opposite page.

How to use this curve.

Example :

Given : 100.10³ daN of 9 5/8'' casing are suspended below a 9 5/8'' 43,50 # N80 joint.
Required : To determine the collapse resistance corrected for the effect of tension for this joint.

Solution : Tensile load : 100.10³ daN, Metal cross-section area : 810³ mm² (see page C17),

$$\text{Axial tensile stress } T_A = \frac{100.10^3}{8\ 103} = 12,34 \text{ hbars.}$$

Minimum yield strength of N80 grade : 55,1 hbars (see page C1), Tensile stress, percent of

$$\text{minimum yield stress : } \frac{T_A}{R_{tn}} \times 100 = \frac{12,34}{55,1} = 22,4 \%$$

Enter chart at 22,4 % on vertical scale and go horizontally to intersect the portion of the ellipse and proceed vertically to the horizontal scale and read 87,0 %.

Minimum collapse resistance of 9 5/8 - 43,50 # N80 casing : 263 bars (see page C17).

87,0 % of 263 bars = 229 bars which is the collapse resistance of 9 5/8'' - 43,50 # N80 casing corrected for the effect of tension.

Note : Computations may be considerably facilitated by the use of the charts on the following pages.

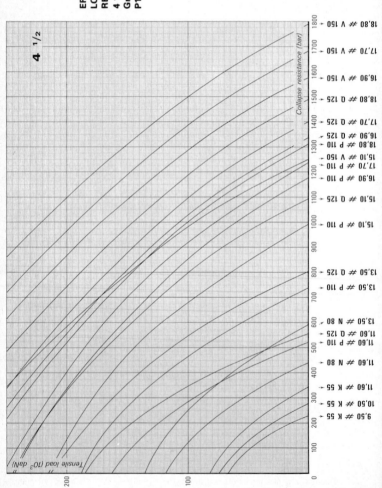

EFFECT OF TENSILE
LOAD ON COLLAPSE
RESISTANCE
4 1/2'' CASING
Grades K55 - N80
P110 - Q125 - V150

EFFECT OF TENSILE LOAD ON COLLAPSE RESISTANCE. 4 1/2" CASING

Grades C75 - C95

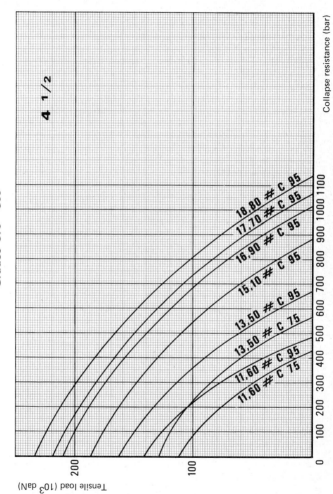

4 ¹/₂

Tensile load (10³ daN)

Collapse resistance (bar)

18,80 # C 95
17,70 # C 95
16,90 # C 95
15,10 # C 95
13,50 # C 95
13,50 # C 75
11,60 # C 95
11,60 # C 75

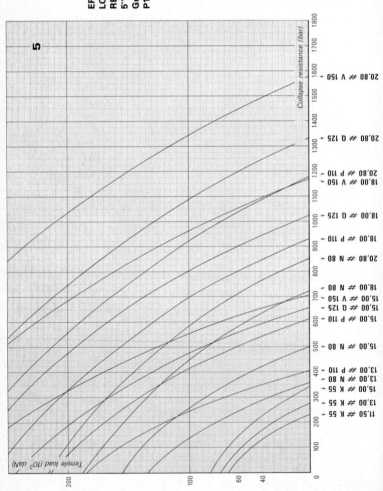

EFFECT OF TENSILE
LOAD ON COLLAPSE
RESISTANCE
5" CASING
Grades K55 - N80
P110 - Q125 - V150

EFFECT OF TENSILE LOAD ON COLLAPSE RESISTANCE. 5" CASING

Grades C75 - C95

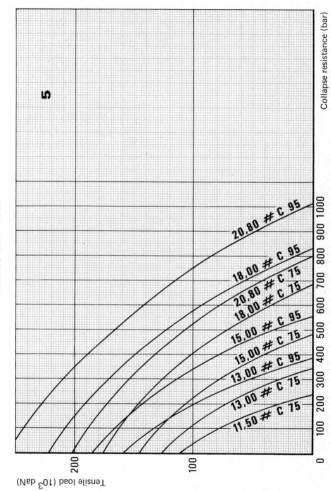

5

Collapse resistance (bar)

Tensile load (10³ daN)

20,80 # C 95
18,00 # C 95
20,80 # C 75
18,00 # C 75
15,00 # C 95
15,00 # C 75
13,00 # C 95
13,00 # C 75
11,50 # C 75

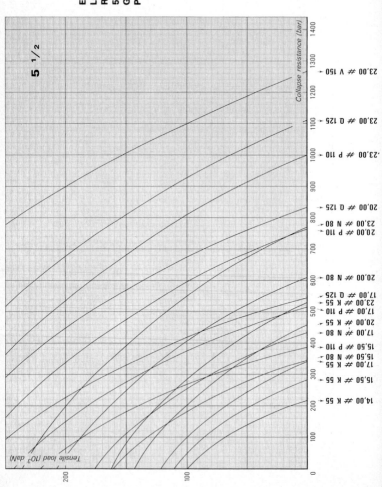

EFFECT OF TENSILE LOAD ON COLLAPSE RESISTANCE
5 1/2" CASING
Grades K55 - N80
P110 - Q125 - V150

5 ¹/₂

Tensile load (10³ daN)

Collapse resistance (bar)

23,00 ≠ V 150
23,00 ≠ Q 125
23,00 ≠ P 110
20,00 ≠ Q 125
23,00 ≠ N 80
20,00 ≠ P 110
20,00 ≠ N 80
17,00 ≠ Q 125
23,00 ≠ K 55
17,00 ≠ P 110
20,00 ≠ K 55
17,00 ≠ N 80
15,50 ≠ P 110
15,50 ≠ N 80
17,00 ≠ K 55
15,50 ≠ K 55
14,00 ≠ K 55

EFFECT OF TENSILE LOAD ON COLLAPSE RESISTANCE. 5 1/2" CASING

Grades C75 - C95

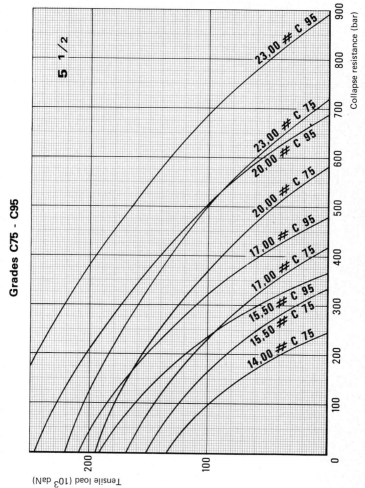

Collapse resistance (bar)

Tensile load (10^3 daN)

5 1/2

23,00 # C 95

23,00 # C 75

20,00 # C 95

20,00 # C 75

17,00 # C 95

17,00 # C 75

15,50 # C 95

15,50 # C 75

14,00 # C 75

EFFECT OF TENSILE
LOAD ON COLLAPSE
RESISTANCE
6 5/8" CASING
Grades K55 - N80
P110 - Q125 - V150

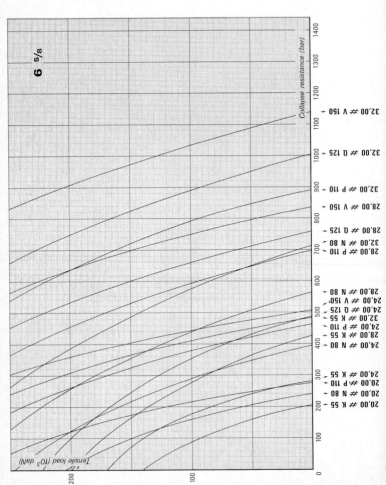

EFFECT OF TENSILE LOAD ON COLLAPSE RESISTANCE. 6 5/8" CASING

Grades C75 - C95

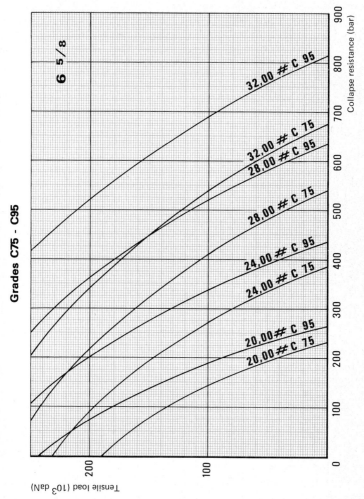

**EFFECT OF TENSILE
LOAD ON COLLAPSE
RESISTANCE
7" CASING
Grades K55 - N80
P110 - Q125 - V150**

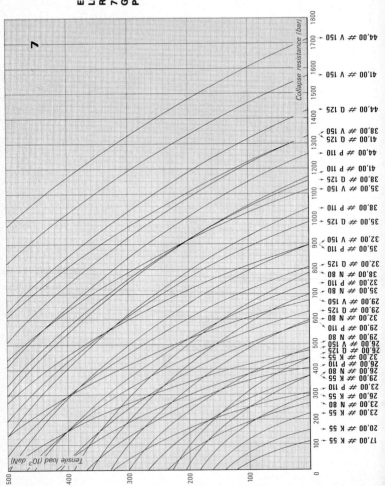

EFFECT OF TENSILE LOAD ON COLLAPSE RESISTANCE. 7" CASING

Grades C75 - C95

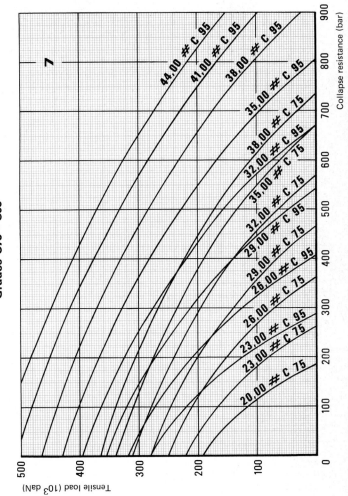

Collapse resistance (bar)

Tensile load (10^3 daN)

44.00 # C 95
41.00 # C 95
38.00 # C 95
35.00 # C 95
38.00 # C 75
32.00 # C 95
35.00 # C 75
32.00 # C 75
29.00 # C 95
29.00 # C 75
26.00 # C 95
26.00 # C 75
26.00 # C 95
23.00 # C 95
23.00 # C 75
20.00 # C 75

7

EFFECT OF TENSILE LOAD ON COLLAPSE RESISTANCE
7 5/8" CASING
Grades K55 - N80 P110 - Q125 - V150

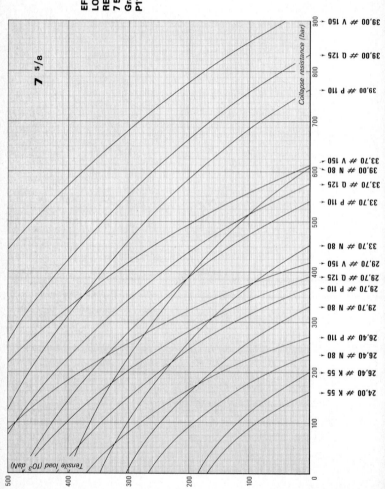

7 5/8

Tensile load (10³ daN)

Collapse resistance (bar)

→ 24,00 ## K 55
→ 26,40 ## K 55
→ 26,40 ## N 80
→ 26,40 ## P 110
→ 29,70 ## N 80
→ 29,70 ## P 110
→ 29,70 ## Q 125
→ 29,70 ## V 150
→ 33,70 ## N 80
→ 33,70 ## P 110
→ 33,70 ## Q 125
→ 39,00 ## N 80
→ 33,70 ## V 150
→ 39,00 ## P 110
→ 39,00 ## Q 125
→ 39,00 ## V 150

EFFECT OF TENSILE LOAD ON COLLAPSE RESISTANCE. 7 5/8" CASING

Grades C75 - C95

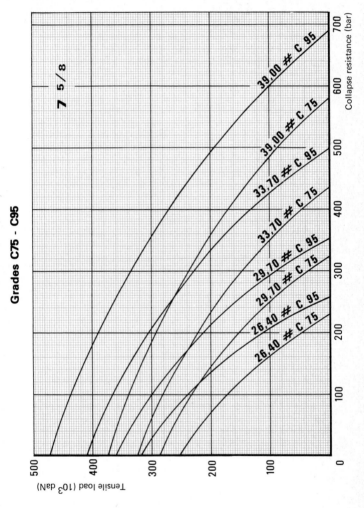

7 5/8

Collapse resistance (bar)

Tensile load (10³ daN)

39,00 # C 95
39,00 # C 75
33,70 # C 95
33,70 # C 75
29,70 # C 95
29,70 # C 75
26,40 # C 95
26,40 # C 75

EFFECT OF TENSILE
LOAD ON COLLAPSE
RESISTANCE
8 5/8'' CASING
Grades K55 - N80
P110 - Q125 - V150

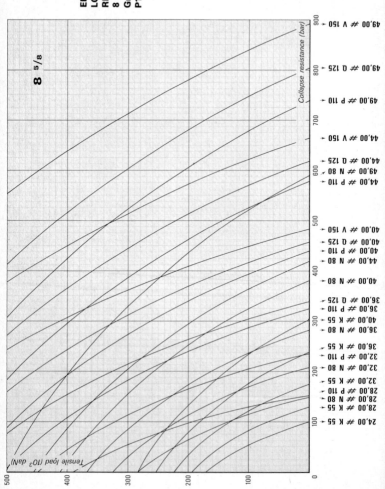

8 ⁵/₈

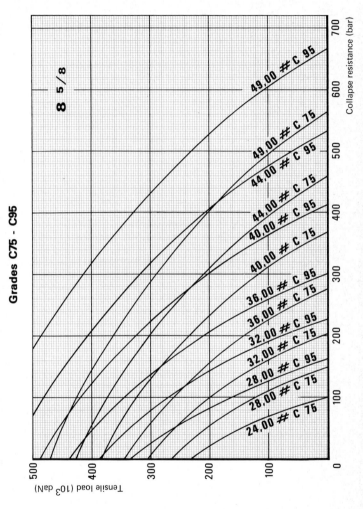

EFFECT OF TENSILE LOAD ON COLLAPSE RESISTANCE. 8 5/8" CASING

Grades C75 - C95

8 5/8

Tensile load (10^3 daN)

Collapse resistance (bar)

49,00 # C 95
49,00 # C 75
44,00 # C 95
44,00 # C 75
40,00 # C 95
40,00 # C 75
36,00 # C 95
36,00 # C 75
32,00 # C 95
32,00 # C 75
28,00 # C 95
28,00 # C 75
24,00 # C 76

EFFECT OF TENSILE LOAD ON COLLAPSE RESISTANCE: 9 5/8" CASING. Grades K55 - N80 - P110 - Q125 - V150

9 5/8

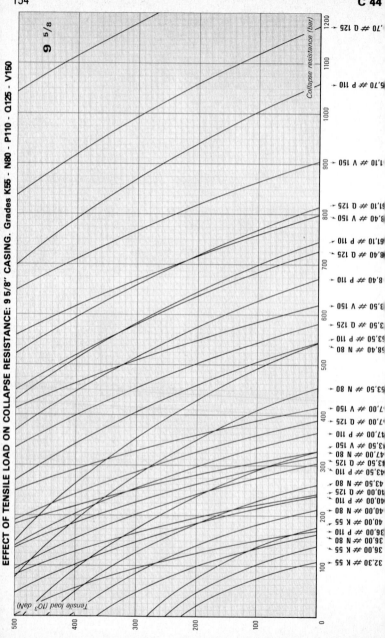

Collapse resistance (bar)

Tensile load (10³ daN)

1,70 ## Q 125
5,70 ## P 110
1,10 ## V 150
51,10 ## Q 125
3,40 ## V 150
61,10 ## P 110
8,40 ## Q 125
8,40 ## P 110
3,50 ## V 150
3,50 ## Q 125
53,50 ## P 110
58,40 ## N 80
53,50 ## N 80
7,00 ## V 150
7,00 ## Q 125
47,00 ## P 110
43,50 ## V 150
47,00 ## N 80
43,50 ## Q 125
43,50 ## P 110
43,50 ## N 80
40,00 ## Q 125
40,00 ## P 110
40,00 ## N 80
40,00 ## K 55
36,00 ## P 110
36,00 ## N 80
36,00 ## K 55
32,30 ## K 55

EFFECT OF TENSILE LOAD ON COLLAPSE RESISTANCE. 9 5/8" CASING

Grades C75 - C95

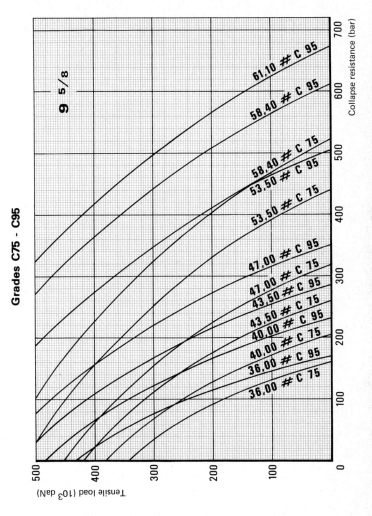

9 5/8

61,10 # C 95
58,40 # C 95
58,40 # C 75
53,50 # C 95
53,50 # C 75
47,00 # C 95
47,00 # C 75
43,50 # C 95
43,50 # C 75
40,00 # C 95
40,00 # C 75
36,00 # C 95
36,00 # C 75

Tensile load (10³ daN)

Collapse resistance (bar)

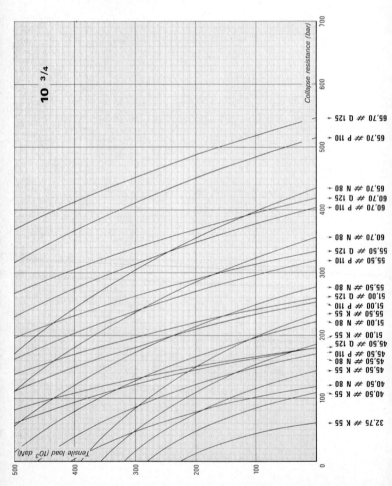

EFFECT OF TENSILE
LOAD ON COLLAPSE
RESISTANCE
10 3/4" CASING
Grades K55 - N80
P110 - Q125

10 ³/₄

Tensile load (10³ daN)

Collapse resistance f (bar)

→ 65,70 # Q 125
→ 65,70 # P 110

→ 65,70 # N 80
→ 60,70 # Q 125
→ 60,70 # P 110

→ 60,70 # N 80
→ 55,50 # Q 125
→ 55,50 # P 110

→ 55,50 # N 80
→ 51,00 # Q 125
→ 51,00 # P 110
→ 55,50 # K 55
→ 51,00 # N 80
→ 51,00 # K 55
→ 45,50 # Q 125
→ 45,50 # P 110
→ 45,50 # N 80
→ 45,50 # K 55
→ 40,50 # N 80
→ 40,50 # K 55

→ 32,75 # K 55

EFFECT OF TENSILE LOAD ON COLLAPSE RESISTANCE. 10 3/4" CASING

Grades C75 - C95

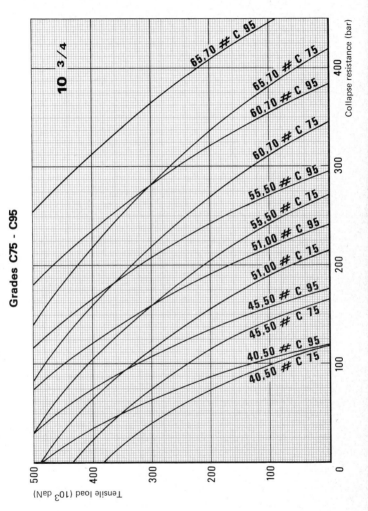

EFFECT OF TENSILE
LOAD ON COLLAPSE
RESISTANCE
11 3/4" CASING
Grades K55 - N80

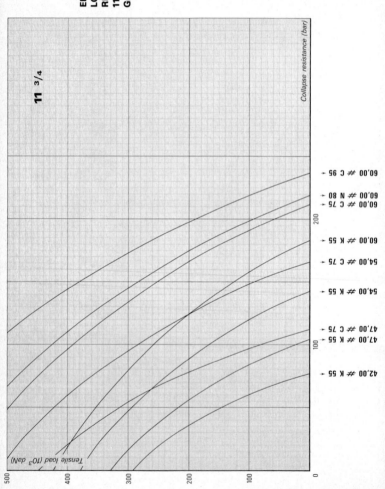

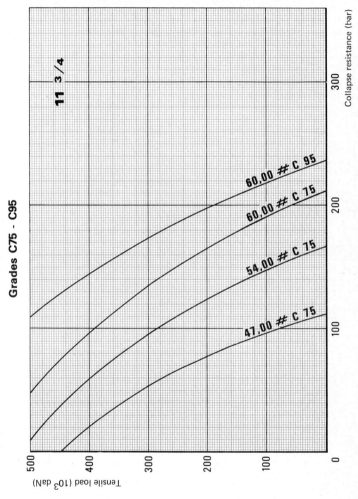

EFFECT OF TENSILE LOAD ON COLLAPSE RESISTANCE. 11 3/4" CASING

Grades C75 - C95

11 3/4

60,00 # C 95

60,00 # C 75

54,00 # C 75

47,00 # C 75

Collapse resistance (bar)

Tensile load (10³ daN)

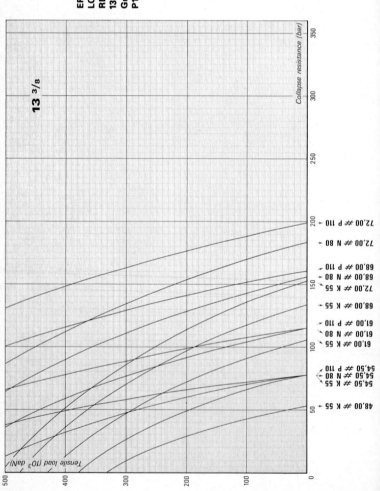

EFFECT OF TENSILE
LOAD ON COLLAPSE
RESISTANCE
13 3/8" CASING
Grades K55 - N80
P110

13 ³/₈

Collapse resistance (bar)

Tensile load (10³ daN)

EFFECT OF TENSILE LOAD ON COLLAPSE RESISTANCE. 13 3/8" CASING

Grades C75 - C95

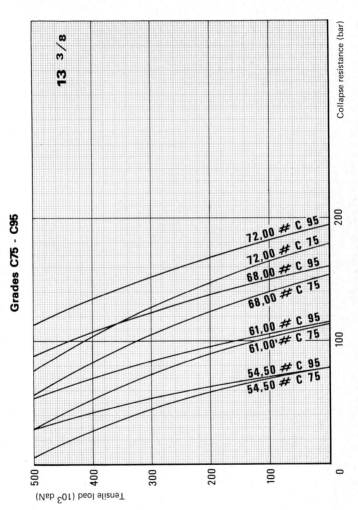

13 3/8

Collapse resistance (bar)

Tensile load (10^3 daN)

72,00 # C 95
72,00 # C 75
68,00 # C 95
68,00 # C 75
61,00 # C 95
61,00 # C 75
54,50 # C 95
54,50 # C 75

PHYSICAL PROPERTIES OF LINE PIPE RISER AND CONDUCTOR PIPE

Outside diameter 22 in (558.8 mm)

Property	Unit	9.52 mm / 0.375 in	12.70 mm / 0.500 in	19.05 mm / 0.750 in
Inside diameter	mm	539.8	533.4	520.7
Inside diameter	in	21.25	21.00	20.50
Cross section area	cm²	164	217	323
Cross section area	in²	25.4	33.6	50.1
Weight	kg/m	129.0	171.0	253.5
Weight	lb/ft	86.6	114.8	170.2
Displacement closed-end	l/m	245.2	245.2	245.2
Displacement closed-end	cu.ft/ft	2.64	2.64	2.64
Capacity	l/m	228.8	223.5	212.9
Capacity	cu.ft/ft	2.46	2.41	2.29
Steel grade B				
Collapse resistance (1)	bar	16.5	40	90
Collapse resistance (1)	psi	239	580	1305
Yield strength (1)	10³ daN	403	533	794
Yield strength (1)	10³ lb	906	1198	1785
Standard test pressures (2)	bar	62	82	123
Standard test pressures (2)	psi	890	1190	1790
Steel grade X52				
Collapse resistance (1)	bar	16.5	40	115
Collapse resistance (1)	psi	239	580	1668
Yield strength (1)	10³ daN	600	795	1183
Yield strength (1)	10³ lb	1349	1787	2659
Standard test pressures (2)	bar	110	147	173
Standard test pressures (2)	psi	1600	2130	2500

Outside diameter 24 in (609.6 mm)

Property	Unit	12.70 mm / 0.500 in	15.88 mm / 0.625 in	19.05 mm / 0.750 in	25.4 mm / 1.000 in
Inside diameter	mm	584.2	577.8	571.5	558.8
Inside diameter	in	23.00	22.75	22.50	22.00
Cross section area	cm²	239	296	354	467
Cross section area	in²	37.0	45.9	54.9	72.4
Weight	kg/m	186.9	232.4	277.4	365.9
Weight	lb/ft	125.5	156.0	186.2	245.6
Displacement closed-end	l/m	291.9	291.9	291.9	291.9
Displacement closed-end	cu.ft/ft	3.14	3.14	3.14	3.14
Capacity	l/m	268.0	262.3	256.5	245.2
Capacity	cu.ft/ft	2.89	2.82	2.76	2.64
Steel grade B					
Collapse resistance (1)	bar	30	51	77	128
Collapse resistance (1)	psi	435	740	1117	1856
Yield strength (1)	10³ daN	588	728	870	1148
Yield strength (1)	10³ lb	1322	1637	1956	2581
Standard test pressures (2)	bar	76	94	113	151
Standard test pressures (2)	psi	1090	1370	1640	2190
Steel grade X52					
Collapse resistance (1)	bar	30	60	94	168
Collapse resistance (1)	psi	435	870	1363	2436
Yield strength (1)	10³ daN	875	1084	1296	1710
Yield strength (1)	10³ lb	1967	2437	2913	3844
Standard test pressures (2)	bar	134	159	159	159
Standard test pressures (2)	psi	1950	2300	2300	2300

Outside diameter 26 in (660.4 mm)

Property	Unit	12.70 mm / 0.500 in	15.88 mm / 0.625 in	19.05 mm / 0.750 in	25.4 mm / 1.000 in
Inside diameter	mm	635.0	628.6	622.3	609.6
Inside diameter	in	25.00	24.75	24.50	24.00
Cross section area	cm²	257	321	383	506
Cross section area	in²	39.8	49.8	59.4	78.4
Weight	kg/m	202.8	252.3	301.3	397.7
Weight	lb/ft	136.2	169.4	202.3	267.0
Displacement closed-end	l/m	342.5	342.5	342.5	342.5
Displacement closed-end	cu.ft/ft	3.69	3.69	3.69	3.69
Capacity	l/m	316.8	310.4	304.2	291.9
Capacity	cu.ft/ft	3.41	3.34	3.27	3.14
Steel grade B					
Collapse resistance (1)	bar	24	41	65	112
Collapse resistance (1)	psi	348	595	943	1624
Yield strength (1)	10³ daN	632	789	941	1244
Yield strength (1)	10³ lb	1421	1774	2115	2797
Standard test pressures (2)	bar	70	87	104	138
Standard test pressures (2)	psi	1010	1260	1510	2000
Steel grade X52					
Collapse resistance (1)	bar	24	47	77	146
Collapse resistance (1)	psi	348	682	1117	2117
Yield strength (1)	10³ daN	941	1175	1402	1853
Yield strength (1)	10³ lb	2115	2641	3152	4166
Standard test pressures (2)	bar	125	138	138	138
Standard test pressures (2)	psi	1800	2000	2000	2000

(1) Collapse resistance and yield strength are calculated by means of formulas from API Bull. 5C3.
(2) Standard test pressures are calculated taking fibre stress equal to 75 % of the specified minimum yield strength for grade B and 90 % for grade X 52 (API Std 5L and 5LX).

PHYSICAL PROPERTIES OF LINE PIPE
RISER AND CONDUCTOR PIPE (continued)

		30				32				36			
Outside diameter	in	30				32				36			
	mm	762.0				812.8				914.4			
Wall thickness	mm	12.70	15.88	19.05	25.4	12.70	15.88	19.05	25.4	12.70	15.88	19.05	25.4
	in	0.500	0.625	0.750	1.000	0.500	0.625	0.750	1.000	0.500	0.625	0.750	1.000
Inside diameter	mm	736.6	730.2	723.9	711.2	787.4	781.0	774.7	762.0	889.0	882.6	876.3	863.6
	in	29.00	28.75	28.50	28.00	31.00	30.75	30.50	30.00	35.00	34.75	34.50	34.00
Cross section area	cm²	300	372	444	588	320	396	476	628	360	448	536	712
	in²	46.5	57.7	68.8	91.1	49.6	61.4	73.8	97.3	55.8	69.4	83.1	110.4
Weight	kg m	234.6	292.1	349.0	461.3	250.6	312.0	372.8	493.1	282.4	351.7	420.6	556.8
	lb ft	157.5	196.1	234.0	309.7	168.2	209.4	250.3	331.1	189.6	236.1	282.4	373.8
Displacement closed-end	l m	456.0	456.0	456.0	456.0	518.8	518.8	518.8	518.8	656.8	656.8	656.8	656.8
	cu.ft ft	4.91	4.91	4.91	4.91	5.58	5.58	5.58	5.58	7.07	7.07	7.07	7.07
Capacity	l m	426.0	418.8	411.6	397.2	486.8	479.2	471.2	456.0	620.8	612.0	603.2	585.6
	cu.ft ft	4.59	4.51	4.43	4.28	5.24	5.16	5.07	4.91	6.68	6.59	6.49	6.30
Steel grade		B X52	B X52	B X52	B X52	B X52	B X52	B X52	B X52	B X52	B X52	B X52	B X52
Collapse resistance (1)	bar	15 15	30 30	46 49	87 109	13 13	25 25	44 44	77 94	9 9	18 18	30 30	60 69
	psi	218 218	435 435	667 711	1262 1581	189 189	363 363	638 638	1120 1363	131 131	261 261	435 435	870 1000
Yield strength (1)	10³ daN	642 1077	796 1335	1070 1594	1417 2110	771 1149	954 1422	1147 1709	1513 2255	868 1292	1080 1608	1292 1924	1716 2556
	10³ lb	1443 2421	1789 3000	2405 3583	3185 4743	1733 2583	2145 3197	2578 3842	3401 5069	1951 2904	2428 3615	2904 4325	3858 5753
Standard test pressures (2)	bar	61 108	76 134	90 162	121 207	57 100	71 127	84 151	113 201	50 89	63 112	76 134	101 180
	psi	880 1560	1090 1950	1310 2340	1750 3000	820 1460	1030 1830	1230 2190	1640 2920	730 1300	910 1620	1090 1950	1460 2600

(1) Collapse resistance and yield strength are calculated by means of formulas from API Bull. 5C3.
(2) Standard test pressures are calculated taking fibre stress equal to 75 % of the specified minimum yield strength for Grade B and 90 % for Grade X52 (API Std 5L and 5LX).

PHYSICAL PROPERTIES OF LINE PIPE
RISER AND CONDUCTOR PIPE (continued)

Outside diameter 40 in / 1016.0 mm

		Wall 12.70 mm / 0.500 in		Wall 15.88 mm / 0.625 in		Wall 19.05 mm / 0.750 in		Wall 25.4 mm / 1.000 in	
		B	X52	B	X52	B	X52	B	X52
Inside diameter	mm	990.6		984.2		977.9		965.2	
	in	39.00		38.75		38.50		38.00	
Cross section area	cm²	400		499		596		790	
	in²	62.0		77.3		92.4		122.5	
Weight	kg m	314.2		391.5		468.3		620.4	
	lb ft	210.9		262.8		314.4		416.5	
Displacement closed-end	l m	810.7		810.7		810.7		810.7	
	cu ft ft	8.73		8.73		8.73		8.73	
Capacity	l m	770.7		760.8		751.1		731.7	
	cu ft ft	8.30		8.19		8.09		7.88	
Collapse resistance (1)	bar	6.5	6.5	13	13	22	22	48	53
	psi	94	94	189	189	319	319	696	769
Yield strength (1)	10³ daN	964	1436	1203	1791	1436	2140	1904	2836
	10³ lb	2167	3228	2704	4026	3228	4810	4280	6375
Standard test pressures (2)	bar	45	80	57	101	68	122	90	162
	psi	660	1170	820	1460	980	1760	1310	2340

Outside diameter 42 in / 1066.8 mm

		Wall 12.70 mm / 0.500 in		Wall 15.88 mm / 0.625 in		Wall 19.05 mm / 0.750 in		Wall 25.4 mm / 1.000 in	
		B	X52	B	X52	B	X52	B	X52
Inside diameter	mm	1041.4		1035.0		1028.7		1016.0	
	in	41.00		40.75		40.50		40.00	
Cross section area	cm²	420		525		627		831	
	in²	65.1		81.4		97.2		128.8	
Weight	kg m	330.1		411.4		492.2		652.2	
	lb ft	221.6		276.2		330.4		437.9	
Displacement closed-end	l m	893.8		893.8		893.8		893.8	
	cu ft ft	9.63		9.63		9.63		9.63	
Capacity	l m	851.8		841.3		831.1		810.7	
	cu ft ft	9.17		9.06		8.95		8.73	
Collapse resistance (1)	bar	5.6	5.6	11	11	19	19	40	46
	psi	81	81	160	160	276	276	580	667
Yield strength (1)	10³ daN	1012	1508	1265	1884	1511	2251	2003	2983
	10³ lb	2275	3390	2844	4235	3397	5060	4503	6706
Standard test pressures (2)	bar	43	77	54	96	65	115	86	154
	psi	620	1110	780	1390	940	1670	1250	2230

(1) Collapse resistance and yield strength are calculated by means of formulas from API Bull. 5C3.
(2) Standard test pressures are calculated taking fibre stress equal to 75 % of the specified minimum yield strength for Grade B and 90 % for Grade X52 (API Std 5L and 5LX).

API TUBING LIST

Outside diameter		Nominal weight	Grade					Wall thickness		Type of ends (1)		
in	mm	lb/ft	H	J	75	N	P	in	mm	Non-upset	External upset	Integral joint
1,050	26,7	1,14 (2)	●	●	●	●		0,113	2,87	x		
		1,20	●	●	●	●		0,113	2,87		x	
1,315	33,4	1,70	●	●	●	●		0,133	3,38	x		
		1,72	●	●	●	●		0,133	3,38			x
		1,80	●	●	●	●		0,133	3,38		x	
1,660	42,2	2,10	●	●				0,125	3,18			x
		2,30	●	●	●	●		0,140	3,56	x		
		2,33	●	●	●	●		0,140	3,56			x
		2,40	●	●	●	●		0,140	3,56		x	
1,900	48,3	2,40	●	●				0,125	3,18			x
		2,75	●	●	●	●		0,145	3,68	x		
		2,76	●	●	●	●		0,145	3,68			x
		2,90	●	●	●	●		0,145	3,68		x	
2,063	52,4	3,25	●	●	●	●		0,156	3,96			x
2 3/8	60,3	4,00	●	●	●	●		0,167	4,24	x		
		4,60	●	●	●	●	●	0,190	4,83	x		
		4,70	●	●	●	●	●	0,190	4,83		x	
		5,80			●	●	●	0,254	6,45	x		
		5,95			●	●	●	0,254	6,45		x	
2 7/8	73,0	6,40	●	●	●	●	●	0,217	5,51	x		
		6,50	●	●	●	●	●	0,217	5,51		x	
		8,60			●	●	●	0,308	7,82	x		
		8,70			●	●	●	0,308	7,82		x	
3 1/2	88,9	7,70	●	●	●	●		0,216	5,49	x		
		9,20	●	●	●	●	●	0,254	6,45	x		
		9,30	●	●	●	●	●	0,254	6,45		x	
		10,20	●	●	●	●		0,289	7,34	x		
		12,70			●	●	●	0,375	9,52	x		
		12,95			●	●	●	0,375	9,52		x	
4	101,6	9,50	●	●	●	●		0,226	5,74	x		
		11,00	●	●	●	●		0,262	6,65		x	
4 1/2	114,3	12,60	●	●	●	●		0,271	6,88	x		
		12,75	●	●	●	●		0,271	6,88		x	

(1) Non-upset tubing is available with regular couplings or special bevel couplings. External upset tubing is available with regular, special bevel or special clearance couplings.
(2) For information purposes only, non API.

API TUBING THREAD

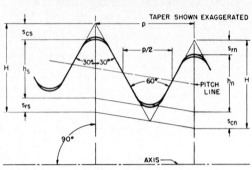

Taper 6,25 %
(0,0625 inch per inch)
(0,750 inch per foot)

Thread element	10 threads per inch p = 2,540	8 threads per inch p = 3,175
H = 0,866 p	2,200 mm	2,750 mm
h = 0,626 p − 0,178	1,412 mm	1,810 mm
tb = 0,120 p + 0,051	0,356 mm	0,432 mm
ts = 0,120 p + 0,127	0,432 mm	0,508 mm

NUMBER OF THREADS PER INCH			
Outside diameter (in)	Non-upset tubing	External upset tubing	Integral joint tubing
1,050	10	10	—
1,315	10	10	10
1,660	10	10	10
1,900	10	10	10
2,063	—	—	10
2 3 8	10	8	—
2 7 8	10	8	—
3 1 2	10	8	—
4	8	8	—
4 1/2	8	8	—

DIMENSIONS AND WEIGHTS OF TUBING
(API Stds 5A, 5AC, 5AX ; API Bul 5C2 and Vallourec notice No 2306)

OD (in) (mm)	Nominal weight (lb/ft)	Thickness (mm)	Inside diameter ID (mm)	Cross section area (mm²)	Drift diameter (mm)	Non-upset: outside diam of coupling (mm)	Non-upset: weight with coupling (kg/m)	Ext upset: coupling OD Reg (mm)	Ext upset: coupling OD Spe (mm)	Ext upset wt Regular short (kg/m)	Ext upset wt Regular long (kg/m)	Ext upset wt Special short (kg/m)	Ext upset wt Special long (kg/m)	Integral joint OD (mm)	Integral joint ID (mm)	Integral joint weight (kg/m)	VAM OD heavy (mm)	VAM OD regular (mm)	VAM OD special (mm)	VAM wt heavy (kg/m)	VAM wt regular (kg/m)	VAM wt special (kg/m)
1.050 (26.7)	1.14	2.87	20.9	215	18.5	33.4	1.70	—	—	—	—	—	—	—	—	—	—	—	—	—	—	—
	1.20	2.87	20.9	215	18.5	—	—	42.2	—	1.79	—	—	—	—	—	—	—	—	—	—	—	—
1.315 (33.4)	1.70	3.38	26.6	319	24.3	42.2	2.53	—	—	—	—	—	—	—	—	—	—	—	—	—	—	—
	1.72	3.38	26.6	319	24.3	—	—	—	—	—	—	—	—	39.4	24.6	2.52	—	—	—	—	—	—
	1.80	3.38	26.6	319	24.3	—	—	48.3	—	2.60	—	—	—	—	—	—	—	—	—	—	—	—
1.660 (42.2)	2.10	3.18	35.8	389	32.7	—	—	—	—	—	—	—	—	47.8	33.0	3.07	—	—	—	—	—	—
	2.30	3.56	35.1	431	32.7	52.2	3.44	—	—	—	—	—	—	—	—	—	—	—	—	—	—	—
	2.33	3.56	35.1	431	32.7	—	—	—	—	—	—	—	—	47.8	33.0	3.39	—	—	—	—	—	—
	2.40	3.56	35.1	431	32.7	—	—	55.9	—	3.50	—	—	—	—	—	—	—	—	—	—	—	—
1.900 (48.3)	2.40	3.18	41.9	450	38.5	—	—	—	—	—	—	—	—	53.6	38.9	3.54	—	—	—	—	—	—
	2.75	3.68	40.9	516	38.5	55.9	4.09	—	—	—	—	—	—	—	—	—	—	—	—	—	—	—
	2.76	3.68	40.9	516	38.5	—	—	—	—	—	—	—	—	53.6	38.9	4.06	—	—	—	—	—	—
	2.90	3.68	40.9*	516	38.5	—	—	63.5	—	4.20	—	—	—	—	—	—	—	—	—	—	—	—
2.063 (52.4)	3.25	3.96	44.5	603	42.1	—	—	—	—	—	—	—	—	59.1	42.5	4.75	—	—	—	—	—	—
2 3/8 (60.3)	4.00	4.24	51.8	747	49.5	73.0	5.98	—	—	—	—	—	—	—	—	—	—	—	—	—	—	—
	4.60	4.83	50.7	841	48.3	73.0	6.71	—	—	—	—	—	—	—	—	—	73.0	68.5	66.5	6.75	6.67	6.64
	4.70	4.83	50.7	841	48.3	—	—	77.8	73.9	6.89	6.94	6.80	6.83	—	—	—	—	—	—	—	—	—
	5.10*	5.54	49.2	952	46.8	—	—	—	—	—	—	—	—	—	—	—	73.0	68.5	66.5	7.61	7.53	7.49
	5.80	6.45	47.4	1092	45.0	73.0	8.66	—	—	—	—	—	—	—	—	—	73.0	70.5	68.5	8.70	8.66	8.62
	5.95	6.45	47.4	1092	45.0	—	—	77.8	73.9	8.83	8.87	8.75	8.77	—	—	—	—	—	—	—	—	—
	6.30	7.12	46.1	1188	43.7	—	—	—	—	—	—	—	—	—	—	—	73.0	72.5	—	9.44	9.43	—
	7.30	8.53	43.2	1389	40.9	—	—	—	—	—	—	—	—	—	—	—	73.0	72.5	—	10.98	10.97	—

* These nominal weights are not API).

DIMENSIONS AND WEIGHTS OF TUBING (continued)
(API Stds 5A, 5AC, 5AX ; API Bul 5C2 and Vallourec notice No. 2306)

Outside diameter (OD) (in) (mm)	Pipe Body Nominal weight (lb/ft)	Thickness (mm)	Inside diameter (ID) (mm)	Cross section area (mm²)	Drift diameter (mm)	Non-upset OD of coupling (mm)	Non-upset Weight with coupling (kg/m)	Ext. upset OD of coupling Reg. (mm)	Ext. upset OD of coupling Spe. (mm)	EU Weight Regular short (kg/m)	EU Weight Regular long (kg/m)	EU Weight Special short (kg/m)	EU Weight Special long (kg/m)	Integral OD (mm)	Integral ID (mm)	Integral Weight (kg/m)	VAM OD heavy (mm)	VAM OD regular (mm)	VAM OD special (mm)	VAM Weight heavy (kg/m)	VAM Weight regular (kg/m)	VAM Weight special (kg/m)
2 7/8 (73.0)	6.40	5.51	62.0	1169	59.6	88.9	9.41	93.2	87.9	9.58	9.66	9.45	9.48	—	—	—	88.9	81.2	80.0	9.28	9.25	9.23
	6.50	5.51	62.0	1169	59.6	—	—	—	—	—	—	—	—	—	—	—	—	—	—	—	—	—
	7.70*	7.01	59.0	1454	56.6	—	—	—	—	—	—	—	—	—	—	—	88.9	84.5	82.9	9.92	9.81	9.77
	8.60	7.82	57.4	1603	55.0	88.8	12.75	93.2	87.9	12.93	13.01	12.80	12.83	—	—	—	88.9	84.5	82.9	12.80	12.69	12.65
	8.70	7.82	57.4	1603	55.0	—	—	—	—	—	—	—	—	—	—	—	88.9	87.7	—	14.64	14.53	—
	9.70	9.19	54.6	1844	52.3	—	—	—	—	—	—	—	—	—	—	—	88.9	87.7	—	16.06	16.03	—
	10.70	10.28	52.5	2028	50.1	—	—	—	—	—	—	—	—	—	—	—	—	—	—	—	—	—
3 1/2 (88.9)	7.70	5.49	77.9	1438	74.8	108.0	11.68	—	—	—	—	—	—	—	—	—	108.0	96.6	96.4	11.75	11.36	13.18
	9.20	6.45	76.0	1671	74.8	108.0	13.48	—	—	—	—	—	—	—	—	—	108.0	98.1	—	13.57	13.23	—
	9.30	6.45	76.0	1671	72.8	—	—	114.3	106.2	13.80	13.91	13.51	13.57	—	—	—	—	—	—	—	—	—
	10.20	7.34	74.2	1881	71.0	108.0	15.11	—	—	—	—	—	—	—	—	—	108.0	99.5	97.9	15.20	14.90	14.85
	12.70	9.52	69.8	2375	66.7	108.0	18.93	114.3	106.2	19.24	19.35	18.96	18.99	—	—	—	108.0	102.5	100.5	19.02	18.83	18.76
	12.95	9.52	69.8*	2375	66.7	—	—	—	—	—	—	—	—	—	—	—	—	—	—	—	—	—
	13.70*	10.50	67.9	2600	64.7	—	—	—	—	—	—	—	—	—	—	—	108.0	105.1	102.3	20.61	20.51	20.41
	14.70*	11.43	66.1	2777	62.9	—	—	—	—	—	—	—	—	—	—	—	108.0	105.1	102.3	22.12	22.01	21.91
	15.80*	12.40	64.1	2917	60.9	—	—	—	—	—	—	—	—	—	—	—	108.0	—	104.9	23.67	—	23.56
4 (101.6)	9.50	5.74	90.1	1729	86.9	120.6	14.02	—	—	—	—	—	—	—	—	—	120.6	109.9	—	14.14	13.69	15.70
	10.90*	6.65	88.3	1985	85.1	—	—	—	—	—	—	—	—	—	—	—	120.6	110.9	—	16.12	15.76	—
	11.00	6.65	88.3	1985	85.1	—	—	127.0	—	16.36	—	—	—	—	—	—	—	—	—	—	—	—
	13.00*	8.38	84.8	2455	81.7	—	—	—	—	—	—	—	—	—	—	—	120.6	113.5	—	19.71	19.41	19.30
	14.80*	9.65	82.3	2788	79.1	—	—	—	—	—	—	—	—	—	—	—	120.6	117.2	—	22.29	22.14	—
	16.50*	10.92	79.8	3112	76.6	—	—	—	—	—	—	—	—	—	—	—	120.6	117.2	—	24.78	24.64	—
4 1/2 (114.3)	12.60	6.88	100.5	2323	97.4	132.1	18.66	—	—	—	—	—	—	—	—	—	132.1	123.5	122.0	18.83	18.39	18.32
	12.75	6.88	100.5	2323	97.4	—	—	141.3	—	19.20	—	—	—	—	—	—	—	—	—	—	—	—

* These nominal weights are not API.

PERFORMANCE PROPERTIES OF TUBING
(API Bul 5C2 and Vallourec notice)

Outside diameter (OD) (in)(mm)	Nominal weight Non-upset (lb/ft)	Upset (lb/ft)	Integral (lb/ft)	VAM (lb/ft)	Wall thickness (mm)	Grade	Pipe body Collapse resistance (bar)(1)	Pipe body Internal yield pressure (bar)(2)	Pipe body yield strength (10³ daN)	Round threaded coup. Yield strength(1) Non-upset (10³ daN)	Round threaded coup. Yield strength(1) Upset (10³ daN)	Round threaded coup. Make-up torque (m-daN)	Integral joint Yield strength (10³ daN)	Integral joint Make-up torque (m-daN)	Heavy weight (10³ daN)	VAM coupling
1.050 (2.67)	1.14	1.20	—	—	2.87	H40	530	519	5.93	2.83	5.93	62	—	—		
						K55	728	714	8.14	3.89	8.14	80	—	—		
						C75	994	974	11.10	5.30	11.10	105	—	—		
						N80	1060	1039	11.85	5.63	11.85	110	—	—		
						P105*	1391	1245	15.56	7.43	15.56		—	—		
1.315 (33.4)	1.70	1.80	1.72	—	3.38	H40	501	488	8.80	4.89	8.80	60	6.96	42		
						K55	690	671	12.10	6.71	12.10	77	9.57	54		
						C75	940	915	16.50	9.15	16.50	100	14.04	70		
						N80	1003	976	17.60	9.75	17.60	107	13.93	76		
						P105*	1316		23.09	12.80	23.09		—	—		
1.660 (4.22)	—	—	2.10	—	3.18	H40	384	363	10.75	—	—	—	9.76	51		
						K55	528	500	14.77	—	—	—	13.32	67		
	2.30	2.40	2.33	—	3.56	H40	426	407	11.93	6.92	11.93	72	9.76	51		
						K55	586	560	16.37	9.51	16.37	93	13.32	67		
						C75	798	763	22.33	13.03	22.33	120	18.14	88		
						N80	852	814	23.82	13.83	23.82	130	19.32	93		
						P105*	1118	1068	31.27	18.15	31.27		—	—		
1.900 (48.3)	—	—	2.40	—	3.18	H40	339	317	12.42	—	—	—	11.72	61		
						K55	458	436	17.08	—	—	—	16.08	78		
	2.75	2.90	2.76	—	3.68	H40	389	368	14.24	8.51	14.24	90	11.72	61		
						K55	534	507	19.58	11.69	19.58	120	16.08	78		
						C75	729	691	26.70	15.94	26.70	155	21.97	103		
						N80	778	736	28.49	17.00	28.49	165	23.44	110		
						P105*	985	967	37.39	22.32	37.39		—	—		
2.063 (52.4)	—	—	3.25	—	3.96	H40	386	365		—	—	—	15.59	77		
						K55	530	502	22.56	—	—	—	21.38	100		
						C75	723	684	30.41	—	—	—	29.22	130		
						N80	771	730	32.37	—	—	—	31.19	140		
						P105*				—	—	—				

* Non API grades or nominal weights. (1) Based on nominal weights. (2) Based on 87,5 % of nominal thickness.

(1) Based on nominal thickness.

PERFORMANCE PROPERTIES OF TUBING (continued)
(API Bul 5C2 and Vallourec notice)

Outside diameter (OD) (in) (mm)	Nominal weight Non-upset (lb/ft)	Nominal weight Upset (lb/ft)	Nominal weight Integral (lb/ft)	Nominal weight VAM (lb/ft)	Wall thickness (mm)	Grade	Pipe body Collapse resistance (bar)(1)	Pipe body Internal yield pressure (bar)(2)	Round threaded coup. Yield strength Non-upset (10³ daN)(1)	Round threaded coup. Yield strength Upset (10³ daN)(1)	Round threaded coup. Make-up torque (m-daN)	Integral joint Yield strength (10³ daN)	Integral joint Make-up torque (m-daN)	VAM coupling Heavy weight (10³ daN)	VAM coupling Yield strength Regular same grade (10³ daN)	VAM coupling Yield strength Regular higher grade (10³ daN)	VAM coupling Yield strength Spe clearance same grade (10³ daN)	VAM coupling Yield strength Spe clearance higher grade (10³ daN)	VAM coupling Make-up torque (m-daN)
2 3/8 (60.3)	4 00	—	—	—	4.24	H40	361	339	20 63	—	—	—	—	—	—	—	—	—	—
						K55	496	467	28 37	—	63	—	—	—	—	—	—	—	—
						C75	656	636	38 69	—	82	—	—	—	—	—	—	—	—
						N80	688	678	41 27	—	108	—	—	—	—	—	—	—	—
						P105*	848	892	54 4	—	114	—	—	—	—	—	—	—	—
	4 60	4 70	—	4 60	4.83	H40	406	386	23 23	23 23	—	—	—	23 23	23 23	23 23	19 84	23 23	—
						K55	558	531	31 94	31 94	130	—	—	31 94	31 94	31 94	27 29	31 94	190
						C75	761	724	43 56	43 56	175	—	—	43 56	43 56	46 47	38 17	46 47	270
						N80	812	772	46 47	46 47	230	—	—	46 47	51 60	61 00	39 69	61 00	270
						095*	964	917	55 18	55 18	244	—	—	55 18	55 18	—	47 09	—	270
						P105	1066	1014	61 00	61 00	307	—	—	61 00	67 98	—	51 85	—	290
	5 10*	5 20*	—	5 10*	5.54	K55*	630	609	36 18	36 18	—	—	—	36 18	35 31	35 31	27 46	35 31	217
						C75*	860	831	49 34	49 34	—	—	—	49 34	48 07	52 70	37 27	52 70	270
						N80*	920	886	52 70	52 70	—	—	—	52 70	51 60	67 98	39 40	61 76	270
						095*	1090	1052	62 50	62 50	—	—	—	62 50	61 32	—	47 09	—	270
						P105*	1210	1163	69 00	69 00	—	—	—	69 00	67 98	—	51 90	—	318
	5 80	5 95	—	5 80	6.45	K55*	724	710	41 45	41 45	285	—	—	41 45	41 45	41 45	35 36	41 45	270
						C75	988	968	55 97	55 97	303	—	—	55 97	55 97	60 34	48 22	60 34	270
						N80*	1054	1032	60 34	60 34	380	—	—	60 34	60 34	79 16	60 34	79 16	270
						095*	1251	1227	71 60	71 60	—	—	—	71 60	71 60	—	61 10	—	366
						P105	1383	1355	79 16	79 16	—	—	—	79 16	79 16	—	68 06	—	366
	—	6 30*	—	6 30*	7.12	C75*	1076	1067	61 52	61 52	—	—	—	61 52	61 52	61 52	—	—	420
						N80*	1147	1138	65 64	65 64	—	—	—	65 64	65 64	65 64	—	—	420
						095*	1362	1351	78 00	78 00	—	—	—	78 00	78 00	86 20	—	—	420
						P105*	1506	1494	86 20	86 20	—	—	—	86 20	86 20	—	—	—	420
	—	—	—	7 30*	8.53	C75*	1256	1280	71 90	71 90	—	—	—	71 90	71 90	71 90	—	—	420
						N80*	1340	1366	76 70	76 70	—	—	—	76 70	76 70	76 70	—	—	420
						095*	1591	1622	91 08	91 08	—	—	—	91 08	91 08	—	—	—	468
						P105*	1759	1792	100 66	100 66	—	—	—	100 66	100 66	—	—	—	468

* Non API grades or nominal weights.   (1) Based on nominal thickness.   (2) Based on 87.5 % of nominal thickness.

PERFORMANCE PROPERTIES OF TUBING (continued)
(API Bul 5C2 and Vallourec notice)

Outside diameter (OD) (in) (mm)	Non-upset (lb/ft)	Upset (lb/ft)	Integral (lb/ft)	VAM (lb/ft)	Wall thickness (mm)	Grade	Pipe body Collapse resistance (bar)(1)	Pipe body Internal yield pressure (bar)(2)	Round threaded coup. Yield strength Non-upset (10³daN)	Round threaded coup. Yield strength Upset (10³daN)	Round threaded coup. Make-up torque (m-daN)	Integral joint Yield strength (10³daN)	Integral joint Make-up torque (m-daN)	VAM Heavy weight (10³daN)	VAM Regular same grade (10³daN)	VAM Regular higher grade (10³daN)	VAM Spe same grade (10³daN)	VAM Spe higher grade (10³daN)	VAM Make-up torque (m-daN)
2 7/8 (73.0)	6.40	6.50	—	6.40	5.51	H40	385	364	23.50	32.27	170	—	—	32.27	32.27	32.27	28.21	32.27	320
						K55	529	501	32.32	44.39	225	—	—	44.39	44.39	44.39	38.80	44.39	365
						C75	722	683	44.07	60.54	290	—	—	60.54	60.54	—	52.92	52.97	365
						N80	770	729	47.02	64.57	310	—	—	64.57	64.57	64.57	56.43	64.57	365
						095*	891	865	55.84	76.67	—	—	—	76.67	76.67	—	67.00	—	365
						P105*	966	956	61.73	84.74	—	—	—	84.74	84.74	84.74	74.07	84.74	365
	7.90*	—	—	7.70*	7.01	K55*	658	637	42.16	55.20	390	—	—	55.20	55.20	55.20	38.75	52.97	365
						C75*	898	869	58.86	75.30	—	—	—	75.30	75.30	—	72.11	—	440
						N80*	957	927	62.76	80.31	—	—	—	80.31	80.31	80.31	76.51	80.31	440
						095*	1137	1100	74.52	95.35	—	—	—	95.35	95.35	—	91.23	83.50	440
						P105*	1257	1216	82.37	105.38	—	—	—	105.38	105.38	105.38	74.07	—	440
	8.60	8.70	—	8.60	7.82	C75	989	969	66.52	82.97	385	—	—	82.97	82.97	82.97	72.11	88.50	440
						N80	1055	1034	78.94	88.50	—	—	—	88.50	88.50	88.50	76.89	—	440
						095*	1253	1228	84.27	105.40	—	—	—	105.40	105.40	—	91.23	—	440
						P105	1385	1358	93.14	116.15	—	—	—	116.15	116.15	116.15	100.94	116.15	440
	9.90*	—	—	9.70*	9.19	C75*	1138	1139	79.05	95.46	—	—	—	95.46	95.46	—	—	—	515
						N80*	1214	1215	84.32	101.84	—	—	—	101.84	101.84	101.84	—	—	515
						095*	1442	1443	100.12	120.92	—	—	—	120.92	120.92	—	—	—	560
						P105*	1594	1595	110.67	133.65	—	—	—	133.65	133.65	133.65	—	—	560
	10.85*	—	—	10.70*	10.28	C75*	1252	1275	88.53	104.96	—	—	—	104.96	104.96	—	—	—	560
						N80*	1335	1360	94.42	111.97	—	—	—	111.97	111.97	111.97	—	—	560
						095*	1586	1615	112.13	133.97	—	—	—	133.97	133.97	—	—	—	595
						P105*	1752	1785	123.93	146.97	515	—	—	146.97	146.97	146.97	—	—	595
3 1/2 (88.9)	7.70	7.80*	—	7.70	5.49	H40	319	298	28.97	39.70	125	—	—	39.70	39.70	—	—	—	390
						K55	412	410	39.85	54.60	164	—	—	54.60	54.60	—	—	—	390
						C75	520	558	54.35	74.45	216	—	—	74.45	74.45	—	—	—	440
						N80	542	596	57.87	79.32	230	—	—	79.32	79.32	79.32	—	79.32	440
						095*	610	707	68.83	94.28	—	—	—	94.28	94.28	—	—	—	515
						P105*	651	782	76.08	104.20	—	—	—	104.20	104.20	104.20	—	104.20	515

* Non API grades or nominal weights.
(1) Based on nominal thickness. (2) Based on 87.5 % of nominal thickness.

PERFORMANCE PROPERTIES OF TUBING (continued)
(API Bul 5C2 and Vallourec notice)

Outside diameter (OD) (in)(mm)	Nominal weight Non-upset (lb/ft)	Upset (lb/ft)	Integral (lb/ft)	VAM (lb/ft)	Wall thickness (mm)	Grade	Pipe body Collapse resistance (bar)(1)	Pipe body Internal yield pressure (bar)(2)	Pipe body Yield strength (10¹ daN)	Round threaded coup Non-upset (10¹ daN)	Round threaded coup Upset (10¹ daN)	Round threaded coup Make-up torque (m-daN)	Integral joint Yield strength (10¹ daN)	Integral joint Make-up torque (m-daN)	VAM coupling Heavy weight (10¹ daN)	Regular same grade (10¹ daN)	Regular higher grade (10¹ daN)	Spe clearance same grade (10¹ daN)	Spe clearance higher grade (10¹ daN)	VAM coupling Make-up torque (m-daN)
3 1 2 (88.9)	9.20	9.30*	—	9.20	6.45	H40	371	350	46 10	35 42	46 10	230	—	—	46 10	46 10	46 10	39 95	46 10	393
						K55	511	481	63 46	48 71	63 46	310	—	—	63 46	63 46	63 46	54 93	63 46	440
						C75	692	656	86 51	66 43	86 51	405	—	—	86 51	86 51	—	74 91	—	440
						N80	726	701	92 30	70 85	92 30	430	—	—	92 30	92 30	92 30	80 00	92 30	515
						095*	831	832	109 60	84 14	109 60	—	—	—	109 60	109 60	—	94 89	—	515
						P105	900	919	121 13	93 00	121 13	550	—	—	121 13	121 13	121 13	104 88	121 13	515
	10.20	10.30*	—	10.20	7.34	H40	418	398	51 95	41 40	51 95	175	—	—	51 95	51 95	51 95	44 97	51 95	515
						K55	575	548	71 42	56 68	71 42	230	—	—	71 42	71 42	71 42	61 85	71 42	637
						C75	783	747	97 35	77 29	97 35	305	—	—	97 35	97 35	—	84 35	—	637
						N80	836	797	103 88	82 43	103 88	325	—	—	103 88	103 88	103 88	90 00	103 88	690
						095*	987	947	123 36	97 90	123 36	—	—	—	123 36	123 36	—	106 84	—	690
						P105	1097	1046	136 34	108 30	136 34	—	—	—	136 34	136 34	136 34	118 08	136 34	690
	12.70	12.95	—	12.70	9.52	K55	726	711	90 20	75 46	90 20	545	—	—	90 20	90 20	90 20	79 93	90 20	690
						C75	989	969	122 97	103 00	122 97	580	—	—	122 97	122 97	122 97	106 27	122 97	786
						N80	1055	1034	131 18	109 76	131 18	785	—	—	131 18	131 18	131 18	113 36	131 18	786
						095*	1253	1227	155 77	130 32	155 77	—	—	—	155 77	155 77	155 77	134 60	155 77	854
						P105	1385	1358	172 17	144 04	172 17	—	—	—	172 17	172 17	172 17	148 80	172 17	854
	13.85*	14.10*	—	13.70*	10.50	K55	794	789	98 74	84 00	98 74	—	—	—	98 74	98 74	98 74	81 92	98 74	786
						C75	1083	1076	134 63	114 84	134 63	—	—	—	134 63	134 63	134 63	120 66	134 63	786
						N80	1155	1147	143 61	122 17	143 61	—	—	—	143 61	143 61	143 61	128 71	143 61	786
						095*	1372	1362	170 54	145 08	170 54	—	—	—	170 54	170 54	170 54	153 03	170 54	976
						P105	1576	1506	188 48	160 35	188 48	—	—	—	188 48	188 48	188 48	168 73	188 48	976
	14.90*	15.20*	—	14.70*	11.43	K55	848	853	105 42	90 66	105 42	—	—	—	105 42	105 42	105 42	84 36	105 42	786
						C75	1156	1163	143 76	123 62	143 76	—	—	—	143 76	143 76	143 76	123 60	143 76	908
						N80	1234	1240	153 35	132 00	153 35	—	—	—	153 35	153 35	153 35	128 03	153 35	908
						095*	1465	1473	182 12	156 67	182 12	—	—	—	182 12	182 12	182 12	158 73	182 12	1030
						P105	1619	1628	201 23	173 06	201 23	—	—	—	201 23	201 23	201 23	—	201 23	1030
	15.50*	15.80*	—	15.80*	12.40	K55	910	925	111 47	96 98	111 47	—	—	—	111 47	111 47	—	101 04	111 47	786
						C75	1241	1262	152 00	132 25	152 00	—	—	—	152 00	152 00	—	144 20	162 00	908
						N80	1324	1346	162 00	141 24	162 00	—	—	—	162 00	162 00	162 00	154 50	162 00	908
						095*	1572	1598	192 53	167 50	192 53	—	—	—	192 53	192 53	192 53	183 44	192 53	1030
						P105	1738	1766	212 75	185 13	212 75	—	—	—	212 75	212 75	212 75	202 08	212 75	1030

* Non API grades or nominal weights   (1) Based on nominal thickness   (2) Based on 87.5 % of nominal thickness

PERFORMANCE PROPERTIES OF TUBING (continued)
(API Bul 5C2 and Vallourec notice)

Outside diameter (OD) (in)(mm)	Nominal weight Non-upset (lb/ft)	Upset (lb/ft)	Integral (lb/ft)	VAM (lb/ft)	Wall thickness (mm)	Grade	Pipe body Collapse resistance (bar)(1)	Internal yield pressure (bar)(2)	Round threaded coup. Yield strength (1) Non-upset (10³ daN)	Upset (10³ daN)	Make-up torque (m-daN)	Integral joint Yield strength (10³ daN)	Make-up torque (m-daN)	VAM coupling Heavy weight (10³ daN)	Yield strength (1) Regular same grade (10³ daN)	Regular yield grade (10³ daN)	Spe clearance same grade (10³ daN)	Spe clearance higher grade (10³ daN)	Make-up torque (m-daN)
4 (101.6)	9.50	9.70*	—	9.50	5.74	H40	280	273	32.07	47.73	110	—	—	47.73	47.73	—	48.81	—	340
						K55	352	375	44.00	65.65	165	—	—	65.65	65.65	—	67.12	—	490
						C75	438	512	60.13	89.50	220	—	—	89.50	89.50	95.47	91.53	109.64	490
						N80	454	545	64.14	95.47	235	—	—	95.47	95.47	—	97.63	—	490
						095*	503	648	76.00	113.00		—	—	113.00	113.00	125.00	116.00	125.00	490
						P105*	532	716	84.00	125.00		—	—	125.00	125.00	—	128.00	—	
	10.90*	11.00	—	10.90	6.65	H40	338	316	39.14	54.82	260	—	—	54.82	54.82	—	48.49	—	465
						K55	454	434	53.33	75.37	345	—	—	75.37	75.37	—	67.12	—	540
						C75	580	593	73.41	102.78	460	—	—	102.78	102.78	109.64	88.49	93.19	540
						N80	606	632	79.20	109.64	615	—	—	109.64	109.64	—	120.66	—	540
						095*	687	751	93.00	130.00		—	—	130.00	130.00	144.00	153.00	135.50	540
						P105*	738	830	103.00	144.00		—	—	144.00	144.00	—	169.00	178.00	540
	13.20*	13.40*	—	13.00	8.38	K55*	574	548	71.66	93.19		—	—	93.19	93.19	—	88.49	—	600
						C75*	783	747	97.72	127.10		—	—	127.10	127.10	—	120.66	154.00	640
						N80*	835	796	104.20	135.50		—	—	135.50	135.50	—	128.70	—	640
						095*	992	946	123.80	161.00		—	—	161.00	161.00	—	153.00	202.00	660
						P105*	1096	1045	137.00	178.00		—	—	178.00	178.00	—	169.00	—	660
	14.92*	15.10*	—	14.80	9.65	C75*	889	860	115.00	144.35		—	—	144.35	144.35	—	—	154.00	785
						N80*	948	917	123.00	154.00		—	—	154.00	154.00	—	—	—	785
						095*	1125	1089	145.50	183.00		—	—	183.00	183.00	—	—	202.00	785
						P105*	1245	1204	161.00	202.00		—	—	202.00	202.00	—	—	—	785
	16.60*	16.80*	—	16.50	10.92	C75*	992	973	131.72	161.00		—	—	161.00	161.00	—	—	172.00	810
						N80*	1058	1038	140.50	172.00		—	—	172.00	172.00	—	—	—	810
						095*	1257	1232	166.80	204.00		—	—	204.00	204.00	—	—	225.00	880
						P105*	1389	1362	184.00	225.00		—	—	225.00	225.00	—	—	—	880
4 1/2 (114.3)	12.60*	12.75	—	12.60	6.88	H40	310	291	46.48	64.14	290	—	—	64.14	64.14	64.14	56.26	64.14	635
						K55	395	400	63.90	88.20	385	—	—	88.20	88.20	88.20	77.22	88.20	635
						C75	497	545	87.16	120.28	510	—	—	120.28	120.28	—	105.30	128.00	635
						N80	517	581	93.00	128.00	545	—	—	128.00	128.00	128.00	112.00	—	635
						095*	579	690	110.00	152.00		—	—	152.00	152.00	152.00	133.00	167.00	635
						P105*	617	763	122.00	167.00		—	—	167.00	167.00	167.00	147.00	—	635

* Non API grades or nominal weights.   (1) Based on nominal thickness.   (2) Based on 87.5 % of nominal thickness.

API LINE PIPE THREAD

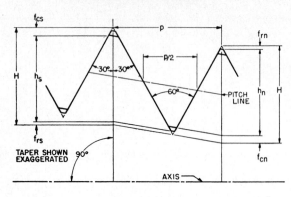

Taper : 6,25 %
(0,0625 inch per inch)
(0,750 inch per foot)

Thread element	27 threads per inch p = 0,940	18 threads per inch p = 1,412	14 threads per inch p = 1,814	11 1/2 threads per inch p = 2,210	8 threads per inch p = 3,175
H = 0,866 p	0,815	1,222	1,572	1,913	2,748
h = 0,760 p	0,714	1,072	1,379	1,679	2,413
tb = 0,033 p	0,030	0,046	0,061	0,074	0,104
ts = 0,073 p	0,069	0,104	0,132	0,160	0,231

DIMENSIONS, WEIGHTS AND TEST PRESSURES
OF THREADED LINE PIPE
(API Std 5L)

Nominal size (in)	Nominal weight (lb/ft)	Outside diameter (OD) (mm)	Thickness (mm)	Inside diameter (ID) (mm)	Weight with coupling (kg/m)	Outside diameter of coupling (mm)	Number of threads per (in)	Test pressure		
								Longitu-dinal welded (bar)	Grade A (bar)	Grade B (bar)
1 1/8	0,25	10,3	1,73	6,8	0,37	14,3	27	48	48	48
1/4	0,43	13,7	2,24	9,2	0,64	18,3	18	48	48	48
3/8	0,57	17,1	2,31	12,5	0,85	22,2	18	48	48	48
1/2	0,86	21,3	2,77	15,8	1,28	27,0	14	48	48	48
3/4	1,14	26,7	2,87	21,0	1,69	33,4	14	48	48	48
1	1,70	33,4	3,38	26,6	2,51	40,0	11 1/2	48	48	48
1 1/4	2,30	42,2	3,56	35,1	3,41	52,2	11 1/2	69	69	76
1 1/2	2,75	48,3	3,68	40,9	4,07	55,9	11 1/2	69	69	76
2	3,75	60,3	3,91	52,5	5,49	73,0	11 1/2	69	69	76
2 1/2	5,90	73,0	5,16	62,7	8,70	85,7	8	69	69	76
3	7,70	88,9	5,49	77,9	11,37	101,6	8	69	69	76
3 1/2	9,25	101,6	5,74	90,1	13,71	117,5	8	83	83	90
4	11,00	114,3	6,02	102,3	16,27	132,1	8	83	83	90
5	15,00	141,3	6,55	128,2	22,03	159,9	8	—	83	90
6	19,45	168,3	7,11	154,1	28,59	187,7	8	—	83	90
8	25,55	219,1	7,04	205,0	37,46	244,5	8	—	80	93
8	29,35	219,1	8,18	202,7	43,17	244,5	8	—	92	108
10	32,75	273,0	7,09	258,8	47,38	298,4	8	—	64	75
10	35,75	273,0	7,80	257,4	51,87	298,4	8	—	71	83
10	41,85	273,0	9,27	254,5	61,08	298,4	8	—	84	99
12	45,45	323,8	8,38	307,0	66,68	335,6	8	—	64	75
12	51,15	323,8	9,52	304,8	75,22	335,6	8	—	73	85
14 D	57,00	355,6	9,52	336,6	82,39	381,0	8	—	66	77
16 D	65,30	406,4	9,52	837,4	94,57	431,8	8	—	58	68
18 D	73,00	457,2	9,52	438,2	106,75	482,6	8	—	52	61
20 D	81,00	508,0	9,52	489,0	118,98	533,4	8	—	47	54

Note : For plain end pipe, refer to API Stds 5L, 5LU, 5LX and to '' Formulaire du Producteur '' (Technip edition in French).

COUPLING DIMENSIONS
FOR THREADED LINE PIPE

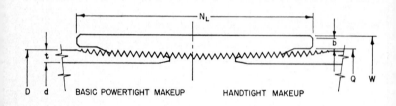

BASIC POWERTIGHT MAKEUP   HANDTIGHT MAKEUP

Nominal size (in)	Outside diameter of coupling W (mm)	Length N_L (mm)	Diameter of recess Q (mm)	Width of bearing face b (mm)	Calculated coupling weight (kg)
1/8	14,3	27,0	11,9	0,8	0,02
1/4	18,3	41,3	15,3	0,8	0,04
3/8	22,2	41,3	18,8	0,8	0,06
1/2	27,0	54,0	22,9	1,6	0,11
3/4	33,4	54,0	28,0	1,6	0,15
1	40,0	66,7	35,0	2,4	0,25
1 1/4	52,2	69,8	43,8	2,4	0,47
1 1/2	55,9	69,8	49,9	2,4	0,41
2	73,0	73,0	62,7	3,2	0,84
2 1/2	85,7	104,8	75,4	4,8	1,48
3	101,6	108,0	91,3	4,8	1,86
3 1/2	117,5	111,1	104,0	4,8	2,69
4	132,1	114,3	116,7	6,4	3,45
5	159,9	117,5	143,7	6,4	4,53
6	187,7	123,8	170,7	6,4	5,87
8	244,5	133,4	221,5	6,4	10,52
10	298,4	146,0	275,4	9,5	14,32
12	355,6	155,6	326,2	9,5	22,37
14 D	381,0	161,9	358,0	9,5	20,81
16 D	431,8	171,4	408,8	9,5	25,35
18 D	482,6	181,0	459,6	9,5	30,20
20 D	533,4	193,7	510,4	9,5	36,03

DIMENSIONS, WEIGHTS OF PIPE " GAS SERIES "

Nominal size (in)	Obsolete size	Outside diameter (OD) (mm)	Number of threads per inch	Light " gas tubing "			Regular " gas tubing "			Heavy " gas tubing "		
				Thickness (mm)	Weight (kg/m)		Thickness (mm)	Weight (kg/m)		Thickness (mm)	Weight (kg/m)	
					plain end	thread and coupling		plain end	thread and coupling		plain end	thread and coupling
1/8	5-10	10.2	28	1,8	0,369	0,372	2,0	0,407	0,410	2.65	0,493	0,496
1/4	8-13	13.5	19	2,0	0,573	0,577	2,35	0,650	0,654	2.9	0,769	0,773
3/8	12-17	17.2	19	2,0	0,747	0,753	2,35	0,852	0,858	2.9	1,02	1,03
1/2	15-21	21.3	14	2,35	1,10	1,11	2,65	1,22	1,23	3.25	1,45	1,46
3/4	20-27	26.9	14	2,35	1,41	1,42	2,65	1,58	1,59	3.25	1,90	1,91
1	26-34	33.7	11	2,90	2,21	2,23	3,25	2,44	2,46	4.05	2,97	2,99
1 1/4	33-42	42.4	11	2,90	2,84	2,87	3,25	3,14	3,17	4.05	3,84	3,87
1 1/2	40-49	48.3	11	2,90	3,26	3,80	3,25	3,61	3,65	4.05	4,43	4,47
2	50-60	60.3	11	3,25	4,56	4,63	3,65	5,10	5,17	4.5	6,17	6,24
2 1/2	66-76	76.1	11	3,25	5,80	5,92	3,65	6,51	6,63	4.5	7,90	8,02
3	80-90	88.9	11	3,25	6,81	6,98	4,05	8,47	8,64	4.85	10,1	10,3
3 1/2	90-102	101.6	11	3,65	8,74	8,92	4,05	9,72	9,90	4.85	11,6	11,8
4	102-114	114.3	11	3,65	9,89	10,20	4,50	12,40	12,40	5.4	14,4	14,7
5	—	139.7	11	3,65	—	—	4,50	15,00	15,50	5.4	17,8	18,3

Taper 6,25 % (0,0625 inch per inch) (3/4 of an inch per foot).

D CAPACITIES AND ANNULAR VOLUMES

Summary

CLEARANCE BETWEEN STANDARD BITS AND CASING SIZES

Outside diameter (in) (mm)	Nominal weight (lb/ft)	Thickness (mm)	Inside diameter (mm)	Drift (mm)	Bit size immediately below drift (in)	(mm)	Clearance between bit and casing sizes (mm)
4 1/2" (114,3)	18,80*	10,92	92,46	89,3	—	—	—
	17,70*	10,20	93,90	90,7	—	—	—
	16,90*	9,65	95,00	91,8	—	—	—
	15,10	8,56	97,20	94,0	—	—	—
	13,50	7,37	99,6	96,4	3 3/4	95,2	4,4
	11,60	6,35	101,6	98,4	3 3/4 (3)	95,2	6,4
	10,50	5,69	102,9	99,75	3 7/8	98,4	4,5
	9,50	5,21	103,9	100,7	3 7/8	98,4	5,5
5" (127,0)	20,80*	10,72	105,6	102,4	3 7/8	98,4	7,2
	18,00	9,19	108,6	105,4	4 1/8	104,8	3,8
	15,00	7,52	112,0	108,8	4 1/4	107,9	4,1
	13,00	6,43	114,1	111,0	4 1/4	107,9	6,2
	11,50	5,59	115,8	112,6	4 1/4	107,9	7,9
5 1/2 (139,7)	23,00	10,54	118,6	115,4	4 1/2	114,3	4,3
	20,00	9,17	121,4	118,2	4 5/8	117,5	3,9
	17,00	7,72	124,3	121,1	4 3/4	120,6	3,7
	15,50	6,98	125,7	122,6	4 3/4	120,6	5,1
	14,00	6,20	127,3	124,1	4 7/8	123,8	3,5
6 5/8" (168,3)	32,00	12,06	144,2	141,0	4 7/8	123,8	20,4
	28,00	10,59	147,1	143,9	5 5/8	142,9	4,2
	24,00	8,94	150,4	147,2	5 3/4 (2)	146,0	4,4
	20,00	7,32	153,7	150,5	5 7/8	149,2	4,5
7" (177,8)	44,00*	16,25	145,3	142,1	4 7/8	123,8	21,5
	41,00*	14,98	147,8	144,7	5 5/8	142,9	4,9
	38,00	13,72	150,4	147,2	5 3/4 (2)	146,0	4,4
	35,00	12,65	152,5	149,3	5 7/8	149,2	3,3
	32,00	11,51	154,8	151,6	5 7/8	149,2	5,6
	29,00	10,36	157,1	153,9	6	152,4	4,7
	26,00	9,19	159,4	156,2	6 1/8	155,6	3,8
	23,00	8,05	161,7	158,5	6 1/8	155,6	6,1
	20,00	6,91	164,0	160,8	6 1/4	158,7	5,3
	17,00	5,87	166,1	162,9	6 1/4	158,7	7,4
7 5/8" (193,7)	39,00	12,70	168,3	165,1	6 1/4 (3)	158,7	9,6
	33,70	10,92	171,9	168,7	6 5/8	168,3	3,6
	29,70	9,52	174,7	171,4	6 5/8 (3)	168,3	6,4
	26,40	8,33	177,0	173,8	6 3/4	171,4	5,6
	24,00	7,62	178,5	175,3	6 3/4	171,4	7,1
8 5/8" (219,1)	49,00	14,15	190,8	187,6	7 3/8	187,3	3,5
	44,00	12,70	193,7	190,5	7 3/8	187,3	6,4
	40,00	11,43	196,2	193,0	7 3/8	187,3	8,9
	36,00	10,16	198,8	195,6	7 5/8	193,7	5,1
	32,00	8,94	201,2	198,0	7 5/8	193,7	7,5
	28,00	7,72	203,7	200,5	7 7/8	200,0	3,7
	24,00	6,71	205,7	202,5	7 7/8	200,0	5,7

* Non API.

CLEARANCE BETWEEN STANDARD BITS
AND CASING SIZES (continued)

CASING DIMENSIONS				Drift (mm)	Bit size immediately below drift		Clearance between bit and casing sizes (mm)
Outside diameter (in) (mm)	Nominal weight (lb/ft)	Thickness (mm)	Inside diameter (mm)		(in)	(mm)	
9 5/8 (244,5)	71,80*	19,05	206,4	202,4	7 7/8	200.0	6,4
	61,10*	15,87	212,7	208,8	7 7/8	200.0	12,7
	58,40*	15,11	214,3	210,3	7 7/8	200.0	14,3
	53,50	13,84	216,8	212,8	8 3/8	212.7	4,1
	47,00	11,99	220,5	216,5	8 1/2	215.9	4,6
	43,50	11,05	222,7	218,4	8 1/2	215.9	6,8
	40,00	10,03	224,4	220,4	8 5/8	219.1	5,3
	36,00	8,94	226,6	222,6	8 3/4	222.2	4,4
	32,30	7,92	228,7	224,7	8 3/4	222.2	6,5
10 3/4" (273,0)	65,70	15,11	242,8	238,9	9	228.6	14,2
	60,70	13,84	245,4	241,4	9	228.6	16,7
	55,50	12,57	247,9	243,9	9	228.6	19,3
	51,00	11,43	250,1	246,2	9 5/8	244.5	5,6
	45,50	10,16	252,7	248,8	9 5/8	244.5	8,2
	40,50	8,89	255,2	251,3	9 7/8	250.8	4,4
	32,75	7,09	258,8	254,9	9 7/8	250.8	8,0
11 3/4" (298,4)	60,00	12,42	273,6	269,6	9 7/8	250.8	22,8
	54,00	11,05	276,3	272,4	10 5/8	269.9	6,4
	47,00	9,52	279,4	275,4	10 5/8	269.9	9,5
	42,00	8,46	281,5	277,6	10 5/8	269.9	11,6
13 3/8" (339,7)	72,00	13,06	313,6	309,7	12	304.8	8,8
	68,00	12,19	315,3	311,4	12 1/4	311.1	4,2
	61,00	10,92	317,9	313,9	12 1/4	311.1	6,8
	54,50	9,65	320,4	316,5	12 1/4	311.1	9,3
	48,00	8,38	322,9	319,0	12 1/4	311.1	11,8
16" (406,4)	84,00	12,57	381,3	376,5	14 3/4	374.6	6,7
	75,00	11,13	384,1	379,4	14 3/4	374.6	9,5
	65,00	9,52	387,4	382,6	15	381.0	6,4
18 5/8" (473,1)	87,5	11,05	451,0	446,2	17 1/2	444.5	6,5
20" (508,0)	133,00	16,13	475,7	470,9	18 1/2	469.9	5,8
	106,50	12,70	482,6	477,8	18 1/2	469.9	12,7
	94,00	11,13	485,7	481,0	18 1/2	469.9	15,8

* Non API.

ANNULAR VOLUME BETWEEN CASING AND OPEN HOLE
(litres per metre)

Size of open hole (in)	Casing outside diameter nominal size (in)															
0 (2)	0 (1)	4 1/2	5	5 1/2	6 5/8	7	7 5/8	8 5/8	9 5/8	10 3/4	11 3/4	13 3/8	16	18 5/8	20	30
	l/m	10.30	12.69	15.36	22.34	24.88	29.58	37.87	47.10	58.74	70.16	90.80	130.0	176.2	203.0	455.8
5 7/8	17.49	7.19	4.80														
6	18.24	7.94	5.55	x													
6 1/8	19.01	8.71	6.32	3.65													
6 1/4	19.79	9.49	7.10	4.43													
6 5/8	22.24	11.94	9.55	6.88													
6 3/4	23.09	12.79	10.40	7.73													
7 3/8	27.56	17.26	14.87	12.20	x	x											
7 7/8	31.42	21.12	18.73	16.06	9.08	6.54											
8 3/8	35.54	25.24	22.85	20.18	13.20	10.66											
8 1/2	36.61	26.31	23.92	21.25	14.27	11.73	x										
8 5/8	37.69	27.39	25.00	22.33	15.35	12.81	8.11										
8 3/4	38.79	28.49	26.10	23.43	16.45	13.91	9.21										
9	41.04	30.74	28.35	25.68	18.70	16.16	11.46										
9 5/8	46.94	36.64	34.25	31.58	24.60	22.06	17.36	x									
9 7/8	49.41	39.11	36.72	34.05	27.07	24.53	19.83	11.54									
10 5/8	57.20	46.90	44.51	41.84	34.86	32.32	27.62	19.33	x	x							
12	72.97	62.67	60.28	57.61	50.63	48.09	43.39	35.10	25.87	14.23							
12 1/4	76.04	65.74	63.35	60.68	53.70	51.16	46.46	38.17	28.94	17.30	x						
14 3/4	110.24	99.94	97.55	94.88	87.90	85.36	80.66	72.37	63.14	51.50	40.08	x					
15	114.01	103.71	101.32	98.65	91.67	89.13	84.43	76.14	66.91	55.27	43.85	23.21	x				
17 1/2	155.2	144.9	142.5	139.8	132.9	130.3	125.6	117.3	108.1	96.46	85.04	64.40	25.2				
20	202.7	192	190.0	187.3	180.4	177.8	173.1	164.8	155.6	144.0	132.5	111.9	72.7	x	x		
24	291.9	281	279.2	276.6	269.6	267.0	262.3	254.0	244.8	233.2	221.7	201.1	161.9	115.7	88.9		
26	342.4	332.1	329.7	327.0	320.1	317.5	312.8	304.5	295.3	283.7	272.2	251.6	212.4	166.2	139.4	x
36	656.4	646.1	643.7	641.0	634.1	631.5	626.8	618.5	609.3	597.7	586.2	565.6	526.4	480.2	453.4	200.6

(1) The zero vertical column gives the capacity of the open hole (nominal size) in litres per metre.
(2) The zero horizontal line gives the total displacement of casing with coupling in litres per metre.
x Casing coupling outside diameter is bigger than open hole size.

ANNULAR VOLUMES BETWEEN TWO STRINGS OF CASING (litres per metre)

Outside diameter of outer string	0 (2)	0 (1) l/m	____ Nominal size of inner string ____ 4 1/2	5	5 1/2	6 5/8	7	7 5/8	8 5/8	9 5/8	10 3/4	11 3/4	13 3/8	16	18 5/8	20	30
(displacement l/m →)			10.30	12.69	15.36	22.34	24.88	29.58	37.87	47.10	58.74	70.16	90.80	130.0	176.2	203.0	455.8
6 5/8	20.00	18.55															
	24.00	17.77															
	28.00	16.99															
7	32.00	16.33	6.03	x	x	x											
	17.00	21.66	11.36	8.97	6.30	x											
	20.00	21.12	10.82	8.43	5.76	x											
	23.00	20.53	10.23	7.84	5.17	x											
	26.00	19.95	9.65	7.26	4.59	x											
	29.00	19.38	9.08	6.69	4.02	x											
	32.00	18.82	8.52	6.13	x	x											
	35.00	18.27	7.97	5.58	x	x											
	38.00	17.77	7.47	5.08	x	x											
	41.00	17.21	6.91	4.52	x	x											
	44.00	16.64	6.34	3.95	x	x											
7 5/8	24.00	25.02	14.72	12.33	9.66	x											
	26.40	24.61	14.31	11.92	9.25	x											
	29.70	23.97	13.67	11.28	8.61	x											
	33.70	23.21	12.91	10.52	7.85	x											
	39.00	22.25	11.95	9.56	6.89	x											
8 5/8	24.00	33.23	22.93	20.54	17.87	10.89	8.35	x									
	28.00	32.59	22.29	19.90	17.23	10.25	7.71	x									
	32.00	31.79	21.49	19.10	16.43	9.45	6.91	x									
	36.00	31.04	20.74	18.35	15.68	8.70	6.16	x									
	40.00	30.23	19.93	17.54	14.87	7.89	x	x									
	44.00	29.47	19.17	16.78	14.11	7.13	x										
	49.00	28.59	18.29	15.90	13.23	6.25	x										
9 5/8	32.30	41.08	30.78	28.39	25.72	18.74	16.20	11.50	x								
	36.00	40.33	30.03	27.64	24.97	17.99	15.45	10.75	x								
	40.00	39.55	29.25	26.86	24.19	17.21	14.67	9.97	x								
	43.50	38.95	28.65	26.26	23.59	16.61	14.07	9.37	x								
	47.00	38.19	27.89	25.50	22.83	15.85	13.31	8.61	x								
	53.50	36.92	26.62	24.23	21.56	14.58	12.04	x									

(1) The zero vertical column gives the capacity of the casing in litres per metre.
(2) The zero horizontal line gives the total displacement of casing with coupling in litres per metre.
x The outside diameter of the coupling is bigger than the inside diameter of the outer casing.

ANNULAR VOLUMES BETWEEN TWO STRINGS OF CASING (continued) (litres per metre)

Nominal size of inner string

Outside diameter and weight of outer string	0 (2)	0 (1) l/m	4 1/2	5	5 1/2	6 5/8	7	7 5/8	8 5/8	9 5/8	10 3/4	11 3/4	13 3/8	16	18 5/8	20	30
0			10.30	12.69	15.36	22.34	24.88	29.58	37.87	47.10	58.74	70.16	90.80	130.0	176.2	203.0	455.8
9 5/8 58.40		36.14	25.84	23.45	20.78	13.80	11.26	x	x								
61.10		35.53	25.23	22.84	20.17	13.19	10.65	x	x								
71.80		33.46	23.16	20.77	18.10	11.12	8.58	x	x								
10 3/4 32.75		52.60	42.30	39.91	37.24	30.26	27.72	23.02	14.73	x							
40.50		51.15	40.85	38.46	35.79	28.81	26.27	21.57	13.28	x							
45.50		50.15	39.85	37.46	34.79	27.81	25.27	20.57	12.28	x							
51.00		49.13	38.83	36.44	33.77	26.79	24.25	19.55	11.26	x							
55.50		48.27	37.97	35.58	32.91	25.93	23.39	18.69	x	x							
60.70		47.26	36.96	34.57	31.90	24.92	22.38	17.68	x	x							
65.70		46.30	36.00	33.61	30.94	23.96	21.42	16.72	x	x							
11 3/4 42.00		62.24	51.94	49.55	46.88	39.90	37.36	32.66	24.37	15.14	x						
47.00		61.31	51.01	48.62	45.95	38.97	36.43	31.73	23.44	14.21	x						
54.00		59.96	49.66	47.27	44.60	37.62	35.08	30.38	22.09	12.86	x						
60.00		58.79	48.49	46.10	43.43	36.45	33.91	29.21	20.92	x	x						
13 3/8 48.00		81.89	71.59	69.20	66.53	59.55	57.01	52.31	44.02	34.79	23.15	x					
54.50		80.63	70.33	67.94	65.27	58.29	55.75	51.05	42.76	33.53	21.89	x					
61.00		79.37	69.07	66.68	64.01	57.03	54.49	49.79	41.50	32.27	20.63	x					
68.00		78.08	67.78	65.39	62.72	55.74	53.20	48.50	40.21	30.98	19.34	x					
72.00		77.24	66.94	64.55	61.88	54.90	52.36	47.66	39.37	30.14	18.50	x					
16 65.00		117.87	107.57	105.18	102.51	95.53	92.99	88.29	80.00	70.77	59.13	47.71	27.07				
75.00		115.87	105.57	103.18	100.51	93.53	90.99	86.29	78.00	68.77	57.13	45.71	25.07				
84.00		114.19	103.89	101.50	98.83	91.85	89.31	84.61	76.32	67.09	55.45	44.03	23.39				
18 5/8 87.50		159.74	149.44	147.05	144.38	137.40	134.86	130.16	121.87	112.64	101.00	89.58	68.94	29.74			
20 94.00		185.28	174.98	172.59	169.92	162.94	160.04	155.70	147.41	138.18	126.54	115.12	94.48	55.28	x		
106.50		182.92	172.62	170.23	167.56	160.58	158.04	153.34	145.05	135.82	124.18	112.76	92.12	52.92	x		
133.00		177.76	167.46	165.07	162.40	155.42	152.88	148.18	139.89	130.66	119.02	107.60	86.96	47.76	x		
30 267.0		407.8	397.5	395.11	392.44	385.5	392.9	378.2	369.9	360.7	349.1	337.6	317.0	277.8	231.6	204.8	x
310.0		397.0	386.7	384.31	381.64	374.7	372.1	367.4	359.1	349.9	338.3	326.8	306.2	267.0	220.8	194.0	x

(1) The zero vertical column gives the capacity of the casing in litres per metre.
(2) The zero horizontal line gives the total displacement of casing with coupling in litres per metre.
x The outside diameter of the coupling is bigger than the inside diameter of the outer casing.

ANNULAR VOLUME BETWEEN DRILL PIPE AND OPEN HOLE
(litres per metre)

		Nominal size of drill pipe							
		0 (1)	2 3/8	2 7/8	3 1/2	4	4 1/2	5	5 1/2	6 5/8
	0 (2)	l/m	3,0*	4,4*	6,5*	8,6*	11,0*	13,3*	16,1*	22,9*
5 7/8		17,5	14,5	13,1	11,0	8,9	x			
6		18,2	15,2	13,8	11,7	9,6	7,2	x		
6 1/8		19,0	16,0	14,6	12,5	10,4	8,0	x		
6 1/4		19,8	16,8	15,4	13,3	11,2	8,8	x		
6 5/8		22,2	19,2·	17,8	15,7	13,6	11,2	8,9	x	
6 3/4		23,1	20,1	18,7	16,6	14,5	12,1	9,8	x	
7 3/8		27,6	24,6	23,2	21,1	19,0	16,6	14,3	11,5	4,7
7 7/8		31,4	28,4	27,0	24,9	22,8	20,4	18,1	15,3	8,5
8 3/8		35,5	32,5	31,1	29,0	26,9	24,5	22,2	19,4	12,6
8 1/2		36,6	33,6	32,2	30,1	28,0	25,6	23,3	20,5	13,7
8 5/8		37,7	34,7	33,3	31,2	29,1	26,7	24,4	21,6	14,8
8 3/4		38,8	35,8	34,4	32,3	30,2	27,8	25,5	22,7	15,9
9		41,0	38,0	36,6	34,5	32,4	30,0	27,7	24,9	18,1
9 5/8		46,9	43,9	42,5	40,4	38,3	35,9	33,6	30,8	24,0
9 7/8		49,4	46,4	45,0	42,9	40,8	38,4	36,1	33,3	26,5
10 5/8		57,2	54,2	52,8	50,7	48,6	46,2	43,9	41,1	34,3
12		73,0	70,0	68,6	66,5	64,4	62,0	59,7	56,9	50,1
12 1/4		76,0	73,0	71,6	69,5	67,4	65,0	62,7	59,9	53,1
14 3/4		110,2	107,2	105,8	103,7	101,6	99,2	96,9	94,1	87,3
15		114,0	111,0	109,6	107,5	105,4	103,0	100,7	97,9	91,1
17 1/2		155,2	152,2	150,8	148,7	146,6	144,2	141,9	139,1	132,3
20		202,7	199,7	198,3	196,2	194,1	191,7	189,4	186,6	179,8
24		291,9	288,9	287,5	285,4	283,3	280,9	278,6	275,8	269,0
26		342,4	339,4	338,0	335,9	333,8	331,4	329,1	326,3	319,5
36		656,4	653,4	652,0	649,9	647,8	645,4	643,1	640,3	633,5

Nominal size of open hole (in)

(1) The zero vertical column gives the capacity of open hole in litres per metre.
(2) The zero horizontal line gives the total displacement of drill pipe with tool joint in litres per metre.
* Drill pipe displacement is averaged taking into account the various sizes of the tool joints.
x Tool joint OD is bigger than open hole size.

ANNULAR VOLUME BETWEEN DRILL COLLAR AND OPEN HOLE
(litres per metre)

Outside diameter of drill collar (in)

Nominal size of open hole (in) — 0 (2)	0 (1) l/m	4 3/4	5	5 3/4	6	6 1/4	6 1/2	6 3/4	7	7 3/4	8	8 1/4	9	9 1/2	9 3/4	10	11 1/4	14
(displacement, l/m)		11.4	12.7	16.8	18.2	19.8	21.4	23.1	24.8	30.4	32.4	34.5	41.0	45.7	48.2	50.7	64.1	99.3
5 7/8	17.5	6.1	4.8	0.7	x	x												
6	18.2	6.8	5.5	1.4	x	x												
6 1/8	19.0	7.6	6.3	2.2	0.8													
6 1/4	19.8	8.4	7.1	3.0	1.6													
6 5/8	22.2	10.8	9.5	5.4	4.0	2.4	0.8	x	x									
6 3/4	23.1	11.7	10.4	6.3	4.9	3.3	1.7	x	x									
7 3/8	27.6	16.2	14.9	10.8	9.4	7.8	6.2	4.5	2.8	x								
7 7/8	31.4	20.0	18.7	14.6	13.2	11.6	10.0	8.3	6.6	1.0	x							
8 3/8	35.5	24.1	22.8	18.7	17.3	15.7	14.1	12.4	10.7	5.1	3.1	1.1						
8 1/2	36.6	25.2	23.9	19.8	18.4	16.8	15.2	13.5	11.8	6.2	4.2	2.2	x					
8 5/8	37.7	26.3	25.0	20.9	19.5	17.9	16.3	14.6	12.9	7.3	5.3	3.3	x					
8 3/4	38.8	27.4	26.1	22.0	20.6	19.0	17.4	15.7	14.0	8.4	6.4	4.3	x					
9	41.0	29.6	28.3	24.2	22.8	21.2	19.6	17.9	16.2	10.6	8.6	6.5	x					
9 5/8	46.9	35.5	34.2	30.1	28.7	27.1	25.5	23.8	22.1	16.5	14.5	12.4	5.9	1.2	x			
9 7/8	49.4	38.0	36.7	32.6	31.2	29.6	28.0	26.3	24.6	19.0	17.0	14.9	8.4	3.7	1.2	x		
10 5/8	57.2	45.8	44.5	40.4	39.0	37.4	35.8	34.1	32.4	26.8	24.8	22.7	16.2	11.5	9.0	6.5	x	
12	73.0	61.6	60.3	56.2	54.8	53.2	51.6	49.9	48.2	42.6	40.6	38.5	32.0	27.3	24.8	22.3	8.9	x
12 1/4	76.0	64.6	63.3	59.2	57.8	56.2	54.6	52.9	51.2	45.6	43.6	41.5	35.0	30.3	27.8	25.3	11.9	x
14 3/4	110.2	98.8	97.5	93.4	92.0	90.4	88.8	87.1	85.4	79.8	77.8	75.7	69.2	64.5	62.0	59.5	46.1	10.9
15	114.0	102.6	101.3	97.2	95.8	94.2	92.6	90.9	89.2	83.6	81.6	79.5	73.0	68.3	65.8	63.3	49.9	14.7
17 1/2	155.2	143.8	142.5	138.4	137.0	135.4	133.8	132.1	130.4	124.8	122.8	120.7	114.2	109.5	107.0	104.5	91.1	55.9
20	202.7	191.3	190.0	185.9	184.5	182.9	181.3	179.6	177.9	172.3	170.3	168.2	161.7	157.0	154.5	152.0	138.6	103.4
24	291.9	280.5	279.2	275.1	273.7	272.1	270.5	268.8	267.1	261.5	259.5	257.4	250.9	246.2	243.7	241.2	227.8	192.6
26	342.4	331.0	329.7	325.6	324.2	322.6	321.0	319.3	317.6	312.0	310.0	307.9	301.4	296.7	294.2	291.7	278.3	243.1
36	656.4	645.0	643.7	639.6	638.2	636.6	635.0	633.3	631.6	626.0	624.0	621.9	615.4	610.7	608.2	605.7	592.3	557.1

(1) The zero vertical column gives the capacity of open hole in litres per metre.
(2) The zero horizontal line gives the total displacement of drill collar in litres per metre.

ANNULAR VOLUME BETWEEN CASING AND DRILL PIPE
(litres per metre)

OD	Weight	0 (1) l/m	1,050 0,6	1,315 0,9	1,660 1,4	1,900 1,8	2 3/8 3,0	2 7/8 4,4	3 1/2 6,5	4 8,6	4 1/2 11,0	5 13,3	5 1/2 16,1	6 5/8 22,9
4 1/2	9,50	8,5	7,9	7,6	7,1	6,7	5,5	x						
	10,50	8,3	7,7	7,4	6,9	6,5	5,3	x						
	11,60	8,1	7,5	7,2	6,7	6,3	5,1	x						
	13,50	7,8	7,2	6,9	6,4	6,0	4,8	x						
	15,10	7,4	6,8	6,5	6,0	5,6	4,4	x						
	16,90	7,1	6,5	6,2	5,7	5,3	4,1	x						
	17,70	7,0	6,4	6,1	5,6	5,2	4	x						
	18,80	6,7	6,1	5,8	5,3	4,9	3,7	x						
5	11,50	10,5	9,9	9,6	9,1	8,7	7,5	6,1	x					
	13,00	10,2	9,6	9,3	8,8	8,4	7,2	5,8	x					
	15,00	9,9	9,3	9,0	8,5	8,1	6,9	5,5	x					
	18,00	9,3	8,7	8,4	7,9	7,5	6,3	x						
	20,80	8,8	8,2	7,9	7,4	7,0	5,8	x						
5 1/2	14,00	12,7	12,1	11,8	11,3	10,9	9,7	8,3	x					
	15,50	12,4	11,8	11,5	11,0	10,6	9,4	8,0	x					
	17,00	12,1	11,5	11,2	10,7	10,3	9,1	7,7	x					
	20,00	11,6	11,0	10,7	10,2	9,8	8,6	7,2	x					
	23,00	11,1	10,5	10,2	9,7	9,3	8,1	6,7	x					
6 5/8	20,00	18,6	18,0	17,7	17,2	16,8	15,6	14,2	12,1	x				
	24,00	17,8	17,2	16,9	16,4	16,0	14,8	13,4	11,3	x				
	28,00	17,0	16,4	16,1	15,6	15,2	14,0	12,6	10,5	x				
	32,00	16,3	15,7	15,4	14,9	14,5	13,3	11,9	9,8	x				
7	17,00	21,7	21,1	20,8	20,3	19,9	18,7	17,3	15,2	13,1	x			
	20,00	21,1	20,5	20,2	19,7	19,3	18,1	16,7	14,6	12,5	x			
	23,00	20,5	19,9	19,6	19,1	18,7	17,5	16,1	14,0	11,9	x			
	26,00	20,0	19,4	19,1	18,6	18,2	17,0	15,6	13,5	11,4	x			
	29,00	19,4	18,8	18,5	18,0	17,6	16,4	15,0	12,9	10,8	x			
	32,00	18,8	18,2	17,9	17,4	17,0	15,8	14,4	12,3	x				
	35,00	18,3	17,7	17,4	16,9	16,5	15,3	13,9	11,8	x				
	38,00	17,8	17,2	16,9	16,4	16,0	14,8	13,4	11,3	x				
	41,00	17,2	16,6	16,3	15,8	15,4	14,2	12,8	10,7	x				
	44,00	16,6	16,0	15,7	15,2	14,8	13,6	12,2	10,1	x				
7 5/8	24,00	25,0	24,4	24,1	23,6	23,2	22,0	20,6	18,5	16,4	14,0	x		
	26,40	24,6	24,0	23,7	23,2	22,8	21,6	20,2	18,1	16,0	13,6	x		
	29,70	24,0	23,4	23,1	22,6	22,2	21,0	19,6	17,5	15,4	13,0	x		
	33,70	23,2	22,6	22,3	21,8	21,4	20,2	18,8	16,7	14,6	12,2	x		
	39,00	22,3	21,7	21,4	20,9	20,5	19,3	17,9	15,8	13,7	x			
8 5/8	24,00	33,2	32,6	32,3	31,8	31,4	30,2	28,8	26,7	24,6	22,2	19,9	17,1	x
	28,00	32,6	32,0	31,7	31,2	30,8	29,6	28,2	26,1	24,0	21,6	19,3	16,5	x
	32,00	31,8	31,2	29,9	29,4	29,0	28,8	27,4	25,3	23,2	20,8	18,5	15,7	x
	36,00	31,0	30,4	30,1	29,6	29,2	28,0	26,6	24,5	22,4	20,0	17,7	14,9	X

Outside diameter and weight of casing (in and lb/ft)

(1) The zero vertical column gives the capacity of the casing in litres per metre.
(2) The zero horizontal line gives the total displacement of drill pipe with tool joint in litres per metre.
* The outside diameter of the tool joint is bigger than the inside diameter of the casing.

ANNULAR VOLUME BETWEEN CASING AND DRILL PIPE
(litres per metre) (continued)

			0 (1)	1,050	1,315	1,660	1,900	2 3/8	2 7/8	3 1/2	4	4 1/2	5	5 1/2	6 5/8
			0 (2)	l/m	0,6	0,9	1,4	1,8	3,0	4,4	6,5	8,6	11,0	13,3	16,1	22,9

Nominal size of drill pipe (in) — Outside diameter and weight of casing (in and lb/ft):

casing	weight	0 (1)	1,050	1,315	1,660	1,900	2 3/8	2 7/8	3 1/2	4	4 1/2	5	5 1/2	6 5/8
8 5/8	40,00	30,2	29,6	29,3	28,8	28,4	27,2	25,8	23,7	21,6	19,2	16,9	14,1	x
	44,00	29,5	28,9	28,6	28,1	27,7	26,5	25,1	23,0	20,9	18,5	16,2	13,4	x
	49,00	28,6	28,0	27,7	27,2	26,8	25,6	24,2	22,1	20,0	17,6	15,3	12,5	x
9 5/8	32,30	41,1	40,5	40,2	39,7	39,3	38,1	36,7	34,6	32,5	30,1	27,8	25,0	18,2
	36,00	40,3	39,7	39,4	38,9	38,5	37,3	35,9	33,8	31,7	29,3	27,0	24,2	17,4
	40,00	39,6	39,0	38,7	38,2	37,8	36,6	35,2	33,1	31,0	28,6	26,3	23,5	16,7
	43,50	39,0	38,4	38,1	37,6	37,2	36,0	34,6	32,5	30,4	28,0	25,7	22,9	16,1
	47,00	38,2	37,6	37,3	36,8	36,4	35,2	33,8	31,7	29,6	27,2	24,9	22,1	15,3
	53,50	36,9	36,3	36,0	35,5	35,1	33,9	32,5	30,4	28,3	25,9	23,6	20,8	14,0
	58,40	36,1	35,5	35,2	34,7	34,3	33,1	31,7	29,6	27,5	25,1	22,8	20,0	13,2
	61,10	35,5	34,9	34,6	34,1	33,7	32,5	31,1	29,0	26,9	24,5	22,2	19,4	12,6
	71,80	33,5	32,9	32,6	32,1	31,7	30,5	29,1	27,0	24,9	22,5	20,2	17,8	x
10 3/4	32,75	52,6	52,0	51,7	51,2	50,8	49,6	48,2	46,1	44,0	41,6	39,3	36,5	29,7
	40,50	51,2	50,6	50,3	49,8	49,4	48,2	46,8	44,7	42,6	40,2	37,9	35,1	28,3
	45,50	50,2	49,6	49,3	48,8	48,4	47,2	45,8	43,7	41,6	39,2	36,9	34,1	27,3
	51,00	49,1	48,5	48,2	47,7	47,3	46,1	44,7	42,6	40,5	38,1	35,8	33,0	26,2
	55,50	48,3	47,7	47,4	46,9	46,5	45,3	43,9	41,8	39,7	37,3	35,0	32,2	25,4
	60,70	47,3	46,7	46,4	45,9	45,5	44,3	42,9	40,8	38,7	36,3	34,0	31,2	24,4
	65,70	46,3	45,7	45,4	44,9	44,5	43,3	41,9	39,8	37,7	35,3	33,0	30,2	23,4
11 3/4	42,00	62,2	61,6	61,3	60,8	60,4	59,2	57,8	55,7	53,6	51,2	48,9	46,1	39,3
	47,00	61,3	60,7	60,4	59,9	59,5	58,3	56,9	54,8	52,7	50,3	48,0	45,2	38,4
	54,00	60,0	59,4	59,1	58,6	58,2	57,0	55,6	53,5	51,4	49,0	46,7	43,9	37,1
	60,00	58,8	58,2	57,9	57,4	57,0	55,8	54,4	52,3	50,2	47,8	45,5	42,7	35,9
13 3/8	48,00	81,9	81,3	81,0	80,5	80,1	78,9	77,5	75,4	73,3	70,9	68,6	65,8	59,0
	54,50	80,6	80,0	79,7	79,2	78,8	77,6	76,2	74,1	72,0	69,6	67,3	64,5	57,7
	61,00	79,4	78,8	78,5	78,0	77,6	76,4	75,0	72,9	70,8	68,4	66,1	63,3	56,5
	68,00	78,1	77,5	77,2	76,7	76,3	75,1	73,7	71,6	69,5	67,1	64,8	62,0	55,2
	72,00	77,2	76,6	76,3	75,8	75,4	74,2	72,8	70,7	68,6	66,2	63,9	61,1	54,3
16	65,00	117,9	117,3	117,0	116,5	116,1	114,9	113,5	111,4	109,3	106,9	104,6	101,8	95,0
	75,00	115,9	115,3	115,0	114,5	114,1	112,9	111,5	109,4	107,3	104,9	102,6	99,8	93,0
	84,00	114,2	113,6	113,3	112,8	112,4	111,2	109,8	107,7	105,6	103,2	100,9	98,1	91,3
18 5/8	87,50	159,7	159,1	158,8	158,3	157,9	156,7	155,3	153,2	151,1	148,7	146,4	143,6	136,8
20	94,00	185,3	184,7	184,4	183,9	183,5	182,3	180,9	178,8	176,7	174,3	172,0	169,2	162,4
	106,50	182,9	182,3	182,0	181,5	181,1	179,9	178,5	176,4	174,3	171,9	169,6	166,8	160,0
	133,00	177,8	177,2	176,9	176,4	176,0	174,8	173,4	171,3	169,2	166,8	164,5	161,7	154,9
30	267,0	407,8	407,2	406,9	406,4	406,0	404,8	403,4	401,3	399,2	396,8	394,5	391,7	384,9
	310,0	397,0	396,4	396,1	395,6	395,2	394,0	392,6	390,5	388,4	386,0	383,7	380,9	374,1

(1) The zero vertical column gives the capacity of the casing in litres per metre.
(2) The zero horizontal line gives the total displacement of drill pipe with tool joint in litres per metre.
x The outside diameter of the tool is bigger than the inside diameter of the casing.

ANNULAR VOLUME BETWEEN CASING AND TUBING
(litres per metre)

			Outside diameter of tubing (in)									
			0 (1)	1,050	1,315	1,660	1,900	2,063	2 3/8	2 7/8	3 1/2	4	4 1/2
		0 (2)	l/m	0,56	0,88	1,41	1,83	2,16	2,87	4,21	6,25	8,15	10,32
Outside diameter and weight of casing (in and lb/ft)	4 1/2	9,50	8,48	7,92	7,60	7,07	6,65	6,32	5,61	4,27	x		
		10,50	8,32	7,76	7,44	6,91	6,49	6,16	5,45	4,11	x		
		11,60	8,11	7,55	7,23	6,70	6,28	5,95	5,24	3,90	x		
		13,50	7,79	7,23	6,91	6,38	5,96	5,63	4,92	3,58	x		
		15,10	7,42	6,86	6,54	6,01	5,59	5,26	4,55	3,21	x		
		16,90	7,10	6,54	6,22	5,69	5,27	4,84	4,23	x			
		17,70	6,95	6,39	6,07	5,54	5,12	4,79	4,08	x			
		18,80	6,73	6,17	5,85	5,32	4,90	4,57	3,86	x			
	5	11,50	10,53	9,97	9,65	9,12	8,70	8,37	7,66	6,32	x	x	
		13,00	10,22	9,66	9,34	8,81	8,39	8,06	7,35	6,01	x	x	
		15,00	9,85	9,29	8,97	8,44	8,02	7,69	6,98	5,64	x		
		18,00	9,26	8,70	8,38	7,85	7,43	7,10	6,39	5,05	x		
		20,80	8,75	8,19	7,87	7,34	6,92	6,59	5,88	4,54	x		
	5 1/2	14,00	12,73	12,17	11,85	11,32	10,90	10,57	9,86	8,52	6,48	x	
		15,50	12,41	11,85	11,53	11,00	10,58	10,25	9,54	8,20	6,16	x	
		17,00	12,13	11,57	11,25	10,72	10,30	9,97	9,26	7,92	5,88	x	
		20,00	11,58	11,02	10,70	10,17	9,75	9,42	8,71	7,37	5,33	x	
		23,00	11,05	10,49	10,17	9,64	9,22	8,89	8,18	6,84	4,80	x	
	6 5/8	20,00	18,55	17,99	17,67	17,14	16,72	16,39	15,68	14,34	12,30	10,40	8,23
		24,00	17,77	17,21	16,89	16,36	15,94	15,61	14,90	13,56	11,52	9,62	7,45
		28,00	16,99	16,43	16,11	15,58	15,16	14,83	14,12	12,78	10,74	8,84	6,67
		32,00	16,33	15,77	15,45	14,92	14,50	14,17	13,46	12,12	10,08	8,18	x
	7	17,00	21,66	21,10	20,78	20,25	19,83	19,50	18,79	17,45	15,41	13,51	11,34
		20,00	21,12	20,56	20,24	19,71	19,29	18,96	18,25	16,91	14,87	12,97	10,80
		23,00	20,53	19,97	19,65	19,12	18,70	18,37	17,66	16,32	14,28	12,38	10,21
		26,00	19,95	19,39	19,07	18,54	18,12	17,79	17,08	15,74	13,70	11,80	9,63
		29,00	19,38	18,82	18,50	17,97	17,55	17,22	16,51	15,17	13,13	11,23	9,06
		32,00	18,82	18,26	17,94	17,41	16,99	16,66	15,95	14,61	12,57	10,67	8,50
		35,00	18,27	17,71	17,39	16,86	16,44	16,11	15,40	14,06	12,02	10,12	7,95
		38,00	17,77	17,21	16,89	16,36	15,94	15,61	14,90	13,56	11,52	9,62	7,45
		41,00	17,21	16,65	16,33	15,80	15,38	15,05	14,34	13,00	10,96	9,06	6,89
		44,00	16,64	16,08	15,76	15,23	14,81	14,48	13,77	12,43	10,39	8,49	6,32

(1) The zero vertical column gives the capacity of the tubing in litres per metre.
(2) The zero horizontal line gives the total displacement of tubing with coupling in litres per metre.

ANNULAR VOLUME BETWEEN CASING AND TUBING
(litres per metre) (continued)

				Outside diameter of tubing (in)								
			0 (1)	1,050	1,315	1,660	1,900	2,063	2 3/8	2 7/8	3 1/2	4	4 1/2
		0 (2)	l/m	0,56	0,88	1,41	1,83	2,16	2,87	4,21	6,25	8,15	10,32
		24,00	25,02	24,46	24,14	23,61	23,19	22,86	22,15	20,81	18,77	16,87	14,70
		26,40	24,61	24,05	23,73	23,20	22,78	22,45	21,74	20,40	18,36	16,46	14,29
	7 5/8	29,70	23,97	23,41	23,09	22,56	22,14	21,81	21,10	19,76	17,72	15,82	13,65
		33,70	23,21	22,75	22,33	21,80	21,38	21,05	20,34	19,00	16,96	15,06	12,89
		39,00	22,25	21,79	21,37	20,84	20,42	20,09	19,38	18,04	16,00	14,10	11,93
		24,00	33,23	32,77	32,35	31,82	31,40	31,07	30,36	29,02	26,98	25,08	22,91
		28,00	32,59	32,03	31,71	31,18	30,76	30,43	29,72	28,38	26,34	24,44	22,27
	8 5/8	32,00	31,79	31,23	30,91	30,38	29,96	29,63	28,92	27,58	25,54	23,64	21,47
		36,00	31,04	30,48	30,16	29,63	29,21	28,88	28,17	26,83	24,79	22,89	20,72
		40,00	30,23	29,67	29,35	28,82	28,40	28,07	27,36	26,02	23,98	22,08	19,91
		44,00	29,47	28,91	28,59	28,06	27,64	27,31	26,60	25,26	23,22	21,32	19,15
		49,00	28,59	28,03	27,71	27,18	26,76	26,43	25,72	24,38	22,34	20,44	18,27
		32,30	41,08	40,62	40,20	39,67	39,25	38,92	38,21	36,87	34,83	32,93	30,76
		36,00	40,33	39,77	39,45	38,92	38,50	38,17	37,46	36,12	34,08	32,18	30,01
		40,00	39,55	38,99	38,67	38,14	37,72	37,39	36,68	35,34	33,30	31,40	29,23
Outside diameter and weight of the casing (in and lb/ft)	9 5/8	43,50	38,95	38,39	38,07	37,54	37,12	36,79	36,08	34,74	32,70	30,80	28,63
		47,00	38,19	37,73	37,31	36,78	36,36	36,03	35,32	33,98	31,94	30,04	27,87
		53,50	36,92	36,36	36,04	35,51	35,09	34,76	34,05	32,71	30,67	28,77	26,60
		58,40	36,14	35,58	35,26	34,73	34,31	33,98	33,27	31,93	29,89	28,00	25,82
		61,10	35,53	34,97	34,65	34,12	33,70	33,37	32,66	31,32	29,28	27,38	25,21
		71,80	33,46	32,90	32,58	32,05	31,63	31,30	30,59	29,25	27,21	25,31	23,14

(1) The zero vertical column gives the capacity of the tubing in litres per metre.
(2) The zero horizontal line gives the total displacement of tubing with coupling in litres per metre.

CASING CAPACITIES (1) (litres per metre)

OD and total displacement (in-l/m)	Nominal weight (lb/ft)	Thickness (mm)	Capacity (l/m)
4 1/2 (10.26)	9.50	5.21	8.48
	10.50	5.69	8.32
	11.60	6.35	8.11
	13.50	7.37	7.79
	15.10	8.56	7.42
	16.90	9.65	7.10
	17.70	10.20	6.95
	18.80	10.92	6.73
5 (12.67)	11.50	5.59	19.53
	13.00	6.43	10.22
	15.00	7.52	9.85
	18.00	9.19	9.26
	20.80	10.72	8.75
5 1/2 (15.33)	14.00	6.20	12.73
	15.50	6.98	12.41
	17.00	7.72	12.13
	20.00	9.17	11.58
	23.00	10.54	11.05
6 5/8 (22.24)	20.00	7.32	18.55
	24.00	8.94	17.77
	28.00	10.59	16.99
	32.00	12.06	16.33
7 (24.83)	17.00	5.87	21.67
	20.00	6.91	21.12
	23.00	8.05	20.54
	26.00	9.19	19.96
	29.00	10.36	19.38
	32.00	11.51	18.82
	35.00	12.65	18.27
	38.00	13.72	17.77
	41.00	14.98	17.21
	44.00	16.25	16.64

OD and total displacement (in-l/m)	Nominal weight (lb/ft)	Thickness (mm)	Capacity (l/m)
7 5/8 (29.46)	24.00	7.62	25.02
	26.40	8.33	24.61
	29.70	9.52	23.97
	33.70	10.92	23.21
	39.00	12.70	22.25
8 5/8 (37.69)	24.00	6.71	33.23
	28.00	7.72	32.59
	32.00	8.94	31.79
	36.00	10.16	31.04
	40.00	11.43	30.23
	44.00	12.70	29.47
	49.00	14.15	28.59
9 5/8 (46.94)	32.30	7.92	41.08
	36.00	8.94	40.33
	40.00	10.03	39.55
	43.50	11.05	38.95
	47.00	11.99	38.19
	53.50	13.84	36.92
	58.40	15.11	36.14
	61.10	15.87	35.53
	71.80	19.05	33.46
10 3/4 (58.56)	32.75	7.09	52.60
	40.50	8.89	51.15
	45.50	10.16	50.15
	51.00	11.43	49.13
	55.50	12.57	48.27
	60.70	13.84	47.26
	65.70	15.11	46.30

OD and total displacement (in-l/m)	Nominal weight (lb/ft)	Thickness (mm)	Capacity (l/m)
11 3/4 (69.96)	42.00	8.46	62.24
	47.00	9.52	61.31
	54.00	11.05	59.96
	60.00	12.42	58.79
13 3/8 (90.65)	48.00	8.38	81.89
	54.50	9.65	80.63
	61.00	10.92	79.37
	68.00	12.19	78.08
	72.00	13.06	77.24
16 (129.72)	65.00	9.52	117.87
	75.00	11.13	115.87
	84.00	12.57	114.19
18 5/8 (175.75)	87.50	11.05	159.74
20 (202.68)	94.00	11.13	185.3
	106.50	12.70	182.9
	133.00	16.13	177.8
30 (455.8)	267.00	20.60	397.0
	310.00	25.40	407.8

(1) These figures do not include any allowance for couplings - See in section C for capacities with couplings.

TUBING CAPACITIES (1) (litres per metre)

Outside diameter (OD) (in-mm)	Nominal weight (lb/ft)	Thickness (mm)	Inside diameter (ID) (mm)	Total displacement (l/m)	Capacity (l/m)	Metal displacement (l/m)	Linear metres per cubic metre (m)
1,050 (26,7)	1,14-1,20	2,87	20,9	0,56	0,34	0,22	2941,2
1,315 (33,4)	1,70-1,72-1,80	3,38	26,6	0,89	0,56	0,33	1785,7
1,660 (42,2)	2,10	3,18	35,8	1,41	1,01	0,40	990,1
	2,30-2,33-2,40	3,56	35,1	1,41	0,97	0,44	1030,9
1,900 (48,3)	2,40	3,18	41,9	1,83	1,38	0,45	724,6
	2,75-2,76-2,90	3,68	40,9	1,83	1,31	0,52	763,6
2,063 (52,4)	3,25	3,96	44,5	2,16	1,56	0,60	641,0
2 3/8 (60,3)	4,00	4,24	51,8	2,88	2,11	0,77	473,9
	4,60-4,70	4,83	50,7	2,88	2,02	0,86	495,0
	5,10	5,54	49,2	2,88	1,91	0,97	523,5
	5,80-5,95	6,45	47,4	2,88	1,77	1,11	565,0
	6,30	7,12	46,1	2,88	1,67	1,21	598,8
	7,30	8,53	43,3	2,88	1,47	1,41	680,2
2 7/8 (73,0)	6,40-6,50	5,51	62,0	4,21	3,02	1,19	331,1
	7,70	7,01	59,0	4,21	2,73	1,48	366,3
	8,60-8,70	7,82	57,4	4,21	2,59	1,62	386,1
	9,70	9,19	54,6	4,21	2,34	1,87	427,3
	10,70	10,28	52,5	4,21	2,16	2,05	463,0
3 1/2 (88,9)	7,70	5,49	77,9	6,26	4,77	1,49	209,6
	9,20-9,30	6,45	76,0	6,26	4,54	1,72	220,3
	10,20	7,34	74,2	6,26	4,33	1,93	230,9
	12,70-12,95	9,52	69,8	6,26	3,83	2,43	261,1
	13,70	10,50	67,9	6,26	3,60	2,66	277,8
	14,70	11,43	66,0	6,26	3,43	2,83	291,5
	15,80	12,40	64,1	6,26	3,29	2,97	304,0
4 (101,6)	9,50	5,74	90,1	8,16	6,38	1,78	156,7
	10,90-11,00	6,65	88,3	8,16	6,12	2,04	163,4
	13,00	8,38	84,8	8,16	5,65	2,51	177,0
	14,80	9,65	82,3	8,16	5,31	2,85	188,3
	16,50	10,92	79,8	8,16	4,99	3,17	200,4
4 1/2 (114,3)	12,60-12,75	6,88	100,5	10,35	7,94	2,41	125,9

(1) Capacities with couplings. Total displacements of tubings with couplings are an average.

CAPACITIES OF DRILL PIPES (1) (litres per metre)

Nominal size (in-mm)	Nominal weight (lb/ft)	Thickness (mm)	Inside diameter (mm)	Total displacement (l/m)	Capacity (l/m)	Metal displacement (l/m)	Lineal metre per cubic metre (m)
1,050 (26,7)	1,55	3,91	18,9	0,560	0,280	0,280	3571,4
1,315 (33,4)	2,30	4,55	24,3	0,88	0,46	0,42	2155,2
1,660 (42,2)	3,29	5,03	32,1	1,41	0,81	0,60	1234,6
1,900 (48,3)	4,19	5,56	37,2	1,83	1,09	0,74	921,7
2 3/8 (60,3)	6,65	7,11	46,1	2,86	1,67	1,19	598,8
2 7 8 (73,0)	10,40	9,19	54,6	4,19	2,34	1,85	427,4
3 1/2 (88,9)	9,50	6,45	76,0	6,21	4,54	1,67	220,3
	13,30	9,35	70,2	6,21	3,87	2,34	258,4
	15,50	11,40	66,1	6,21	3,43	2,78	291,5
4 (101,6)	11,85	6,65	88,3	8,11	6,13	1,98	163,1
	14,00	8,38	84,8	8,11	5,65	2,46	177,0
4 1/2 (114,3)	13,75	6,88	100,5	10,26	7,94	2,32	125,9
	16,60	8,56	97,2	10,26	7,42	2,84	134,8
	20,00	10,92	92,5	10,26	6,72	3,54	148,8
5 (127,0)	16,25	7,52	112,0	12,67	9,85	2,82	101,5
	19,50	9,19	108,6	12,67	9,27	3,40	107,9
	25,60	12,70	101,6	12,67	8,11	4,56	123,3
5 1/2 (139,7)	21,90	9,17	121,4	15,33	11,57	3,76	86,4
	24,70	10,54	118,6	15,33	11,05	4,28	90,5
6 5/8 (168,3)	25,20	8,38	151,5	22,24	18,03	4,21	55,5

(1) No allowance made for tool joints - The capacities of drill pipes with tool joints are given in section B (characteristics of drill pipes and tool joints).

CAPACITIES OF CYLINDERS (1)
(Diameters from 1″ to 24 7/8″) (litres per metre)

Diameter (in)	Capacity (l/m)	Diameter (in)	Capacity (l/m)	Diameter (in)	Capacity (l/m)	Diameter (in)	Capacity (l/m)	Diameter (in)	Capacity (l/m)	Diameter (in)	Capacity (l/m)
1	0,507	5	12,67	9	41,04	13	85,63	17	146,4	21	223,5
1 1/8	0,641	5 1/8	13,31	9 1/8	42,19	13 1/8	87,29	17 1/8	148,6	21 1/8	226,1
1 1/4	0,792	5 1/4	13,97	9 1/4	43,36	13 1/4	88,96	17 1/4	150,8	21 1/4	228,8
1 3/8	0,958	5 3/8	14,64	9 3/8	44,53	13 3/8	90,65	17 3/8	153,0	21 3/8	231,5
1 1/2	1,140	5 1/2	15,33	9 1/2	45,73	13 1/2	92,35	17 1/2	155,2	21 1/2	234,2
1 5/8	1,338	5 5/8	16,03	9 5/8	46,94	13 5/8	94,07	17 5/8	157,4	21 5/8	237,0
1 3/4	1,552	5 3/4	16,75	9 3/4	48,17	13 3/4	95,80	17 3/4	159,6	21 3/4	239,7
1 7/8	1,781	5 7/8	17,49	9 7/8	49,41	13 7/8	97,55	17 7/8	161,8	21 7/8	242,5
2	2,027	6	18,24	10	50,67	14	99,31	18	164,2	22	245,2
2 1/8	2,288	6 1/8	19,01	10 1/8	51,95	14 1/8	101,10	18 1/8	166,5	22 1/8	248,0
2 1/4	2,565	6 1/4	19,79	10 1/4	53,24	14 1/4	102,89	18 1/4	168,8	22 1/4	250,9
2 3/8	2,858	6 3/8	20,59	10 3/8	54,54	14 3/8	104,71	18 3/8	171,1	22 3/8	253,7
2 1/2	3,167	6 1/2	21,41	10 1/2	55,86	14 1/2	106,54	18 1/2	173,4	22 1/2	256,5
2 5/8	3,492	6 5/8	22,24	10 5/8	57,20	14 5/8	108,38	18 5/8	175,8	22 5/8	259,4
2 3/4	3,832	6 3/4	23,09	10 3/4	58,56	14 3/4	110,24	18 3/4	178,1	22 3/4	262,3
2 7/8	4,188	6 7/8	23,95	10 7/8	59,93	14 7/8	112,12	18 7/8	180,5	22 7/8	265,1
3	4,560	7	24,83	11	61,31	15	114,01	19	182,9	23	268,0
3 1/8	4,948	7 1/8	25,72	11 1/8	62,71	15 1/8	115,92	19 1/8	185,3	23 1/8	271,0
3 1/4	5,352	7 1/4	26,63	11 1/4	64,13	15 1/4	117,84	19 1/4	187,8	23 1/4	273,9
3 3/8	5,772	7 3/8	27,56	11 3/8	65,56	15 3/8	119,78	19 3/8	190,2	23 3/8	276,9
3 1/2	6,207	7 1/2	28,50	11 1/2	67,01	15 1/2	121,74	19 1/2	192,7	23 1/2	279,8
3 5/8	6,658	7 5/8	29,46	11 5/8	68,48	15 5/8	123,71	19 5/8	195,2	23 5/8	282,8
3 3/4	7,126	7 3/4	30,43	11 3/4	69,96	15 3/4	125,70	10 3/4	197,6	23 3/4	285,8
3 7/8	7,609	7 7/8	31,42	11 7/8	71,45	15 7/8	127,70	19 7/8	200,2	23 7/8	288,8
4	8,107	8	32,43	12	72,97	16	129,72	20	202,7	24	291,9
4 1/8	8,622	8 1/8	33,45	12 1/8	74,49	16 1/8	131,75	20 1/8	205,2	24 1/8	294,9
4 1/4	9,152	8 1/4	34,49	12 1/4	76,04	16 1/4	133,80	20 1/4	207,8	24 1/4	298,0
4 3/8	9,699	8 3/8	35,54	12 3/8	77,60	16 3/8	135,87	20 3/8	210,4	24 3/8	301,1
4 1/2	10,261	8 1/2	36,61	12 1/2	79,17	16 1/2	137,95	20 1/2	212,9	24 1/2	304,2
4 5/8	10,839	8 5/8	37,69	12 5/8	80,76	16 5/8	140,05	20 5/8	215,5	24 5/8	307,3
4 3/4	11,433	8 3/4	38,79	12 3/4	82,37	16 3/4	142,16	20 3/4	218,2	24 3/4	310,4
4 7/8	12,042	8 7/8	39,91	12 7/8	83,99	16 7/8	144,29	20 7/8	220,8	24 7/8	313,5

(1) The area of a circle is given in square centimetres (cm²) by multiplying by 10 the capacity in l/m.
Example : The area of a 8 1/2″ diameter circle is : 36,61 × 10 = 366,1 cm².

CAPACITIES OF CYLINDERS (1) (continued)
(diameters from 25″ to 36 7/8″) (litres per metre)

Diameter (in)	Capacity (l/m)	Diameter (in)	Capacity (l/m)	Diameter (in)	Capacity (l/m)
25	316,8	29	426,0	33	552,0
25 1/8	319,9	29 1/8	429,9	33 1/8	556,0
25 1/4	323,2	29 1/4	433,6	33 1/4	560,4
25 3/8	326,2	29 3/8	437,2	33 3/8	564,4
25 1/2	329,6	29 1/2	440,8	33 1/2	568,8
25 5/8	332,8	29 5/8	444,7	33 5/8	572,9
25 3/4	336,0	29 3/4	448,4	33 3/4	577,2
25 7/8	339,2	29 7/8	452,2	33 7/8	581,4
26	342,5	30	456,0	34	585,6
26 1/8	345,9	30 1/8	459,9	34 1/8	590,1
26 1/4	349,2	30 1/4	463,6	34 1/4	594,4
26 3/8	352,5	30 3/8	467,4	34 3/8	598,7
26 1/2	356,0	30 1/2	471,2	34 1/2	603,2
26 5/8	359,2	30 5/8	475,2	34 5/8	607,5
26 3/4	362,8	30 3/4	479,2	34 3/4	612,0
26 7/8	366,0	30 7/8	483,0	34 7/8	616,3
27	369,6	31	486,8	35	620,8
27 1/8	372,8	31 1/8	490,9	35 1/8	625,2
27 1/4	376,4	31 1/4	494,8	35 1/4	629,6
27 3/8	379,7	31 3/8	498,8	35 3/8	634,1
27 1/2	383,2	31 1/2	502,8	35 1/2	638,4
27 5/8	386,7	31 5/8	506,8	35 5/8	643,1
27 3/4	390,4	31 3/4	510,8	35 3/4	647,6
27 7/8	393,7	31 7/8	514,8	35 7/8	652,1
28	397,2	32	518,8	36	656,8
28 1/8	400,8	32 1/8	523,0	36 1/8	661,3
28 1/4	404,4	32 1/4	527,2	36 1/4	666,0
28 3/8	407,9	32 3/8	531,1	36 3/8	670,4
28 1/2	411,6	32 1/2	535,2	36 1/2	675,2
28 5/8	415,2	32 5/8	539,4	36 5/8	679,7
28 3/4	418,8	32 3/4	543,6	36 3/4	684,4
28 7/8	422,4	32 7/8	547,6	36 7/8	689,0

(1) The area of a circle is given in square centimetre (cm^2) by multiplying by 10 the capacity in l/m.
Example : The area of a 28″ diameter circle is : $397,2 \times 10 = 3972 \ cm^2$.

E ROTARY DRILLING EQUIPMENT TURBO-DRILLING

Summary

HOW TO USE ROCK BIT CLASSIFICATION TABLES

I. IADC AND API CLASSIFICATION OF ROCK BITS.

The Board of Directors of the IADC has approved a standard roller bit nomenclature system. It has also been approved as a recommended practice of the API and appears in API RP 7G.

The manufacturer assigns to each bit design three numbers that correspond to a specific block on the form (see next page).

As an example , an insert bit can be designated by all manufacturers by 7-4-7.

The first number designates the series (here 7 = insert bit for hard semi-abrasive and abrasive formations).

The second number 1 through 4 designates the type *i.e.* the formation hardness subclassification from softest to hardest (here 4 = hardest in series 7).

The third number designates the features (here 7 = friction sealed bearing and gage).

II. HOW TO CHOOSE A ROCK BIT.

The formation to be drilled is soft, having low compressive strength and high drillability. We want to use a milled tooth bit with sealed bearings.

The table on page E 2 enables us to find the numerical code of the bit :

> First number : soft and high drillability formation : 1
> Second number : estimaded hardness in series 1 : 2
> Third number : sealed bearing : 4
> Numerical code of the bit 1-2-4.

The table on page E 3 gives a choice between manufacturers :

> Hughes : X3
> Reed   : S12
> Security : S33
> Smith   : SDT
> SMF   : ES3

III. HOW TO KNOW THE DESIGN OF A ROCK BIT.

Let us take an SMF EM8. Which formation is it designed for ? The table on page E 3 indicates that SMF EM8 bit is equivalent to Hughes XC, Reed S23, Security M44L, Smith ST2 and that its numerical code is 2-3-4.

The table on page E 2 shows that it is a milled tooth bit for medium to medium hard formation with grade hardness 3 in series 2. It has sealed bearings.

IV. NOTE. Bit classifications are general and are to be used simply as guides. All bit types will drill effectively in formations other than those specified. It is the responsability of the manufacturer to classify his bits in the table at his own discretion.

TABLE FOR CLASSIFICATION OF ROCK (ROLLER) BITS

| | | | FEATURES | | | | | | | | |
SERIES	FORMATIONS	TYPES	1 Standard	2 "T" gage	3 Gage insert	4 Roller sealed bearings	5 Sealed bearings and gage	6 Friction sealed bearings	7 Friction sealed bearings and gage	8 Other	9 Other
MILLED TOOTH BITS											
1	Soft formations having low compressive strength and high drillability	1	1-1-1	1-1-2	1-1-3	1-1-4	1-1-5	1-1-6	1-1-7	1-1-8	1-1-9
		2	1-2-1	1-2-2	1-2-3	1-2-4	1-2-5	1-2-6	1-2-7	1-2-8	1-2-9
		3	1-3-1	1-3-2	1-3-3	1-3-4	1-3-5	1-3-6	1-3-7	1-3-8	1-3-9
		4	1-4-1	1-4-2	1-4-3	1-4-4	1-4-5	1-4-6	1-4-7	1-4-8	1-4-9
2	Medium to medium hard formations with high compressive strength	1	2-1-1	2-1-2	2-1-3	2-1-4	2-1-5	2-1-6	2-1-7	2-1-8	2-1-9
		2	2-2-1	2-2-2	2-2-3	2-2-4	2-2-5	2-2-6	2-2-7	2-2-8	2-2-9
		3	2-3-1	2-3-2	2-3-3	2-3-4	2-3-5	2-3-6	2-3-7	2-3-8	2-3-9
		4	2-4-1	2-4-2	2-4-3	2-4-4	2-4-5	2-4-6	2-4-7	2-4-8	2-4-9
3	Hard semi-abrasive or abrasive formations	1	3-1-1	3-1-2	3-1-3	3-1-4	3-1-5	3-1-6	3-1-7	3-1-8	3-1-9
		2	3-2-1	3-2-2	3-2-3	3-2-4	3-2-5	3-2-6	3-2-7	3-2-8	3-2-9
		3	3-3-1	3-3-2	3-3-3	3-3-4	3-3-5	3-3-6	3-3-7	3-3-8	3-3-9
		4	3-4-1	3-4-2	3-4-3	3-4-4	3-4-5	3-4-6	3-4-7	3-4-8	3-4-9
INSERT BITS											
5	Soft to medium formations with low compressive strength	1	5-1-1	5-1-2	5-1-3	5-1-4	5-1-5	5-1-6	5-1-7	5-1-8	5-1-9
		2	5-2-1	5-2-2	5-2-3	5-2-4	5-2-5	5-2-6	5-2-7	5-2-8	5-2-9
		3	5-3-1	5-3-2	5-3-3	5-3-4	5-3-5	5-3-6	5-3-7	5-3-8	5-3-9
		4	5-4-1	5-4-2	5-4-3	5-4-4	5-4-5	5-4-6	5-4-7	5-4-8	5-4-9
6	Medium hard formations with high compressive strength	1	6-1-1	6-1-2	6-1-3	6-1-4	6-1-5	6-1-6	6-1-7	6-1-8	6-1-9
		2	6-2-1	6-2-2	6-2-3	6-2-4	6-2-5	6-2-6	6-2-7	6-2-8	6-2-9
		3	6-3-1	6-3-2	6-3-3	6-3-4	6-3-5	6-3-6	6-3-7	6-3-8	6-3-9
		4	6-4-1	6-4-2	6-4-3	6-4-4	6-4-5	6-4-6	6-4-7	6-4-8	6-4-9
7	Hard semi-abrasive and abrasive formations	1	7-1-1	7-1-2	7-1-3	7-1-4	7-1-5	7-1-6	7-1-7	7-1-8	7-1-9
		2	7-2-1	7-2-2	7-2-3	7-2-4	7-2-5	7-2-6	7-2-7	7-2-8	7-2-9
		3	7-3-1	7-3-2	7-3-3	7-3-4	7-3-5	7-3-6	7-3-7	7-3-8	7-3-9
		4	7-4-1	7-4-2	7-4-3	7-4-4	7-4-5	7-4-6	7-4-7	7-4-8	7-4-9
8	Extremely hard and abrasive formations	1	8-1-1	8-1-2	8-1-3	8-1-4	8-1-5	8-1-6	8-1-7	8-1-8	8-1-9
		2	8-2-1	8-2-2	8-2-3	8-2-4	8-2-5	8-2-6	8-2-7	8-2-8	8-2-9
		3	8-3-1	8-3-2	8-3-3	8-3-4	8-3-5	8-3-6	8-3-7	8-3-8	8-3-9
		4	8-4-1	8-4-2	8-4-3	8-4-4	8-4-5	8-4-6	8-4-7	8-4-8	8-4-9

COMPARISON OF ROCK BITS (MILLED TOOTH BITS)

SOFT FORMATIONS HAVING HIGH DRILLABILITY

Numerical code	Hughes	Reed	Security	Smith	SMF
1-1-1	OSC-3A	Y11	S3S	DS	TS2
1-1-2					
1-1-3	X3A	S11	S3SG / S33S	SDS	ES2
1-1-4			S33SF		
1-1-5		F11			
1-1-6					
1-1-7					
1-2-1	OSC-3 / OSC-3C	Y12 / Y12T	S3 / S3T	DT / DTT	TS3
1-2-2			S3TG		
1-2-3			S33		TS3K / ES3 / ES3K
1-2-4	X3	S12		SDT	
1-2-5					
1-2-6	J2	F12			
1-2-7					
1-3-1	OSC/1G / OSC-1C	Y13 / Y13T	S4 / S4T	DG / DGT	TS5
1-3-2	ODG	Y13G	S4TG	DGH	
1-3-3	X1G	S13	S44	SDG	TS5K / ES5 / ES5K
1-3-4	XDG	S13G	S44G	SDGH	
1-3-5	J3		S44F		
1-3-6					
1-3-7					
1-4-1	OSC		S6 / S6T	K2	TS8
1-4-2			S6TG	K2H	
1-4-3			S66		ES8
1-4-4					
1-4-5					
1-4-6	J4	F14			
1-4-7					

MEDIUM TO MEDIUM HARD FORMATIONS

Numerical code	Hughes	Reed	Security	Smith	SMF
2-1-1	OWV-OW4	Y21	M4N	V1	TM2
2-1-2					
2-1-3	ODV-OD4	Y21G	M4NG	V1H	TM2K
2-1-4	XV	S21	M44N	SV1	EM2
2-1-5	XDV	S21G	M44NG	SVH	EM2K
2-1-6		F21	M44NF		
2-1-7					
2-2-1	WO	Y22	M4	V2	TM5
2-2-2					
2-2-3		Y22G	M4G	V2H	TM5K
2-2-4				SV2	EM5
2-2-5				SV2H	EM5K
2-2-6		F22			
2-2-7					
2-3-1	OWC	Y23	M4L	T2	TM8
2-3-2					
2-3-3		Y23G	M4LG	T2H	TM8K
2-3-4	XC	S23	M44L	ST2	EM8
2-3-5		S23G			EM8K
2-3-6			M44LF		
2-3-7					
2-4-1			M5		
2-4-2					
2-4-3					
2-4-4					
2-4-5					
2-4-6					
2-4-7					

Note. Numerical code given by the manufacturers.

COMPARISON OF ROCK BITS
(MILLED TOOTH BITS) (continued)

HARD, SEMI-ABRASIVE OR ABRASIVE FORMATIONS

Numerical code	Hughes	Reed	Security	Smith	SMF
3-1-1	W7-W7C	Y31	H7	L4	
3-1-2			H7T		
3-1-3	WD7	Y31G	H7TG	L4H	
3-1-4	X7	S31	H77	SL4	
3-1-5	XD7	S31G	H77G	SL4H	
3-1-6	J7				
3-1-7	JD7	F31G			
3-2-1	W7R2	Y32	H7U	W4	TH3
3-2-2			H7UG	W4H	TH3K
3-2-3		Y32G	H77U		EH3
3-2-4					EH3K
3-2-5					
3-2-6					
3-2-7					
3-3-1		Y33			
3-3-2					
3-3-3			H7SG		
3-3-4					
3-3-5					
3-3-6					
3-3-7					
3-4-1	WR	Y34	HC	WC	
3-4-2					
3-4-3	WDR	Y34G	HCG	WCH	TH5K
3-4-4	XWR	S34	H77C	SWC	
3-4-5	XDR	S34G	H77S	SWCH	EH5K
3-4-6	J8	F34	H77CF		
3-4-7	JD8				

Note. Numerical code given by the manufacturers.

FOR FUTURE USE

Numerical code	Hughes	Reed	Security	Smith	SMF
4-1-1					
4-1-2					
4-1-3					
4-1-4					
4-1-5					
4-1-6					
4-1-7					
4-2-1					
4-2-2					
4-2-3					
4-2-4					
4-2-5					
4-2-6					
4-2-7					
4-3-1					
4-3-2					
4-3-3					
4-3-4					
4-3-5					
4-3-6					
4-3-7					
4-4-1					
4-4-2					
4-4-3					
4-4-4					
4-4-5					
4-4-6					
4-4-7					

COMPARISON OF ROCK BITS
(INSERT BITS)

SOFT TO MEDIUM FORMATIONS

Numerical code	Hughes	Reed	Security	Smith	SMF
5-1-1					
5-1-2					
5-1-3	J22	FP51			
5-1-4			S84		
5-1-5			S84F		
5-1-6					
5-1-7					
5-2-1					
5-2-2					
5-2-3	J33	S52	S86	2JS	ES6
5-2-4		F52/FP52	S86F	F2	LS6
5-2-5					
5-2-6					
5-2-7					
5-3-1			S8		
5-3-2		S53	S88	3JS	EM6
5-3-3		F53/FP53	S88F	F3	LM6
5-3-4					
5-3-5					
5-3-6					
5-3-7					
5-4-1					
5-4-2					
5-4-3		FP54			
5-4-4					
5-4-5					
5-4-6					
5-4-7					

MEDIUM HARD FORMATIONS

Numerical code	Hughes	Reed	Security	Smith	SMF
6-1-1					
6-1-2					
6-1-3					
6-1-4	X44		M84F	4JS	
6-1-5	J44			F4	
6-1-6					
6-1-7					
6-2-1	44R		S88F	5J	EM9
6-2-2			M8		
6-2-3	X55R			5JS	
6-2-4	J55R	S62	M88		
6-2-5		F62/FP62	M88F-M89TF	F5	LM9
6-2-6					
6-2-7					
6-3-1					
6-3-2	X55	S63			
6-3-3			M89F	F57	
6-3-4	J55	F63/FP63			
6-3-5					
6-3-6					
6-3-7					
6-4-1					
6-4-2					
6-4-3		S64		6JS	
6-4-4					
6-4-5		F64/FP64			
6-4-6					
6-4-7					

Note. Numerical code given by the manufacturers.

COMPARISON OF ROCK BITS
(INSERT BITS) (continued)

HARD SEMI-ABRASIVE AND ABRASIVE FORMATIONS

Numerical code	Hughes	Reed	Security	Smith	SMF
7-1-1					
7-1-2					
7-1-3					
7-1-4	J77				EH6
7-1-5					LH6
7-1-6					
7-1-7					
7-2-1	RG7				
7-2-2	55R RG7X				
7-2-3		Y72	H8		
7-2-4		S72	H88		
7-2-5				6JS	
7-2-6	J88	F72/FP72	H88F	F6	
7-2-7					
7-3-1					
7-3-2		Y73	H8	7J	
7-3-3					
7-3-4		S73	H88	7JS	EH7
7-3-5		F73/FP73		F7	LH7
7-3-6					
7-3-7					
7-4-1	RG1				
7-4-2		Y74	H9		
7-4-3	RG1X				
7-4-4		S74	H99		
7-4-5				F7	
7-4-6		F74/FP74	H99F		
7-4-7					

EXTREMELY HARD AND ABRASIVE FORMATIONS

Numerical code	Hughes	Reed	Security	Smith	SMF
8-1-1					
8-1-2					
8-1-3					EH8
8-1-4					
8-1-5					
8-1-6					
8-1-7					
8-2-1					
8-2-2					
8-2-3					
8-2-4					
8-2-5					
8-2-6					
8-2-7					
8-3-1					
8-3-2					
8-3-3	RG2B RG2BX	Y83	H10	9J	EH9
8-3-4					
8-3-5	J99	S83	H100	9JS	
8-3-6		F83/FP83	H100F	F9	
8-3-7					
8-4-1					
8-4-2					
8-4-3					
8-4-4					
8-4-5					
8-4-6					
8-4-7					

Note. Numerical code given by the manufacturers.

MANUFACTURERS CODE
FOR THE DESIGNATION OF ROCK BITS

FEATURE \ MANUFACTURERS	Hughes	Reed	Security	Smith	SMF
Standard non sealed bearing		prefix Y	number 3 to 7		prefix T
Sealed bearing	letter X	prefix S	digit double	prefix S	prefix E
Gage insert	letter D	suffix G (1)	suffix G	suffix H	suffix K
T gage		suffix T	suffix T	letter T	
Insert bit			number 8 to 10		
Friction bearings	prefix J	prefix F or FP	suffix F	prefix F	prefix L

FORMATION					
Soft			letter S at the beginning		letter S as second letter
Medium			letter M at the beginning		letter M as second letter
Hard			letter H at the beginning		letter H as second letter

(1) All Reed insert bits have a gage protection, their designations may or may not have the suffix G.

GRADING USED BITS

I. TOOTH WEAR

Tooth wear is estimated in eighths (1/8) of the initial height of the tooth. Since tooth wear is not uniform on any row of a given cone, it is advisable to take several readings and report an average figure.

TOOTH WEAR	MILLED TOOTH	INSERT BIT
T1	Tooth height 1/8 gone	1/8 of insert lost or broken
T2	Tooth height 2/8 gone	2/8 of insert lost or broken
T3	Tooth height 3/8 gone	3/8 of insert lost or broken
T4	Tooth height 4/8 gone	4/8 of insert lost or broken
T5	Tooth height 5/8 gone	5/8 of insert lost or broken
T6	Tooth height 6/8 gone	6/8 of insert lost or broken
T7	Tooth height 7/8 gone	7/8 of insert lost or broken
T8	Tooth height all gone	All of insert lost or broken

Note : if any one row has a majority of teeth broken, add the letters '' BT ''.

Example : height of the teeth of a new bit = 18 mm. Average height of the teeth on a worn bit : 6 mm. Wear 18 — 6 = 12 mm.
To express this wear in eighths, calculate :
$\frac{12}{18} \times 8 = 5,33 \approx 5$ *i.e.* nearly 5/8 gone : teeth wear : T5.

II. BEARING CONDITION (check the worst cone)

The measurement of the bearing wear is very subjective. It is recommended to try estimate it in eighths of the life of the bearing.
Since mechanical aids are not available, it is necessary to « eyeball » the bearing wear and estimate rotating hours left in bearing. Knowing the rotating hours of the bit at the bottom of the well it is possible to calculate the ratio :

$$\frac{\text{estimated rotating hours left in bearing}}{\text{rotating hours at the bottom}} = \frac{T_e}{T_r}$$

and express this ratio in eighths of the life of the bearing

$\frac{T_e}{T_r}$	Condition
7	B1 bearing life gone 1/8
3	B2 bearing life gone 2/8
1,66	B3 bearing life gone 3/8
1	B4 bearing life gone 4/8
0,6	B5 bearing life gone 5/8
0,33	B6 bearing life gone 6/8
0,15	B7 bearing life gone 7/8
0	B8 bearing life gone 8/8

GRADING USED BITS (continued)

Example : A roller bit is pulled out of the hole after 12 h of rotation at the bottom (T_r = 12). The driller estimates that the worst cone could rotate 4 more hours before being completely worn out (T_e = 4)

$$\frac{T_e}{T_r} = \frac{4}{12} = 0,33 \; i.e. \; B6$$

Note : When a cone is locked the bearing wear is B8.

III. GAGE WEAR

When the bit pulled out of the hole is in gage use the terminology letter I.

When the bit pulled out of the hole is out of gage use the terminology letter O followed by the amount of gage in fractions of an inch.

To measure the amount of gage wear on a used bit, set the ring gage on two cones and measure the distance between the ring gage and the third cone in fractions of an inch.

IV. HOW TO REPORT BIT WEAR.

Example # 1 : T2 - B4 - I means :

Teeth 2/8 gone or inserts 2/8 lost or broken, 4/8 of bearing life gone, bit in gage.

Example # 2 : T6 BT - B6 - O 1/2 means :

Teeth 6/8 gone and teeth broken or inserts lost or broken, 6/8 of bearing life gone, bit out of gage 1/2 inch.

DRILLING FACTORS

Weight on bit and rotary speed

Formation classification	Weight per inch of bit diameter		Rotary speed rpm
	Pounds per inch	daN per inch	
Soft	3370 to 6750 4050 to 7800	1500 to 3000 1800 to 3500	250 to 100 180 to 60
Medium	4500 to 9000	2000 to 4000	120 to 40
Hard milled tooth bit	5600 to 11250	2500 to 5000	70 to 35
Hard insert bit	2250 to 5600 4500 to 9000	1000 to 2500 2000 to 4000	70 to 35 65 to 35
Hard friction bearing bit	4500 to 6750	2000 to 3000	60 to 35

Mud circulation rate

The mud circulation rate must be fast enough to remove the cuttings from the bottom of the hole and carry them to the surface.

The upward velocity of the mud in the annulus between the drill pipe and the wall can vary from 25 to 60 m/min (65 to 200 ft/min) depending on the drillability of the formations and the size of the hole.

Pump output and annular velocities for different sizes of hole

Size of drilling bit (in)	Pump output normally used		Annular velocity of mud (1)	
	l/min	gal/min	m/min	ft/min
17 1/2	3000 to 4000	800 to 1050	21 to 28	69 to 92
15	2800 to 3500	740 to 925	28 to 35	92 to 115
12 1/4	2200 to 2600	580 to 690	35 to 41	115 to 130
9 7/8	1500 to 1900	400 to 520	41 to 52	130 to 170
8 1/2	1200 to 1600	320 to 420	50 to 67	164 to 220
7 7/8	1200 to 1600	320 to 420	51 to 69	169 to 226
6 3/4	800 to 1000	210 to 260	53 to 67	173 to 220
6	600 to 800	160 to 210	51 to 68	169 to 223

(1) Annular velocity of mud may be slightly different depending on the size of the drill pipe.

Calculation of pump output to obtain a given annular velocity

The graph on the next page gives the method of finding the pump output in l/min knowing the annular volume between the drill pipe and the wall and the annular velocity chosen.

Example : Drilling 12 1/4'' hole with 5'' drill pipe. Annular velocity chosen 40 m/mm. Figure page D6 gives an annular volume of 62,7 l/m between 5'' drill pipe and 12 1/4'' hole. The graph gives a pump output of 2500 l/min.

PUMP OUTPUT - ANNULAR VELOCITY
ANNULAR VOLUME

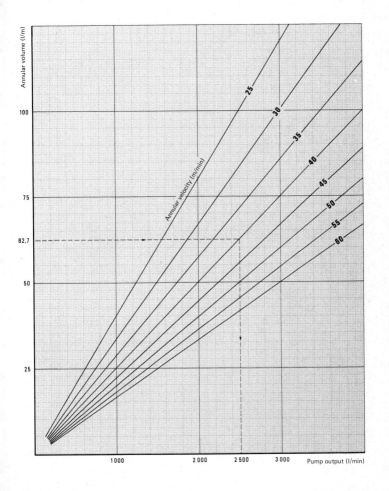

STRETCH OF SUSPENDED DRILL PIPE

Stretch due to its own weight :

$$A_a = 7,85 \frac{L^2}{2E}$$

L = length of string (m),
E = modulus of elasticity = 21 000 hbars.

$$\boxed{A_a \text{ in metres } = 1,87.10^{-7} \ L^2}$$

Shrinkage due to buoyancy in mud :

$$A_b = - \frac{d_b L^2}{E} (1 - y)$$

d_b = mud specific gravity,
L = length of string (m),
E = modulus of elasticity = 21 000 hbars,
y = Poisson's ratio = 0,3 for steel.

$$\boxed{A_b \text{ in metres } = - 0,334.10^{-7} \ d_b \ L^2}$$

Stretch due to temperature :

$$\boxed{A_t = 11,8.10^{-6} \ Ldt}$$

L = length of string (m),
dt = average temperature of the mud.

Total stretch :

$$A = A_a + A_b + A_t$$

$$\boxed{A = L^2 \ 10^{-7} \ (1,87 - 0,334 \ d_b) + 11,8.10^{-6} \ L \ dt}$$

CRITICAL ROTARY SPEEDS

Two types of vibration may occur. The pipe between each tool joint may vibrate in nodes, as a violin string. This critical speed may be approximately predicted by the formula.

$$RPM = \frac{4\,760\,000}{l^2}\,(D^2 + d^2)^{1/2}$$

where :
RPM = revolutions per minute,
l = length of one pipe, inches,
D = outside diameter of pipe, inches,
d = inside diameter of pipe, inches.

or : $RPM = \dfrac{1200}{l^2}\,(D^2 + d^2)^{1/2}$

where :
l = length of one pipe, metre,
D = outside diameter, cm,
d = inside diameter, cm.

Pipe size (in)	Approx. critical rotary speed RIPM
2 3/8	95←(110)→125
2 7/8	110←(130)→150
3 1/2	135←(160)→185
4	160←(185)→210
4 1/2	180←(210)→240
5	200←(235)→270
5 1/2	220←(260)→300

The critical speed predicted by this formula, number in brackets, is probably accurate within 15 % (number at left and number at right).

The second type of vibration is of the spring pendulum type and may be approximately predicted by the following formula :

$$RPM = \frac{258\,000}{L\ (ft)} \quad or \quad RPM = \frac{78\,600}{L\ (m)}$$

where L = total length of string in feet or in metres. Secondary and higher harmonic vibrations will occur at 4, 9, 16, 25, 36, etc... times the speed in this equation.

Vibrations of the spring pendulum type are probably less significant than the mode type. Each higher harmonic of the spring pendulum type vibration is also less significant.

Particular care should be taken to avoid operating under those conditions which would be the critical speed for both types of vibration because the combination would be particularly bad.

COINCIDENCE OF WELL DEPTH AND ROTARY SPEEDS FOR TWO TYPES OF VIBRATION

Drill pipe OD (in)	Speed RPM	Vibration coincidence depth							
		m	ft	m	ft	m	ft	m	ft
2 3/8	110	700	2300	2 800	9400	6 300	21000	—	—
2 7/8	130	600	1960	2 400	8000	5 400	18000	9 600	32000
3 1/2	160	500	1600	2 000	6600	4 500	14800	8 000	26000
4	185	425	1390	1 700	5600	3 800	12700	6 800	22000
4 1/2	210	370	1200	1 500	5000	3 300	11200	6 000	19700
5	235	330	1100	1 300	4500	3 000	10000	4 200	17500
5 1/2	260	300	980	1 200	4000	2 700	9000	3 800	15900

BUOYANCY FACTOR
Steel density = 7,85 kg/l

DRILLING MUD DENSITY			FACTOR	DRILLING MUD DENSITY			FACTOR
kg/l	lb/gal	lb/cu.ft		kg/l	lb/gal	lb/cu.ft	
1,00	8,35	62,4	0,873	1,62	13,52	101,2	0,793
1,02	8,51	63,7	0,869	1,64	13,68	102,4	0,791
1,04	8,68	64,9	0,867	1,66	13,85	103,7	0,789
1,06	8,85	66,2	0,864	1,68	14,02	104,9	0,786
1,08	9,01	67,4	0,862	1,70	14,18	106,2	0,783
1,10	9,18	68,7	0,859	1,72	14,35	107,4	0,781
1,12	9,35	69,9	0,857	1,74	14,52	108,7	0,779
1,14	9,51	71,2	0,854	1,76	14,68	109,9	0,776
1,16	9,68	72,4	0,852	1,78	14,85	111,2	0,773
1,18	9,85	73,7	0,849	1,80	15,02	112,4	0,771
1,20	10,01	74,9	0,847	1,82	15,18	113,7	0,768
1,22	10,18	76,2	0,844	1,84	15,35	114,9	0,765
1,24	10,35	77,4	0,842	1,86	15,53	116,2	0,763
1,26	10,52	78,7	0,839	1,88	15,69	117,4	0,760
1,28	10,68	79,9	0,837	1,90	15,86	118,7	0,758
1,30	10,85	81,2	0,834	1,92	16,02	119,9	0,755
1,32	11,02	82,4	0,832	1,94	16,18	121,2	0,752
1,34	11,18	83,7	0,829	1,96	16,36	122,4	0,749
1,36	11,35	84,9	0,827	1,98	16,53	123,7	0,747
1,38	11,52	86,2	0,824	2,00	16,69	124,9	0,745
1,40	11,68	87,4	0,822	2,02	16,86	126,2	0,742
1,42	11,85	88,7	0,819	2,04	17,02	127,4	0,739
1,44	12,02	89,9	0,817	2,06	17,18	128,7	0,737
1,46	12,18	91,2	0,814	2,08	17,36	129,9	0,734
1,48	12,35	92,4	0,812	2,10	17,53	131,2	0,732
1,50	12,52	93,7	0,809	2,12	17,69	132,4	0,729
1,52	12,68	94,9	0,807	2,14	17,86	133,7	0,727
1,54	12,85	96,2	0,804	2,16	18,02	134,9	0,725
1,56	13,02	97,4	0,801	2,18	18,19	136,2	0,722
1,58	13,18	98,7	0,798	2,20	18,36	137,4	0,719
1,60	13,35	99,9	0,796	2,22	18,54	138,7	0,717

Calculation of the apparent weight
of the drilling string

Apparent weight = weight in air — buoyant force

$$\text{Buoyant force} = \frac{\text{weight in air} \times \text{mud density}}{\text{steel density}}$$

$$\text{Apparent weight} = \text{weight in air} \left[\frac{\text{steel density} - \text{mud density}}{\text{steel density}} \right]$$

Apparent weight = weight in air × buoyancy factor

CALCULATION OF A TAPERED DRILLING STRING

I. MAXIMUM LENGTH OF A SPECIFIC SIZE, GRADE AND INSPECTION CLASS OF A DRILL PIPE

$$L_1 = \frac{T_1 \times 0,9}{S \times P_{t1} \times C_a} - \frac{P_{mt} \times L_{mt}}{P_{t1}} \quad \text{or} \quad \frac{T_1 \times 0,9 - M}{P_{t1} \times C_a} - \frac{P_{mt} \times L_{mt}}{P_{t1}}$$

Where : L_1 = maximum length of drill pipe (m),
T_1 = theoretical tension load from tables (daN),
S = safety factor (generally 1,33),
P_{t1} = mass / metre of drill pipe in air (kg/m),
C_a = buoyancy factor (depends on mud density) (see E 14),
P_{mt} = mass / metre of drill collars (kg/m),
L_{mt} = total length of drill-collars (m),
M = margin of over pull (daN).

Note. $T_1 \times 0,9$ is the maximum allowable design load in tension (daN)

$\dfrac{T_1 \times 0,9}{S}$ is the maximum static allowable load (daN)

0,9 is a constant relating proportional limit to yield strength
M is the difference between the calculated load and the maximum allowable tension load.

If the string is to be a tapered string, *i.e.,* to consist of more than one size, grade or inspection class of drill pipe, the pipe having the lowest load capacity should be placed just above the drill collars and the maximum length is calculated as shown previously. The $P_{mt} \times L_{mt}$ term in the equation is replaced by a term representing the weight in air of the drill collars plus the drill pipe assembly in the lower string.

II. EXAMPLE CALCULATIONS OF A TYPICAL DRILL STRING DESIGN

Design parameters :

Depth and hole size...	4 500 m - 8 1/2''
Mud weight..	1,2
Safety factor or Margin of over pull.............................	1,33 or 30 000 daN
Length of drill collars 6 3/4 × 2 13/16..........................	247 m
Pipe size, weight and grade : 5'' × 19,5 # × grade E, class II......	P_{t1} = 31,26 kg/m
Theoretical tension load from table (B6).........................	139 000 daN

From equation :

$$= \frac{T_1 \times 0,9}{S \times P_{t1} \times C_a} - \frac{P_{mt} \times L_{mt}}{P_{t1}} = \frac{139\,000 \times 0,9}{1,33 \times 31,26 \times 0,847} - \frac{149,4 \times 247}{31,26} = 3552 - 1180 = \boxed{2372 \text{ m}}$$

Maximum depth 2372 + 247 = 2619 m

It is apparent that drill pipe of a higher strength will be required to reach 4 500 m.

Add 5'' × 19,50 # (4 1/2 FH) Grade 95, class I (new) :

Air weight of number 1 drill pipe and drill collars :

$P_1 = L_1 \times P_{t1} + L_{mt} \times P_{mt} = 2372 \times 31,26 + 247 \times 149,4 = 111.050,5$ daN

From equation :

$$L_2 = \frac{223\,000 \times 0,9}{1,33 \times 31,8 \times 0,847} - \frac{111\,050,5}{31,8} = 2\,110 \text{ m}$$

This is more drill pipe than required, so final drill string will consist of the following.

6 3/4 × 2 13/16 drill collars	length 247 m	weight 36 902 kg
5'', 19,50 #, grade E, class II	length 2 372 m	weight 74 149 kg
5'', 19,50 #, grade 95, class I	length 1 881 m	weight 59 816 kg
Total	length 4 500 m	weight 170 867 kg

HOW TO APPLY MAKE-UP TORQUE

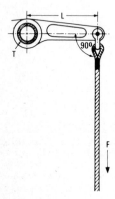

Torque is the measurement of the amount of twist applied to two members as they are screwed together. The product of the tong arm length L (measured from axis to axis) and the line pull force F is the measurement of the torque.

$$C = F \times L$$

when the arm tong and the pulling line are at right angles (90°).

The torque is given in metre - decanewtons (or foot-pounds) when the force is in decanewtons (or pounds) and the length of the tong arm is in metres (or feet).

The tables in this book give the recommended make-up torque for the size of connections in metre-decanewtons.

To find the force to apply on the pull-line divide the recommended make-up torque by the number of metres in the effective length of the tong arm.

Example : Recommended make-up torque for drill collar :
7 3/4 × 2 13/16'' with a 6 5/8 REG connection : 7 900 m daN.
  Length of the tong arm L = 1,17 m.
  The force to be applied on the pull-line will be :

$$F = \frac{C}{L} = \frac{7900}{1,17} = 6752 \text{ daN}$$

Note. A line-pull measuring device should be used in making-up connections. It is important that line-pull be measured when the line is at right angle or 90° to the tong handle.
When applying line-pull to the tongs, apply a long steady pull rather than jerking the line.

HOW TO APPLY MAKE-UP TORQUE (continued)

Torque measuring devices and total line-pull forces required

The torque indicators normally used are graduated in force units (kgf or lbf, tf or ton f). In the future they will be graduated in international system units *i.e.*, daN or 10^3 daN. These measuring devices give a force and not a torque.

It must be remembered that the force to be applied to make-up connections with the recommended torque is :

$$F = \frac{C}{L}$$

Example. For 6 3/4 OD × 2 13/16'' ID drill collars and NC 50 connection, the tables recommend 4 800 m-daN of make-up torque. Let's say the effective tong arm length is 1,22 m (48'' tong) :

$$F = \frac{4800}{1,22} = 3\ 934 \text{ daN}$$

The 3 934 daN of line pull is the total pull required on the end of the tong. This may or may not be the amount of line pull reading on the torque indicator, as this depends on the location of the indicator in the hook-up system. This may or may not be the amount of cathead pull for the same reason.

Following are some examples of hook-ups used to make-up drill collar connections.

First examples :

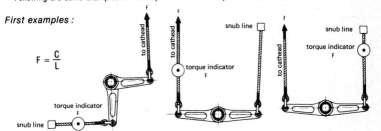

For these three hook-ups the amount of cathead pull and the torque indicator reading are equal to F (3934 daN for example).

Second examples :

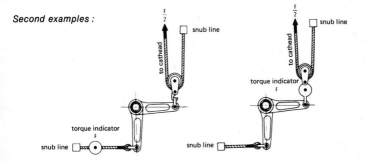

For these two hook-ups the amount of cathead pull is half $\left(\frac{F}{2}\right)$ of the torque indicator reading (F).

HOW TO APPLY MAKE-UP TORQUE (continued)

Torque measuring devices and total line-pull forces required

Third examples :

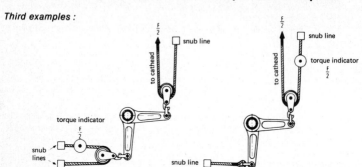

For these two examples the amount of cathead pull and the torque indicator reading are the same and equal to $\frac{F}{2}$, *i.e.* to apply a recommended force of 3934 daN it will be necessary to pull with the cathead until the reading on the torque indicator reaches $\frac{F}{2}$ or $\frac{3934}{2}$ = 1967 daN.

Fourth examples :

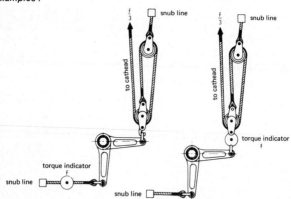

For these two examples the amount of cathead pull is one third $(\frac{F}{3})$ of the torque indicator reading (F).

Note. The third and fourth examples are systems in which high torques can be used while the cathead exerts forces lower than those required to obtain the torques wanted.

RECOMMENDED LUBRICANTS

ITEM	GRADE	TYPE	REMARKS

ENGINE
  Gas-butane : large
            small
  Diesel : general
            G.M.
            Caterpillar
	SAE 40
	SAE 30
	SAE 40
	SAE 30
	SAE 30
		MIL-L-2104A or
		MIL-L-2104B or
		supplement S-1
		Series 3
			1. Grade can be adjusted for extreme weather conditions, but it is better to use correct grade and control engine operating temperature.
			2. Extreme care should be used to prevent contamination of oil with permanent type antifreeze. Crankcase oils contaminated with some antifreeze additives will freeze engines.

GEAR OIL
  Slush pumps
  Rotary
  Swivel
  Pillow block bearings
  Gear reducers
	SAE 90
	SAE 90
	SAE 90
	SAE 90
		lead naphthenate,
		non-corrosive,
		anti-foaming,
		gear lube
			Do not use gear oils which contain EP compounds of chlorine and sulphur.

RUST AND OXIDATION OILS
(R and O oils)
  Draworks - oil bath
  Air compressors-All
  Air line lubricators
	SAE 30
	SAE 30
	SAE 10
		R and O
		straight mineral
			1. Do not use detergent oils ; these will form an emulsion in the presence of water.
			2. Rust and oxidation (R and O) straight mineral oil will protect equipment during stack-out.

GREASE Grease fittings - All	NGLI n° 2 (1)	multipurpose (lithium)	More expensive but only one kind needed on rig.
HYDRAULIC FLUID	Use manufacturer's recommendations		
OPEN CHAIN SYSTEM	used oil		

(1) NLGI : National Lubricating Grease Institute.

OILS AND GREASES - CORRELATION BETWEEN MANUFACTURER'S TRADE NAMES

TYPE	AGIP	BP	ELF	ESSO	FINA	MOBIL	SHELL	TOTAL
DIESEL ENGINE OIL								
MIL-2104A		ENERGOL HD	ELAN	ESTOR HDX	DELTA	DELVAC 1130	ROTELLA S	HDO
S1	ALPHA	DIESEL S1	DISAL HD1	ESTOR SDX	DILANO	DELVAC 1230	ROTELLA T	HDI-A
S2	GAMMA	DIESEL S2	DISAL HD2	ESTOR D3	SOLNA S3	DELVAC 1330	RIMULA	HDI-B et HD2
S3	SIGMA S3	DIESEL S3	DISAL HD3	MULTIGRADE S1	MULTIGRADE	DELVAC 20W 40	ROTELLA M-MULTI	HD3 et HD3-C
Multigrade S1 or MIL-L-210A4		VANELLUS	MULTI-PERFO	MULTIGRADE S2		UNIVERSAL 20 W 40	ROTELLA T-MULTI	HD1-M
Multigrade S2 or MIL-L-2104B		BP-TOU						HD2-M
COMPOUND OIL								
Mineral		GEAR-OIL	GIRELF	GEAR-OIL-ST	PERSANE	MOBILUBE-C	DENTAX	BOITE-PONT
Mineral with Pb naphthenate	ROTRA EP	ENERGOL GR-EP	REDUCT-ELF	PEN-O-LED-EP	SATURNA-EP	MOBIL COMPOUND	MACOMA	TORILIS
EP-MIL-L-2105A (1)		GEAR-OIL EP	TRANSELF-EP	GEAR-OIL-GP	PONTONIC-N	MOBILUBE-GX	SPIRAX-EP	TOTAL-EP
EP-MIL-L-2105B (1)		HYDROGEAR-EP	TRANSELF TYPE B	GEAR-OIL-GX	PONTONIC-EP	MOBILUBE-HD	SPIRAX-HD	TOTAL-EP-TYPE B
Multipurpose (2)		BP-TOU	PRESTIGRADE	UNIFARM	UNIVERSAL	UNIVERSAL	SUPERTRACTOR-HP	HD1-M
Industrial hydraulic	OSO	ENERGOL-HLP	ACANTIS	NUTOH		MOBIL FLUID 422	TELLUS	AZOLLA
Hydraulic high V.I.		ENERGOL-SHF	HYDRELF	OLEOFLUID-EP	HYDROFLUID EP-34	DTE-13	TELLUS-T	EQUIVIS
HYDRAULIC COUPLING AND TORQUE CONVERTOR OIL								
Spec. GM Type A Suffix A		ATF	RENAULT MATIC	ATF	PURFIMATIC	ATF-200Y	DONAX T6	FLUIDE-A
Spec. GM DEXRON		AUTRAN-DX	ELF MATIC-G	ATF-DEXRON	DEXRON-ATF	ATF-220	ATF-DEXRON	DEXRON
Spec. Ford		TFB-W1		TORQUE FLUID 35		ATF-210	DONAX-T7	ATF-33
		ENERGOL-CS		AQUILA	CIRKAN	DTE	TALPA	CORTIS
COMPRESSOR OIL								
ALL PURPOSES GREASE								
Lithium	GR MU2	LS2 MULTIPURPOSE	ELF-MULTI	BEACON-L2	MARSON-L2	MOBIGREASE MP	RETINAX-A	MULTIS
Lithium + others			ELF-MULTI-MOS, (5)	MULTI-GREASE H (3)	MARSON-LM31(5)	MOB-GREASE SUPER (4)	ALVANIA-EP-2	MULTIS-MS (5)
RUST PREVENTIVE OIL		PROTECTIVE-OIL	STOCKAGE	MOTOR-PROTECT	RUSAN	INFILREX-100	ENSIS	STOCKAGE
		ENERGOL-PAR-B	ANTIROUILLE 301	BANOX-324	RUSAN-S27	MOBILARMA 778	ENSISCOMPOUND352	OSYRIS-WS

(1) Mineral oil with sulphuror chlorine compounds (contact manufacturer).
(2) Detergent oil.
(3) Lime and lithine additives.
(4) Ca, Pb and MOS₂ compound additives.
(5) Lithium + MOS.
Note. These recommendations for lubrication are made as a guide. In all cases the lubricant recommendations given by the manufacturer should be used.

TURBO-DRILLING
LONGITUDINAL CROSS-SECTION OF A TURBODRILL

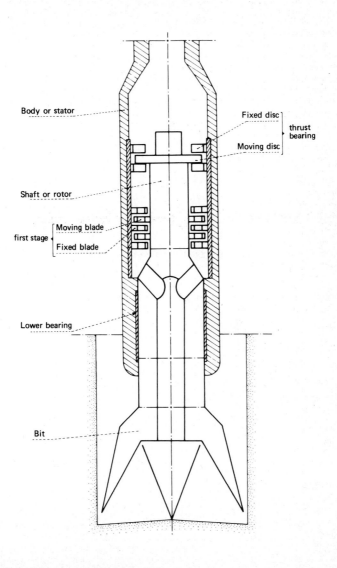

Body or stator

Fixed disc ⎤
          ⎥ thrust
          ⎦ bearing
Moving disc

Shaft or rotor

first stage { Moving blade
            { Fixed blade

Lower bearing

Bit

TURBO-DRILLING (continued)

I. TURBODRILL CHARACTERISTICS

The elements to characterise a turbo drill are :
  outside diameter of the body,
  number of stages,
  number of sections,
  type of blades,
  length and weight.

The hydraulic characteristics are given for a nominal pump output and a mud specific gravity equal to 1. These are the nominal characteristics :
  nominal speed N_n,
  nominal horse power output P_n,
  nominal drive torque C_n,
  nominal pressure drop Δp_n,
  axial thrust P_a.

The nominal horse power output is the maximum horse power output obtained with the nominal pump output.

The efficiency of the turbodrill is equal to the ratio of the mechanical horse power by the hydraulic horse power furnished to the turbodrill :

$$\eta_g = \frac{\text{mechanical horse power}}{\text{hydraulic horse power}} = \frac{\pi CN}{5\Delta pQ}$$

II. VARIATIONS OF NOMINAL CHARACTERISTICS WITH PUMP OUTPUT AND SPECIFIC GRAVITY OF MUD.

The actual values of horse power (P_{n1}), torque (C_{n1}), speed (N_{n1}) and pressure drop (Δp_{n1}) for a given fluid discharge Q_1 may be calculated by the following formulae from nominal values P_n, C_n, N_n, Δp_n at nominal discharge Q.

$$N_{n1} = N_n \times \frac{Q_1}{Q}$$

$$C_{n1} = C_n \times d_1 \left(\frac{Q_1}{Q}\right)^2$$

$$P_{n1} = P_n \times d_1 \left(\frac{Q_1}{Q}\right)^3$$

$$\Delta p_{n1} = \Delta p_n \times d_1 \left(\frac{Q_1}{Q}\right)^2$$

III. TURBODRILL CHECKING

Before running in check :

a. *Rotation.* check the rotation of the shaft with a chain tong.

Never run a turbodrill if the shaft is not rotating freely when applying a torque of some metre-decanewtons.

TURBO-DRILLING (continued)
TYPICAL CHARACTERISTIC CURVES

given specific gravity of circulation fluid d - constant pump output Q

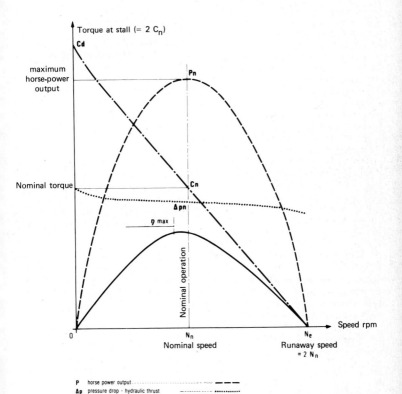

P	horse power output	― ― ―
Δp	pressure drop - hydraulic thrust	•••••••••
η	efficiency	─────
C	torque	―·―·―·

TURBO-DRILLING (continued)

b. *Measuring clearances*

Turbodrills are supplied with data sheets giving initial clearances and permissible wear limits.

Axial clearance is measured at the lower end of turbodrill with a sliding caliper between the end of lower bearing and the shaft shoulder.

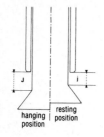

hanging position | resting position

The measurements are taken in both the hanging and the resting position.

The clearance is J-i.

For a new turbodrill the axial clearance is about 1 mm.

The maximum permissible wear given by the manufacturer is around 5 to 6 mm.

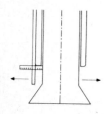

Radial clearance is measured with a rule and a graduated gauge as shown in the drawing at left.

The radical clearance must be 1 or 2 mm as a maximum (check with manufacturer for wear limits)

c. *Adjustment marks.* Turbodrills or turbodrill parts are always delivered ready to use. There are chisel marks across the connections and body. If necessary, these connections should be tightened with rotary tongs until alignment of chisel marks is achieved. On the contrary if connections are overtightened they never should be unscrewed.

d. *Filter.* Do not use turbodrills without the special filter set at the top of the last drill-pipe. This filter is always provided with the turbodrill.

IV. TURBO-DRILLING

In turbodrilling the speed of rotation and the weight on the bit are inter related.

Weight on the bit - Weight on bit is determined by the best penetration rate but it must be lower than the stalling weight (weight for which the turbine stops).

Weight on bit must be kept as constant as possible.

Pump discharge - As a rule discharge should not exceed the maximum allowed for each type of turbodrill.

Remember that the horse power output is proportional to the cube of the flow rate ratio $\frac{Q_1}{Q}$

F WIRE ROPE. CORDAGE LIFTING EQUIPMENT

Summary

API CLASSIFICATION OF WIRE ROPES

CLASSIFICATION 6 × 19	
6 × 19 Warrington	Fibre core
6 × 19 Seale	Fibre core
6 × 19 Seale	Wire rope core
6 × 21 Filler Wire	Fibre core
6 × 25 Filler Wire	Fibre core
6 × 25 Filler Wire	Wire rope core
6 × 25 Seale	Wire rope core
6 × 26 Warrington Seale	Wire rope core

CLASSIFICATION 6 × 37	
6 × 31 Filler Wire Seale	Wire rope core
6 × 31 Warrington Seale	Wire rope core
6 × 36 Filler Wire	Wire rope core
6 × 37 Two-Operation	Fibre core
6 × 37 Filler Wire Seale	Fibre core
6 × 41 Filler Wire	Wire rope core
6 × 41 Filler Wire	Fibre core
6 × 49 Filler Wire Seale	Wire rope core

CLASSIFICATION 8 × 19	
8 × 21 Filler Wire	Wire rope core
8 × 25 Filler Wire	Wire rope core

CLASSIFICATION 18 × 7 et 19 × 7	
18 × 7 Non-rotating	Fibre core
19 × 7 Non-rotating	Wire rope core

CLASSIFICATION 6 × 7	
6 × 7	Fibre core

TYPICAL SIZES AND CONSTRUCTIONS OF WIRE ROPE FOR OIL FIELD SERVICE (API RP 9B)

SERVICE AND WELL DEPTH	WIRE ROPE DIAMETRE (in)	WIRE ROPE DESCRIPTION
SAND LINES		
shallow	1/4 to 1/2 incl.	6 ×7 PS or IPS bright or galv.
intermediate	1/2 to 9/16	PF or NPF, RL, FC
deep	9/16 to 5/16	
DRILLING LINES		
coring and slim hole		6 × 19 Seale or 6 × 25 FW
shallow	7/8 to 1	PS or IPS, PF or NPF, RL, IWRC or FC
intermediate	1 to 1/8	6 × 19 Seale or 6 × 25 FW, EIPS, PF, RL or IWRC
DRILLING LINES (1)		
Large rotary rigs		
shallow	1 to 1 1/8	6 × 19 Seale or 6 × 21 FW
intermediate	1 1/8 to 1 1/4	PS or IPS, PF or NPF, RL, IWC or FC
deep	1 1/4 to 2	6 × 19 Seale or 6 × 21 FW, EIPS, PF, IWRC
WINCH LINES		6 × 31 Seale or 6 × 21 FW, IPS, PF, RL, IWRC or FC
heavy duty	5/8 to 7/8	6 × 31 Seale or 6 × 21 FW, EIPS, PF, RL, IWRC
MOORING LINES	7/8 to 2 incl.	6 × 19 IPS, PF, galv, IWRC or 6 × 19 EIPS PF, IWRC
	1 3/8 to 3 1/2 incl.	6 × 37 IPS, PF, galv, IWRC or 6 × 37 EIPS PF, IWRC.
	3 3/4 to 5'' incl.	6 × 61 IPS, PF, galv, IWRC or 6 × 61 EIPS PF, IWRC
MAST RAISING LINES	1 3/8 and smaller	6 × 25 FW, IPS or EIPS, PF, IWRC
	1 1/2 and larger	6 × 37 IPS or EIPS, PF, IWRC

ABBREVIATIONS :

FW   = Filler wire construction,
PS   = Plow Steel (rope wire having a breaking strength from 157 to 176 hbar 227 700 to 255 300 psi),
IPS   = Improved Plow Steel (breaking strength from 177 to 196 hbar 256 700 to 284 300 psi),
EIPS   = Extra Improved Plow-Steel (breaking strength from 197 to 215 hbar 285 700 to 310 800 psi),
PF   = Preformed,
NPF   = Non-Preformed,
RL   = Right Lay,
LL   = Left Lay,
FC   = Fiber Core,
IWRC = Independent Wire Rope Core.

(1) For drilling lines 1 3/8'' and larger some manufacturers recommend wire ropes type 6 × 26 Warrington Seale IPS or EIPS, IWRC which have better flexibility and better fatigue resistance for a good strain resistance.

API WIRE ROPES

I. SHEAVE DIAMETER FACTOR

Wire rope 6 × 19 Seale   D/d = 34
Wire rope 6 × 21 Filler   D/d = 30   D : tread diameter of sheave,
Wire rope 6 × 25 Filler   D/d = 27   d : nominal rope diameter.
Wire rope 6 × 31   D/d = 25
Wire rope 6 × 37   D/d = 18

II. SAFETY FACTOR

$$F = \frac{T}{P}$$

T = nominal breaking strength of wire rope,
P = calculated total static load.

Minimum safety factor	API	French " Service of Mines "
Rotary drilling line	3	5
Rotary drilling line setting casing	2	3
Pulling on stuck pipe	2	3

III. STRAND CONSTRUCTION

6 × 19 Seale   **6 × 25 Filler**   **6 × 19 Warrington**
wire rope core   **wire rope core**   **fibre core**

Ordinary strand. All the wires of the strand have the same diameter. Each successive wire layer contains 6 wires more (1 + 6 + 12 + 18 + ...).

Seale strand. All layers contain the same number of wires, or, at least, the two outside layers (1 + 9 + 9 + ...).

Seale lay filler wire strand. The inner layers contain filling wires (1 + 6,6 + 12) (shown above).

Warrington strand. The outside layer contains filling wires (1 + 6 + 6,6) (shown above).

Mixed strands :
Filler Wire Seale strand : (1 + 5,5 + 10 + 10),
Warrington Seale strand : (1 + 5 + 5,5 + 10).

IV. DIFFERENT KINDS OF WIRE ROPE TWISTING

right lay   **left lay**   **right lay**   **left lay**
regular lay   **regular lay**   **lang lay**   **lang lay**

In ropes with " regular " lay the wires are twisted in one direction and the strands in the opposite direction.

In ropes with " lang " lay the wire and the strands are twisted in the same direction.

ANTI-ROTATING WIRE ROPES

When a load is handled with one or two lines, it is better to use anti-rotating wire ropes.

In a anti-rotating wire rope the layers are twisted in the opposite directions so that the resulting torque will be as small as possible.

ANTI-ROTATING WIRE ROPE TYPES

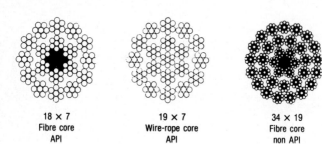

18 × 7	19 × 7	34 × 19
Fibre core	Wire-rope core	Fibre core
API	API	non API

(1) When stringing-up with more than two lines it is not necessary to use anti-rotating wire-rope,
(2) The sheave diameter factor D/d must be 40 (never less than 35),
(3) An anti-rotating wire rope must always be under tension even when idle,
(4) An anti-rotating wire rope must never be untwisted or overtwisted to avoid kinking or bird-caging,
(5) Do not use bull-dog clips, thimbles. Use only sockets sealed with zinc.
(6) The length of the spooling drum must be large enough to have only one layer of wire-rope on it.

Breaking load for anti-rotative wire ropes 18 × 7 and 19 × 7

Nominal diameter (in)	Weight per metre (kg/m)	Breaking load (10³ daN) IPS wire rope core		Nominal diameter (in)	Weight per metre (kg/m)	Breaking load (10³ daN) IPS fibre core	
		minimum	nominal			minimum	nominal
1/2	0,67	8,5	8,8	1/2	0,64	8,5	8,8
9/16	0,86	10,8	11,0	9/16	0,82	10,8	11,0
5/8	1,06	13,3	13,6	5/8	1,01	13,3	13,6
3/4	1,52	18,9	19,4	3/4	1,44	18,9	19,4
7/8	2,07	25,6	26,2	7/8	1,96	25,6	26,2
1	2,71	33,2	34,1	1	2,57	33,2	34,1

BREAKING STRENGTHS FOR WIRE ROPES WITH INDEPENDENT WIRE ROPE CORE (IWRC) 6 × 19 API CLASSIFICATION

Nominal diameter (in)	Weight (kg/m)	BREAKING STRENGTH (10³ daN)									
		Bright or Drawn galvanized (1)				Bright		Galvanized			
		Plow Steel		Improved Plow Steel		Extra improved Plow Steel		Plow Steel		Improved Plow Steel	
		minimum	nominal	minimum	nominal	minimum	nominal	minimum	nominal	minimum	nominal
1/2	0.68	8.7	9	10	10.2	11.5	11.8	7.8	8.1	8.9	9.2
9/16	0.88	11	11.3	12.6	12.9	14.6	14.9	9.9	10.1	11.3	11.6
5/8	1.07	13.5	13.9	15.5	15.9	17.9	18.3	12.1	12.5	14.0	14.3
3/4	1.55	19.2	19.8	22.2	22.8	25.5	26.2	17.3	17.8	19.9	20.5
7/8	2.11	26.1	26.7	30.0	30.8	34.5	35.4	23.5	24.1	27.0	27.7
1	2.75	33.9	34.7	38.9	40.0	44.9	46.0	30.5	31.3	35.1	35.9
1 1/8	3.48	42.6	43.7	48.9	50.3	56.5	57.8	38.3	39.3	44.1	45.2
1 1/4	4.30	52.5	53.7	60.1	61.7	69.4	71.1	47.2	48.4	54.3	55.6
1 3/8	5.21	62.7	64.6	72.5	74.3	83.2	85.4	56.5	58.1	64.9	66.8
1 1/2	6.16	74.7	76.5	85.9	88.0	98.8	101.4	67.2	68.9	77.4	79.2
1 5/8	7.26	86.7	89.0	99.8	102.3	114.5	117.4	78.1	80.1	89.8	92.1
1 3/4	8.44	100.6	103.2	115.4	118.3	132.7	136.1	90.3	92.9	103.8	106.5
1 7/8	9.57					159.9	154.8	103.1	105.7	118.7	121.7
2	10.10					171.7	176.1	116.3	119.3	134.3	137.7

BREAKING STRENGTH FOR WIRE ROPES WITH FIBRE CORE (FC) 6 × 19 API CLASSIFICATION

Nominal diameter (in)	Weight (kg/m)	Bright or Drawn galvanized (1)				Bright		Galvanized			
		Plow Steel		Improved Plow Steel		Extra improved Plow Steel		Plow Steel		Improved Plow Steel	
		minimum	nominal	minimum	nominal	minimum	nominal	minimum	nominal	minimum	nominal
1/2	0.63	8.1	8.3	9.3	9.5			7.3	7.5	8.4	8.6
9/16	0.79	10.2	10.5	11.7	12.0			9.2	9.4	10.5	10.8
5/8	0.98	12.6	12.9	14.5	14.9			11.3	11.6	13.0	13.3
3/4	1.41	18.0	18.4	20.6	21.2			16.1	16.5	18.5	19.0
7/8	1.92	24.3	24.9	27.9	28.6			21.8	22.4	25.1	25.8
1	2.50	31.6	32.4	36.3	37.2			28.5	29.2	32.6	33.5
1 1/8	3.17	39.6	40.7	45.8	46.8			35.6	36.6	41.0	42.1
1 1/4	3.91	48.9	50.0	56.0	57.5			43.9	45.0	50.3	51.7
1 3/8	4.73	58.7	60.1	67.2	69.1			52.9	53.9	60.5	62.2
1 1/2	5.63	69.4	71.2	79.6	81.8			62.3	64.1	71.6	73.7
1 5/8	6.61	81.0	83.1	92.8	95.2			72.9	74.8	83.5	85.7
1 3/4	7.66	93.7	96.1	107.6	110.3			84.3	86.5	96.8	99.3

(1) Zinc coating applied at an intermediate stage of the wire drawing operation.

BREAKING STRENGTH FOR WIRE ROPES 6 × 37 API CLASSIFICATION

Nominal diameter (in)	Weight (kg/m)	INDEPENDENT WIRE ROPE CORE (IWRC)						FIBRE CORE (FB)			
		Bright or drawn galvanized IPS (1)		Bright EIPS (1)		Galvanized IPS (1)		Bright or drawn galvanized IPS (1)		Galvanized IPS (1)	
		minimum	nominal	minimum	nominal	minimum	nominal	minimum	nominal	minimum	nominal
1/2	0.68	9.5	9.8	10.9	11.2	8.5	8.8	8.9	9.1	8.0	8.2
9/16	0.88	12.1	12.4	13.8	14.1	10.9	11.1	11.2	11.5	10.1	10.3
5/8	1.07	14.7	15.1	17.0	17.4	13.3	13.6	13.7	14.1	12.3	12.6
3/4	1.55	21.1	21.6	24.2	24.8	19.0	19.5	19.6	20.1	17.6	18.1
7/8	2.11	28.5	29.3	32.8	33.6	25.7	26.3	26.6	27.2	22.8	24.5
1	2.75	37.1	38.1	42.6	43.7	33.4	34.3	34.5	35.4	31.0	31.8
1 1/8	3.48	46.7	48.0	53.8	55.1	42.1	43.1	43.5	44.6	39.1	40.1
1 1/4	4.30	57.4	58.8	65.8	67.7	51.6	52.9	53.3	54.7	48.0	49.2
1 3/8	5.21	69.1	70.9	79.6	81.6	62.2	63.8	64.3	65.9	57.8	59.3
1 1/2	6.16	82.0	84.1	93.9	96.1	73.8	75.7	76.2	78.2	68.6	70.4
1 5/8	7.26	96.3	98.8	110.2	113.0	86.7	88.9	89.4	91.6	80.4	82.5
1 3/4	8.44	111.0	113.9	126.6	129.9	99.9	102.5	103.2	105.9	92.9	95.3
1 7/8	9.67			145.7	149.5	114.0	116.9				
2	10.10			164.8	169.0	128.8	132.1				
2 1/8	12.43			185.6	190.4	145.2	148.9				
2 1/4	13.93			207.3	212.6	161.6	165.7				
2 3/8	15.48			229.0	234.9	179.6	184.2				
2 1/2	17.26			253.3	259.8	198.3	203.4				
2 5/8	19.05			278.5	285.6	217.8	223.4				
2 3/4	20.83			303.6	311.4	238.1	244.2				
2 7/8	22.77			331.3	339.8	260.0	266.6				
3	24.70			359.1	368.3	281.0	288.2				
3 1/8	26.79			388.6	398.6	303.7	311.4				
3 1/4	29.02			419.0	429.7	327.0	335.5				
3 3/8	31.25			449.3	460.8	352.1	361.1				
3 1/2	33.78			481.3	493.8	377.1	386.7				

(1) See abbreviations page F2.

NOMINAL BREAKING LOADS FOR 6 × 19, 6 × 37
and 6 × 61 CONSTRUCTION MOORING WIRE ROPE (IWRC)
(API Spec 9A)

CONS-TRUCTION CLASSI-FICATION	NOMINAL DIAMETER		APPROXIMATE WEIGHT		NOMINAL BREAKING LOAD			
					Galvanized		Bright	
	mm	in	kg/m	lb/ft	10³daN	10³lb	10³daN	10³lb
6 × 19	26	1	2,75	1,85	41,4	93,1	42,6	95,8
	29	1 1/8	3,48	2,34	52,0	117,0	53,0	119,0
	32	1 1/4	4,30	2,89	64,0	143,8	64,6	145,0
	35	1 3/8	5,21	3,50	76,9	172,8	77,3	174,0
	38	1 1/2	6,19	4,16	91,3	205,2	91,1	205,0
	42	1 5/8	7,26	4,88	106,0	237,6	111,0	250,0
	45	1 3/4	8,44	5,67	123,0	275,4	123,0	287,0
	48	1 7/8	9,67	6,50	139,0	313,2	145,0	327,0
6 × 37	51	2	11,0	7,29	159,0	356,4	164,0	369,0
	54	2 1/8	12,4	8,35	177,0	397,8	184,0	413,0
	57	2 1/4	13,9	9,30	198,0	444,6	205,0	461,0
	61	2 3/8	15,5	10,4	219,0	493,2	235,0	528,0
	64	2 1/2	17,3	11,6	242,0	543,6	258,0	581,0
	67	2 5/8	19,0	12,8	265,0	595,8	283,0	636,0
	70	2 3/4	20,8	14,0	289,0	649,8	309,0	695,0
	74	2 7/8	22,8	15,3	314,0	705,6	345,0	776,0
	77	3	24,7	16,6	340,0	765,0	374,0	841,0
	80	3 1/8	26,8	18,0	367,0	824,4	404,0	907,0
	83	3 1/4	29,0	19,5	394,0	885,6	434,0	977,0
	86	3 3/8	31,3	21,0	424,0	952,2	466,0	1049,0
	90	3 1/2	33,8	22,7	452,0	1015,0	511,0	1148,0
	96	3 3/4	38,7	26,0	506,0	1138,0	562,0	1264,0
6 × 61	103	4	44,0	29,6	571,0	1283,0	634,0	1426,0
	109	4 1/4	49,6	33,3	640,0	1438,0	711,0	1598,0
	115	4 1/2	55,7	37,4	711,0	1598,0	790,0	1776,0
	122	4 3/4	62,1	41,7	785,0	1766,0	873,0	1962,0
	128	5	68,8	46,2	863,0	1944,0	959,0	2156,0

API WIRE ROPES

Efficiency of various fastening devices

(% of breaking strength)

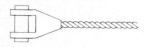

Sockets 100 %

Wire rope clips 80 %

(number of clips varies with wire diameter)

Spliced thimble

3/8'' to 5/8'' wire rope 95 %
3/4'' to 1 1/8'' wire rope 85 %
1 1/4'' to 1 1/2'' wire rope 80 %
1 3/8'' to 2'' wire rope 75 %

Number of clips necessary to obtain 80 % efficiency

Nominal diameter	Number of clips	Length of rope turned back exclusive of eye		Make-up torque for clips	
(in)		cm	in	m-daN	ft-lb
1/8	2	8	3 1/4		
3/16	2	10	3 3/4		
1/4	2	12	4 3/4	2	15
5/16	2	13	5 1/4	4	30
3/8	2	17	6 1/2	6	45
7/16	2	18	7	9	65
1/2	3	29	11 1/2	9	65
9/16	3	30	12	13	95
5/8	3	30	12	13	95
3/4	4	46	18	18	130
7/8	4	48	19	31	225
1	5	66	26	31	225
1 1/8	6	86	34	31	225
1 1/4	6	94	37	49	360
1 3/8	7	112	44	49	360
1 1/2	7	122	48	49	360
1 5/8	7	130	51	58	430
1 3/4	7	135	53	80	590
2	8	180	71	102	750
2 1/4	8	185	73	102	750
2 1/2	9	213	84	102	750
2 3/4	10	254	100	102	750
3	10	269	106	163	1200

Method of attaching clips

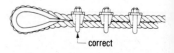

correct

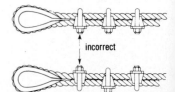

incorrect

Measuring rope diameter

right

wrong

EVALUATION OF ROTARY DRILLING LINE WORK

I. ROUND-TRIP OPERATIONS : Running the string of drill pipe into the hole and pulling the string out of the hole at the depth L :

$$T_m = 0,98 [p. L(L + 1) + 4 L(P + \frac{d}{2})] 10^{-6}$$

where :

T_m = amount of work (10^3 daN-km),

L = depth of hole (m),

l = length of a triple,

d = effective weight of drill collar assembly in mud, minus effective weight of same length of drill pipe in mud (kg),

p = effective weight per metre of drill pipe in mud (kg/m),

P = total weight of travelling block-elevator assembly (kg).

II. DRILLING OPERATIONS

To drill to total depth L_1 :

To drill from depth L_1 to depth L_2 :

$$T_f = 3 T_{m1}$$

$$[T_f] {}^{L_2}_{L_1} = 3 [T_{m2} - T_{m1}]$$

III. CORING OPERATIONS

To core from depth L_1 to depth L_2 :

$$[T_c] {}^{L_2}_{L_1} = 2 [T_{m2} - T_{m1}]$$

EVALUATION OF ROTARY DRILLING LINE WORK
Depths 100 to 500 m

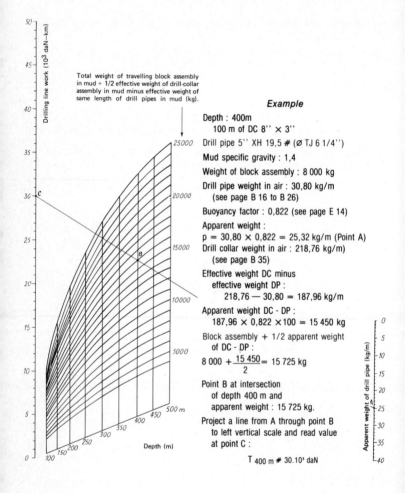

Total weight of travelling block assembly in mud + 1/2 effective weight of drill-collar assembly in mud minus effective weight of same length of drill pipes in mud (kg).

Drilling line work (10³ daN–km)

Example

Depth : 400m
   100 m of DC 8'' × 3''

Drill pipe 5'' XH 19,5 # (Ø TJ 6 1/4'')

Mud specific gravity : 1,4

Weight of block assembly : 8 000 kg

Drill pipe weight in air : 30,80 kg/m
   (see page B 16 to B 26)

Buoyancy factor : 0,822 (see page E 14)

Apparent weight :
p = 30,80 × 0,822 = 25,32 kg/m (Point A)

Drill collar weight in air : 218,76 kg/m)
   (see page B 35)

Effective weight DC minus
   effective weight DP :
   218,76 — 30,80 = 187,96 kg/m

Apparent weight DC - DP :
   187,96 × 0,822 ×100 = 15 450 kg

Block assembly + 1/2 apparent weight
   of DC - DP :

$8\,000 + \dfrac{15\,450}{2} = 15\,725$ kg

Point B at intersection
   of depth 400 m and
   apparent weight : 15 725 kg.

Project a line from A through point B
   to left vertical scale and read value
   at point C :

   T 400 m # 30.10³ daN

Apparent weight of drill pipe (kg/m)

Depth (m)

EVALUATION OF ROTARY DRILLING LINE WORK
Depths 500 to 2 000 m

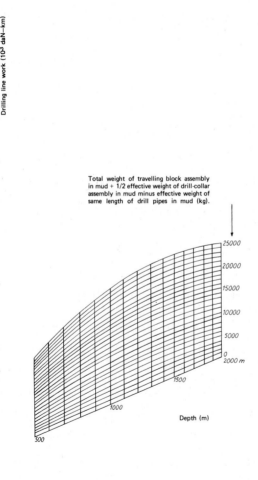

Drilling line work (10^3 daN–km)

Total weight of travelling block assembly in mud + 1/2 effective weight of drill-collar assembly in mud minus effective weight of same length of drill pipes in mud (kg).

25000
20000
15000
10000
5000
0
2000 m

1500

1000

500

Depth (m)

Apparent weight of drill pipe (kg/m)

EVALUATION OF ROTARY DRILLING LINE WORK
Depths 1 500 to 3 500 m

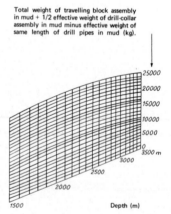

Total weight of travelling block assembly in mud + 1/2 effective weight of drill-collar assembly in mud minus effective weight of same length of drill pipes in mud (kg).

EVALUATION OF ROTARY DRILLING LINE WORK
Depths 3 000 to 6 000 m

Drilling line work. (10^3 daN–km)

1200
1100
1000
900
800
700
600
500
400
300
200
100
50
40
30
20
10
0

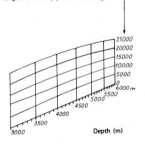

Total weight of travelling block assembly in mud + 1/2 effective weight of drill-collar assembly in mud minus effective weight of same length of drill pipes in mud (kg).

25000
20000
15000
10000
5000
0

6000 m
5500
5000
4500
4000
3500
3000

Depth (m)

Apparent weight of drill pipe (kg/m)

0
5
10
15
20
25
30
35
40

RECOMMENDED CUTOFF LENGTHS FOR ROTARY DRILLING LINES (API RP 9B)

Cutoff length in metres and number of drum laps (1)

DERRICK OR MAST HEIGHT (ft)	DRUM DIAMETER (in)													
	11	13	14	16	18	20	22	24	26	28	30	32	34	36
187										34.6 / 15 1/2 (1)	34.7 / 14 1/2	34.5 / 13 1/2	33.9 / 12 1/2	33.0 / 11 1/2
142-143-147						24.7 / 15 1/2	25.5 / 14 1/2	25.9 / 13 1/2	25.9 / 12 1/2	25.7 / 11 1/2	27.5 / 11 1/2	26.8 / 10 1/2		
133-136-138				22.3 / 17 1/2	22.3 / 15 1/2	23.1 / 14 1/2	21.9 / 12 1/2	23.9 / 12 1/2	23.9 / 11 1/2	25.7 / 11 1/2	25.1 / 10 1/2	24.3 / 9 1/2		
126-129-131				18.5 / 14 1/2	18.0 / 12 1/2	18.4 / 11 1/2	18.4 / 10 1/2	23.9 / 12 1/2	23.9 / 11 1/2	23.5 / 10 1/2	22.7 / 9 1/2	24.3 / 9 1/2		
94-96-100		20.2 / 19 1/2	19.6 / 17 1/2	16.0 / 12 1/2	16.5 / 11 1/2			18.2 / 9 1/2	19.7 / 9 1/2	19.0 / 8 1/2				
87		18.2 / 17 1/2	16.2 / 14 1/2											
66	11.0 / 12 1/2	12.0 / 11 1/2												

(1) In order to change the point of crossover on the drum, where wear and crushing are most severe, the laps to be cut are given in multiples of one-half lap.

CUTOFF PRACTICE FOR DRILLING LINE (continued)
Cumulative work before the first cutoff (1) (API RP 9B)

Derrick or mast height (ft)	Formation hardness	Total work of drilling line before the first cutoff, function of line ∅									
		1''		1 1/8''		1 1/4''		1 3/8''		1 1/2''	
		10³ daN-km	Ton-mile	10³ daN-km	Ton-mile	10³ daN-km	Ton-mile	10³ daN-km	Ton-mile	10³ daN-km	Ton-mile
80 à 87	very hard	716	500								
	hard	716	500								
	medium	716	500								
	soft	859	600								
94 à 100	very hard	716	500	859	600						
	hard	716	500	1003	700						
	medium	716	500	1146	800						
	soft	859	600	1289	900						
126 à 131	very hard			859	600	1432	1000				
	hard			1003	700	1575	1100				
	medium			1146	800	1719	1200				
	soft			1289	900	1862	1300				
133 à 138	very hard			859	600	1432	1000				
	hard			100?	700	1575	1100				
	medium			1146	800	1719	1200				
	soft			1289	900	1862	1300				
142 à 147	very hard					1432	1000	2292	1600		
	hard					1575	1100	2578	1800		
	medium					1719	1200	2864	2000		
	soft					1862	1300	3008	2100		
187 à 189	very hard							2292	1600	2864	2000
	hard							2578	1800	3150	2200
	medium							2864	2000	3437	2400
	soft							3008	2100	3724	2600

(1) All the 10³ daN-km or ton-mile values in the table were assumed using a safety factor of 5. Every body is aware that this safety factor often falls consistently below 5.

In those cases one should refer to the curve at the right. For example if you are operating at a safety factor of 4, the total work found in the table must be multiplied by 0,8.

Example :
Height of mast : 138 ft,
Wire rope diameter : 1 1/4'',
Formation hardness : hard,
Drum diameter : 28 in,
Safety factor : 3.
For a safety factor of 5 the table above gives a total work of 1575 10³ daN-km.
For safety factor of 3 the curve at the right gives a corrective factor of 0,58.
Total work will be 1575 × 0,58 = 914 10³ daN-km.
From table F 14 the total length of the cutoff is 25,7 m or 11 1/2 drum laps.
Note. For the following cutoffs the total work given in the above table must be reduced by 100 ton-mile (143 10³ daN-km) for 1 1/8'' and smaller wire rope diameter and by 200 ton-mile (286 10³ daN-km) for 1 1/4'' and larger wire rope diameter.

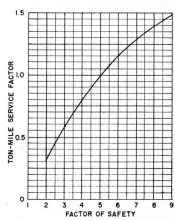

CAPACITIES OF DRUMS AND REELS

1) The length of wire rope, in metres, that can be spooled onto a drum or reel can be figured as below :

$$(A + D) \times A \times B \times K$$

where : $A = \dfrac{H - D}{2}$ (cm)

D = diameter of drum barrel (cm),
H = diameter of drum flanges (cm),
K = factor from table below for size of line selected.

2) When the drum or reel lacks Y (cm) of being filled, the length of wire rope is given by the same formula where :

$$A = \frac{H - D - 2Y}{2}$$

Y = depth not filled on drum in cm.

Nominal diameter of rope (in)	Factor K	Nominal diameter of rope (in)	Factor K	Nominal diameter of rope (in)	Factor K
3/8	0,02939	13/16	0,00658	1 5/8	0,00165
7/16	0,02214	7/8	0,00573	1 3/4	0,00143
1/2	0,01721	1	0,00445	1 7/8	0,00126
9/16	0,01378	1 1/8	0,00355	2	0,00111
5/8	0,01129	1 1/4	0,00283	2 1/8	0,00099
11/16	0,00941	1 3/8	0,00236	2 1/4	0,00089
3/4	0,00796	1 1/2	0,00199	2 3/8	0,00078

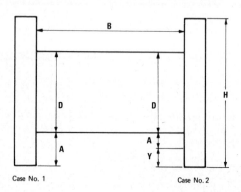

Case No. 1   Case No. 2

CORDAGES

CORDAGE TYPES

Fig. 1. - 3 strands

Fig. 2. - 3 ropes
3 strands each

Fig. 3. - 8 strands plaited 2 × 2
2 × 2 right hand lay
2 × 2 left hand lay

Z = right lay - regular lay,
S = left lay - regular lay.

VARIOUS MATERIALS

Manila : 3 grades :

Superior special
Grade 1
Grade 2

Sisal

Synthetic Fibres :

Nylon (polyamide),	specific weight : 1,14,	melting point : 210-220°C
Terylene (polyester),	specific weight : 1,38,	melting point : 250°C
Polypropylene,	specific weight : 0,91,	melting point : 160-165°C
Polythene-polyethylene,	specific weight : 0,95,	melting point : 130-135°C

Combination : Manila-steel

Note. Safety factor applied to cordages : 10.

CORDAGES
Manila and sisal (English standard BS.2052 and NF G 36-003)

MINIMUM BREAKING STRENGTH (10^3 daN) and MASS (kg/m)

Circumference (in)	Reference diameter (mm)	3 Strands (fig. 1)					9 Strands (fig. 2)					8 Strands plaited (fig. 3)				
		Manila Spec. Sup.	Manila Grade 1	Manila Grade 2	Sisal	Mass (kg/m)	Manila Spec. Sup.	Manila Grade 1	Manila Grade 2	Sisal	Mass (kg/m)	Manila Spec. Sup.	Manila Grade 1	Manila Grade 2	Sisal	Mass (kg/m)
7/8	7	0.38	0.36	0.32	0.32	0.036										
1	8	0.56	0.53	0.47	0.47	0.055										
1 1/4	10	0.74	0.70	0.62	0.62	0.068										
1 1/2	12	1.11	1.05	0.93	0.93	0.105										
1 3/4	14	1.49	1.41	1.26	1.26	0.14										
2	16	2.09	1.99	1.75	1.75	0.19										
2 1/4	18	2.49	2.39	2.09	2.09	0.22										
2 1/2	20	3.34	3.19	2.79	2.79	0.28						3.34	3.19	2.79	2.79	0.29
2 3/4	22	3.99	3.79	3.34	3.34	0.33										
3	24	4.74	4.48	3.98	3.98	0.40						4.74	4.48	3.98	3.98	0.40
3 1/4	26	5.49	5.23	4.63	4.63	0.47										
3 1/2	28	6.23	5.98	5.23	5.23	0.54						6.23	5.98	5.23	5.23	0.55
3 3/4	30	7.23	6.73	5.98	5.98	0.63										
4	32	7.97	7.48	6.73	6.73	0.70						7.98	7.48	6.73	6.73	0.73
4 1/4	34	9.00	8.48	7.48	7.48	0.80										
4 1/2	36	10.0	9.47	8.48	8.48	0.90						10.00	9.46	8.48	8.48	0.91
4 3/4	38	11.0	10.50	9.22	9.22	1.00										
5	40	12.2	11.50	10.20	10.20	1.11	7.73	7.23	6.47	6.47	0.98	12.2	11.5	10.2	10.2	1.14
5 1/2	44	14.7	13.9	12.46	12.46	1.35	9.22	8.62	7.72	7.72	1.16	14.7	13.9	12.5	12.5	1.36
6	48	17.2	16.5	14.4	14.4	1.60	10.7	10.20	8.83	8.83	1.40	17.2	16.5	14.4	14.4	1.64
6 1/2	52	20.2	19.2	17.0	17.0	1.88	12.2	11.5	10.20	10.20	1.65	20.2	19.2	17.0	17.0	1.91
7	56	23.2	22.0	19.4	19.4	2.18	13.9	13.2	11.7	11.7	1.89	23.2	22.0	19.4	19.4	2.23
7 1/2	60	26.4	24.9	22.2	22.2	2.50	15.7	14.9	13.2	13.2	2.20	26.4	24.9	22.2	22.2	2.55
8	64	29.9	28.4	25.2	25.2	2.90	17.7	16.7	14.9	14.9	2.44	29.9	28.4	25.2	25.2	2.91
8 1/2	68	33.4	31.7	28.2	28.2	3.21	19.9	18.7	16.7	16.7	2.81	33.4	31.7	28.2	28.2	3.27
9	72	37.4	35.4	31.4	31.4	3.60	22.0	20.7	18.4	18.4	3.11	37.4	35.4	31.4	31.4	3.68
9 1/2	76	41.4	39.3	34.9	34.9	4.00	23.9	22.7	20.2	20.2	3.48	41.4	39.3	34.9	34.9	4.09
10	80	45.6	43.4	38.4	38.4	4.44	26.7	25.2	22.5	22.5	3.85	45.6	43.3	38.4	38.4	4.55
12	96	65.0	62.1	54.8	54.8	6.39	37.1	35.4	31.4	31.4	5.55	65.0	61.8	54.9	54.9	6.55
14	112	87.5	83.0	73.8	73.8	8.71	50.1	47.4	42.1	42.1	7.57					
16	128	114.8	108.9	96.1	96.1	11.38	64.7	61.3	54.6	54.6	9.95					
18	144	144.2	136.4	121.6	121.6	14.40	81.7	77.2	68.8	68.8	12.51					

CORDAGES (continued)
Synthetic fibres (French standards NF G 36-012, G 36-013, G 36-014, G 36-015 (1)

Circumference (in)	Reference diameter (mm)	Nylon (2)		Terylene (2)		Polypropylene (2) monofilament		Polypropylene multifilament		Polythene	
		10^3daN	kg/m	10^3daN	kg/m	10^3daN	kg/m	10^3daN	kg/m	10^3daN	kg/m
1/2	4	0.31	0.011	0.29	0.015			0.25	0.009	0.20	0.009
5/8	5	0.49	0.016	0.39	0.020	0.54	0.017	0.35	0.013	0.28	0.012
3/4	6	0.74	0.024	0.60	0.029			0.49	0.019	0.39	0.018
7/8	7	1.00	0.032	0.76	0.039	0.94	0.030	0.67	0.025	0.54	0.025
1	8	1.32	0.042	1.01	0.051			0.89	0.033	0.69	0.032
1 1/8	9	1.67	0.053	1.25	0.065	1.40	0.045	1.07	0.041	0.88	0.041
1 1/4	10	2.04	0.065	1.56	0.081			1.33	0.052	1.07	0.050
1 3/8	11	2.45	0.078	1.87	0.097	1.99	0.065	1.61	0.062	1.29	0.060
1 1/2	12	2.94	0.094	2.23	0.116			1.89	0.075	1.51	0.073
1 5/8	13	3.43	0.110	2.67	0.134			2.23	0.087	1.76	0.086
1 3/4	14	4.02	0.128	3.12	0.157	2.74	0.090	2.56	0.101	2.05	0.097
2	16	5.20	0.166	3.98	0.205	3.24	0.115	3.24	0.132	2.74	0.127
2 1/4	18	6.57	0.210	4.98	0.260	4.22	0.148	4.22	0.168	3.38	0.159
2 1/2	20	8.14	0.260	6.23	0.320	5.25	0.180	5.25	0.205	4.19	0.200
2 3/4	22	9.81	0.315	7.47	0.384	5.98	0.220	5.98	0.250	4.98	0.241
3	24	11.77	0.375	8.97	0.460	7.46	0.260	7.46	0.295	5.98	0.286
3 1/2	28	15.5	0.440	12.0	0.630	10.0	0.355	10.0	0.405	7.88	0.391
4	32	19.6	0.665	15.4	0.820	12.5	0.460	12.5	0.527	10.2	0.514
4 1/2	36	24.3	0.84	18.9	1.04	15.9	0.585	15.9	0.668	12.8	0.645
5	40	29.4	1.04	23.4	1.28	18.9	0.720	18.9	0.827	15.3	0.795
6	48	41.2	1.50	32.9	1.85	27.0	1.04	27.0	1.191	22.0	1.145
7	56	54.9	2.03	43.9	2.51	36.8	1.42			29.6	1.568
8	64	70.6	2.65	56.8	3.28	46.6	1.85			37.9	2.045
9	72	88.3	3.36	70.7	4.15	58.9	2.34			47.7	2.582
10	80	107.9	4.15	86.7	5.12	71.6	2.90			58.4	3.10
11	88	128.5	5.02	104.0	6.14	84.8	3.51			70.5	3.70
12	96	151.1	5.98	122.6	7.36	100.0	4.17			86.6	4.40

MINIMUM BREAKING STRENGTH (10^3daN) AND MASS (kg/m)

(1) Standard NF G 36-023 gives the minimum breaking strengths for 8-strand plaited monofilament polypropylene ropes. These breaking strengths are quite the same as those given in the above table for 3 strand ropes.

(2) Also manufactured in 3 strands or in 8 strands plaited 2 × 2 from diameter 32 and above.

CORDAGES (continued)
Normal combination ropes (French standard NF 36-010) (1)

Circum-ference (in)	Approximate diameter (mm)	Number of strands	MINIMUM BREAKING STRENGTH (10³daN)					
			Hemp		Manila		Sisal	
			raw	tarred	raw	tarred	raw	tarred
1 1/4	10	3	1,23	1,10	1,04	0,95	0,99	0,89
1 1/2	12	3 et 4	1,74	1,57	1,47	1,34	1,39	1,26
1 3/4	14	3 et 4	2,35	2,13	1,96	1,81	1,86	1,72
2	16	3 et 4	2,94	2,70	2,45	2,31	2,35	2,16
2 1/4	18	3 et 4	3,83	3,43	3,24	2,94	3,04	2,75
2 1/2	20	3 et 4	4,71	4,27	3,92	3,73	3,73	3,43
2 3/4	22	3 et 4	5,89	5,40	4,91	4,51	4,51	4,12
	25	3 et 4	7,16	6,38	6,08	5,59	5,59	5,20
3 1/2	28	3	8,63	7,75	7,85	7,16	7,36	6,57
3 1/2	28	4	8,24	7,36	7,46	6,77	7,06	6,28
3 3/4	30	3 et 4	9,71	8,73	8,73	8,04	8,24	7,46
4	32	4	10,50	9,42	9,32	8,58	8,53	8,04
4 1/4	34	4	11,77	10,79	10,99	10,01	9,81	9,12
4 1/2	36	4	12,75	11,77	11,77	10,79	10,79	10,10
5	40	4	15,50	13,93	14,72	13,73	13,24	12,26
	45	4	19,13	17,17	18,15	17,17	16,68	15,70
	50	4	22,56	20,60	22,07	19,62	19,62	18,15
	55	4	25,51	23,05	24,53	23,54	22,07	20,60
7 1/2	60	4	30,41	26,98	28,45	26,49	26,00	24,53
	65	4	35,32	31,88	32,37	30,41	29,43	27,47
8 3/4	70	4	40,22	36,30	36,30	33,35	34,34	31,39

(1) The normal combination ropes have 3 strands, without core, or 4 strands twisted together on a fibre core.
Each strand has :
— a fibre core,
— a layer of galvanized wire rope around the core,
— one or two layers of large yarns of hemp, manila or sisal.

CORDAGES (continued)
Special combination ropes (Hall's Barton Ropery C° Ltd)

Circumference (in)	Approximate diameter (mm)	NOMINAL BREAKING LOAD (10³daN) AND MASS (kg/m)									
		4 × 6 construction		6 × 6 construction		6 × 12 construction		6 × 24 construction		6 × 3 × 19 construction Spring lay	
		10³daN	kg/m	10³daN	kg/m	10³daN	kg/m	10³daN	kg/m	10³daN	kg/m
1 1/4	10	1.7	0.156	1.5	0.139	1.8	0.149				
1 1/2	12	2.1	0.186	1.9	0.179	2.3	0.216				
1 3/4	14	2.4	0.223	2.7	0.261	2.7	0.261				
2	16	3.1	0.288	3.2	0.305	3.4	0.342				
2 1/4	18	3.9	0.360	3.9	0.377	4.1	0.402				
2 1/2	20	4.9	0.449	4.8	0.454	4.9	0.489				
2 3/4	22			5.5	0.546	5.6	0.553	3.8	0.377		
3	24			6.5	0.645	6.5	0.658	4.8	0.479	15.4	1.27
3 1/4	26							5.7	0.571	18.1	1.40
3 1/2	28			8.5	0.861	8.7	0.878	6.4	0.621	21.0	1.68
3 3/4	30							7.7	0.769	24.1	1.86
4	32					11.0	1.120	10.3	1.04	27.4	2.16
4 1/4	34									31.0	2.41
4 1/2	36							13.1	1.34	34.9	2.65
4 3/4	38									38.8	3.02
5	40									42.9	3.27
5 1/2	44									52.4	3.99

6 × 3 × 19

6 × 24

6 × 12

6 × 6

4 × 6

SHEAVE GROOVES

Groove radii for new and reconditioned sheaves and gages for worn sheave grooves (API RP 9B)

WIRE ROPE DIAMETER			GROOVE ROOT RADIUS R (1)				GAGES FOR WORN SHEAVE GROOVES					
Nominal diameter	Tolerance on diameter		minimum		maximum		Groove oversize (2)		Gage radius (3) R_G + 0.04 mm + 0.0015 in / -0.00 -0.0000		Marking
(in)	(mm)	(in)	(mm)	(in)	(mm)	(in)	(mm)	(in)	(mm)	(in)	
3/8	+0.8 -0	+1/32 -0	5.2	0.205	5.5	0.215	0.4	1/64	5.0	0.195	API-WORN-3/8-GO
7/16	+0.8 -0	+1/32 -0	6.0	0.235	6.2	0.245	0.4	1/64	5.8	0.227	API-WORN-7/16-GO
1/2	+0.8 -0	+1/32 -0	6.7	0.265	7.0	0.275	0.4	1/64	6.6	0.258	API-WORN-1/2-GO
9/16	+0.8 -0	+1/32 -0	7.6	0.300	7.9	0.310	0.4	1/64	7.3	0.289	API-WORN-9/16-GO
5/8	+0.8 -0	+1/32 -0	8.4	0.330	8.6	0.340	0.4	1/64	8.1	0.320	API-WORN-5/8-GO
3/4	+0.8 -0	+1/32 -0	9.9	0.390	10.2	0.400	0.4	1/64	9.7	0.383	API-WORN-3/4-GO
7/8	+0.8 -0	+1/32 -0	11.7	0.460	12.1	0.475	0.6	1/64	11.4	0.449	API-WORN-7/8-GO
1	+1.2 -0	+3/64 -0	13.3	0.525	13.7	0.540	0.6	3/128	13.0	0.512	API-WORN-1 -GO
1 1/8	+1.2 -0	+3/64 -0	14.9	0.585	15.2	0.600	0.6	3/128	14.6	0.574	API-WORN-1 1/8-GO
1 1/4	+1.2 -0	+1/16 -0	16.6	0.655	17.0	0.670	0.8	3/128	16.3	0.641	API-WORN-1 1/4-GO
1 3/8	+1.6 -0	+1/16 -0	18.3	0.720	18.7	0.735	0.8	1/32	17.9	0.703	API-WORN-1 3/8-GO
1 1/2	+1.6 -0	+1/16 -0	19.8	0.780	20.2	0.795	0.8	1/32	19.5	0.766	API-WORN-1 1/2-GO
1 5/8	+2.4 -0	+3/32 -0	21.8	0.860	22.2	0.875	1.2	3/64	21.2	0.836	API-WORN-1 5/8-GO
1 3/4	+2.4 -0	+3/32 -0	23.5	0.925	23.9	0.940	1.2	3/64	22.8	0.898	API-WORN-1 3/4-GO

(1) Minimum groove root radius is equal to wire rope nominal radius plus 1/2 the positive tolerance with the sum rounded off to the nearest 0.005 in.

(2) Groove oversize equals one half the wire rope oversize tolerance.

(3) Radius $R_G = \dfrac{\text{nominal diameter} + \text{groove oversize}}{2}$

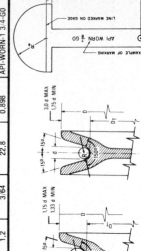

Groove gage

Note. The sheave grooves should have a radius of not less than that of the gage; otherwise a reduction in rope life should be expected.

Sand line sheave

Drilling line and casing line sheave

ELEVATOR LINK ARMS
Remaining capacities of worn link arms

Remaining capacities		Thickness of wear point B							
Link arms		1 3/4	1 5/8	1 1/2	1 3/8	1 1/4	1 1/8		
1 3/4	10³ daN	134	110	91	75	63	51		
	tons	150	123	102	84	71	57		
Link arms		2 1/4	2 1/8	2	1 7/8	1 3/4	1 5/8		
2 1/4	10³ daN	223	187	156	134	116	100		
	tons	250	210	175	150	130	113		
Link arms		2 3/4	2 5/8	2 1/2	2 3/8	2 1/4	2 1/8	2	
2 3/4	10³ daN	312	265	234	206	187	172	156	
	tons	350	298	263	231	210	193	175	
Link arms		3 1/2	3 3/8	3 1/4	3 1/8	3	2 7/8	2 3/4	2 5/8
3 1/2	10³ daN	445	409	378	343	312	285	258	240
	tons	500	460	425	385	350	320	290	270
Link arms		4 3/4	4 5/8	4 1/2	4 3/8	4 1/4	4 1/8	4	3 7/8
4 3/4	10³ daN	668	635	608	581	554	534	514	494
	tons	750	713	683	653	623	600	578	555

Note. The nominal size of elevator link arms is the thickness at point B.

Capacities of new link arms

Nominal size	Capacity	
(in)	tons	10³ daN
1 3/4	150	134
2 1/4	250	223
2 3/4	350	312
3 1/2	500	445
4 3/4	750	668

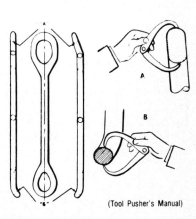

(Tool Pusher's Manual)

DRILL PIPE ELEVATORS
(API Spec 8A)

ELEVATOR SIZE	WELD-ON TOOL JOINTS							
	Taper shoulder				Square shoulder			
	Neck diameter of drill pipe		Elevator bore		Neck diameter of drill pipe		Elevator bore (1)	
	cm	in	cm	in	cm	in	cm	in
2 3/8 IF	65,09	2 9/16	67,47	2 21/32	*	*	*	*
2 7/8 REG	76,20	3	78,58	3 3/32	76,20	3	78,58	3 3/16
2 7/8 IF	80,96	3 3/16	83,34	3 9/32	80,96	3 3/16	85,72	3 3/8
2 7/8 Hydril		no upset	78,58	3 3/32	*	*	*	*
3 1/2 REG et FH	93,66	3 11/16	96,04	3 25/32	92,07	3 5/8	96,84	3 13/16
3 1/2 IF	98,42	3 7/8	100,80	3 31/32	98,42	3 7/8	103,19	4 1/16
3 1/2 Hydril		no upset	96,04	3 25/32	*	*	*	*
4 FH	106,36	4 3/16	108,74	4 9/32	104,77	4 1/8	109,54	4 5/16
4 IF	114,30	4 1/2	121,44	4 25/32	114,30	4 1/2	122,24	4 13/16
4 1/2 REG et FH	119,06	4 11/16	121,44	4 25/32	117,47	4 5/8	122,24	4 13/16
4 1/2 IF	127,00	5	133,35	5 1/4	127,00	5	134,94	5 5/16
5 IEU (2)	130,17	5 1/8	133,35	5 1/4	130,17	5 1/8	134,94	5 5/16
5 1/2 REG et FH	144,46	5 11/16	147,64	5 13/16	144,46	5 11/16	149,22	5 7/8

Note. Elevators with the same bores are the same elevators.
(1) Indicates 1/16 in × 45° chamfer at top of bore, except 3 1/2 IF.
(2) Indicates 5 IEU is a 4 1/2 IF tool joint on 5'' IEU drill pipe.
* Not manufactured.

BRAKE BLOCKS
(API Spec 7)

6 Hole API brake block

API Block No	A (in)	B (in)	C (in)
1	6	1 1/4	3 1/2
2	7	1 1/2	4
3	8	1 3/4	4 1/2
4	9	2	5
5	10	2 1/4	5 1/2
6	11	2 1/2	6
7	12	2 3/4	6 1/2

Brake block thickness

Brake block thickness is not stipulated ; for any given block size, however, several standard thicknesses are provided.
T (in) : 5/8, 3/4, 7/8, 1, 1 1/8, 1 1/4.

4 hole API brake block

API Block No	D (in)	E (in)	F (in)
10	6	1 1/4	3 1/2
11	7	1 1/2	4
12	8	1 1/2	5
13	9	1 1/2	6
14	10	1 1/2	7

Brake block thickness

Brake block thickness is not stipulated ; for any given size, however, several standard thicknesses are provided.
T (in) : 5/8, 3/4, 7/8, 1.

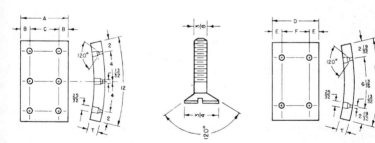

Screws for fastening brake blocks to the brake bands shall be 3/8., 120°, flathead brass machine screw as shown in figure. Screw threads shall be 3/8 - 16 UNC - 2A.

HANDLING APPLIANCES
Shackles (Crosby Group USA)

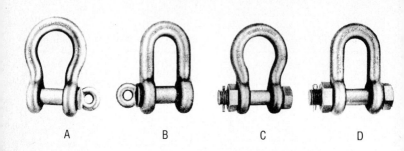

A   B   C   D

Size		Working load		Inside width at pin		Diameter of pin	
in	mm	tons	10³ daN	in	mm	in	mm
1/4*	6,4	1/2	0,44	15/32	11,9	5/16	7,9
5/16*	7,9	3/4	0,67	17/32	13,5	3/8	9,5
3/8*	9,5	1	0,89	21/32	16,7	7/16	11,1
7/16*	11,1	1 1/2	1,34	23/32	18,3	1/2	12,7
1/2	12,7	2	1,78	13/16	20,6	5/8	15,9
5/8	15,9	3 1/4	2,89	1 1/16	27,0	3/4	19,1
3/4	19,1	4 3/4	4,23	1 1/4	31,7	7/8	22,2
7/8	22,2	6 1/2	5,79	1 7/16	36,5	1	25,4
1	25,4	8 1/2	7,57	1 11/16	42,9	1 1/8	28,6
1 1/8	28,6	9 1/2	8,46	1 13/16	46,0	1 1/4	31,7
1 1/4	31,7	12	10,68	2 1/32	51,6	1 3/8	34,9
1 3/8	34,9	13 1/2	12,02	2 1/4	57,1	1 1/2	38,1
1 1/2	38,1	17	15,13	2 3/8	60,3	1 5/8	41,3
1 3/4	44,4	25	22,25	2 7/8	73,0	2	50,8
2	50,8	35	31,15	3 1/4	82,5	2 1/4	57,1
2 1/2	63,5	50	44,50	4 1/8	104,8	2 3/4	69,8
3	76,2	75	66,75	5	127,0	3 1/4	82,5
3 1/2+	88,9	100	89,00	5 3/4	146,0	3 3/4	95,2
4 +	101,6	130	115,70	6 1/2	165,1	4 1/4	107,9

* Model A or B only,
+ Model C or D only.

HANDLING APPLIANCES (continued)
Thimbles and rings (SOFCAB)

Ratios for inside widths :
For wire rope diameter up to 30 mm :
  A = wire rope diameter × 4,5,
  B = wire rope diameter × 3,
  e = steel thickness at the bottom of the groove : wire rope diameter × 0,3.
For wire rope diameter greater than 30 mm :
  A = wire rope diameter × 6,
  B = wire rope diameter × 4,
  e = wire rope diameter × 0,4.

COSSE CŒUR

Dimensions for wire ropes up to 30 mm

Wire rope diameter (mm)	A × B (mm)	e (mm)	Wire rope diameter (mm)	A × B (mm)	e (mm)	Wire rope diameter (mm)	A × B (mm)	e (mm)
5	23 × 15	1,5	12	54 × 36	3,6	24	108 × 72	7,2
6	27 × 18	1,8	14	63 × 42	4,2	26	117 × 78	7,8
7	32 × 21	2,1	16	72 × 48	4,8	28	126 × 84	8,4
8	36 × 24	2,4	18	81 × 54	5,4	30	135 × 90	9,0
9	40 × 27	2,7	20	90 × 60	6,0			
10	45 × 30	3,0	22	99 × 66	6,6			

Rings

Work load given in the table below is calculated for a 35 hbars tensile strength steel and a working rate of r = 10 hbars by the formula :

$$P = \frac{0,8 \ d^3 \ r}{D + 0,3 \ d}$$

Ring thickness d (mm)

Working load (kg)	INSIDE DIAMETER OF RING D (mm)										
	50	75	100	125	150	175	200	225	250	275	300
250	12*	14									
500	15	17*	19	20	21						
1000	19	22	24*	26	27	29	30				
1500		25	27	29*	31	33	34	36	37	38	
2500			32	35	37*	39	40	42	43	45	46
3500				39	41	43*	45	47	49	50	52
4500				43	45	47	49*	51	53	55	56
6000					50	52	55	57*	59	60	62
8000						58	60	63	65*	67	69
9000						60	63	65	67	70*	71
10000						63	65	68	70	72	74*

* Most commonly used rings.

HANDLING APPLIANCES (continued)
Hooks and sockets (SOFCAB)

HOOKS

Working load (kg)	Diameter C (mm)	Length L (mm)	Section area S1 (mm²)	Section area S2 (mm²)	Opening D (mm)	Eye diameter		Length from pin to pin E (mm)	Opening O (mm)	Approx. weight (kg)
						inside B (mm)	outside A (mm)			
250	12	110	18 × 23	15 × 12	24	18	40	63	19	0.200
500	19	146	28 × 17	24 × 16	32	22	52	80	29	0.500
1 000	28	176	35 × 23	32 × 22	45	26	60	91	32	1.100
1 500	33	202	42 × 28	37 × 27	50	30	70	105	35	1.700
2 000	35	223	48 × 32	40 × 30	56	33	78	116	42	2.400
3 000	40	255	62 × 42	46 × 32	62	36	88	134	44	3.800
4 000	43	298	67 × 46	50 × 34	67	40	96	166	52	5.000
5 000	52	327	80 × 54	60 × 40	80	44	104	175	60	7.500
6 000	55	347	88 × 60	65 × 45	85	46	112	183	64	9.000
8 000	58	387	96 × 66	72 × 48	95	48	118	208	72	13.000
10 000	66	453	110 × 75	85 × 56	110	56	134	246	83	18.500

Note. Never use hooks with circular section

Open conical sockets (1)

SOCKETS

Type N°	A (mm)	B (mm)	C (mm)	d (mm)	D (mm)	E (mm)	e (mm)	F (mm)	G (mm)	H (mm)	I (mm)	J (mm)	K (mm)	Breaking strength of wire rope (kg)
1	43	43	22	11	22	6	4,5	44	38	22	9	20	28	8 200
2	59	59	30	15	30	8.5	6,5	60	53	30	13	27	39	16 700
3	67	67	33	17	33	10	7	67	58	34	14	30	43	20 300
4	74	74	37	19	36	10.5	7,5	74	63	38	15	33	48	24 200
5	86	86	43	22	42	12	9	85	75	43	18	39	55	32 800
6	102	102	50	26	50	15	11	104	89	50	22	45	66	48 300
7	114	114	57	29	58	16	12	117	102	59	24	54	74	60 300
8	129	129	65	33	64	18	13	129	112	65	26	60	82	73 700
9	149	149	75	38	74	21	15	152	129	78	30	69	96	98 000
10	178	178	90	45	90	26	19	180	160	90	38	84	113	146 000

(1) The base of the socket should have a slightly reverse conical shape to avoid undue wear when bending.

TENSIONS IN SLINGS

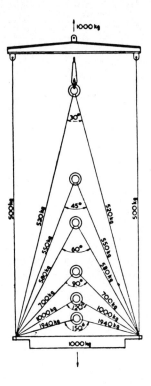

The picture shown above gives the increased tensions caused in the rope during hoisting at a certain angle.

When the angle is greater than 120° the working strengths of the slings should be greater than the load to hoist.

For example : use 1940 kg working load slings to hoist 1000 kg when the angle is 150°.

G POWER. EFFICIENCY FLOW RATE OF MUD PUMPS

Summary

HORSE POWER CORRECTIONS
OF INTERNAL COMBUSTION ENGINES (unsupercharged)

I. WITH ALTITUDE

Altitude (m)	% of power	Altitude (m)	% of power	Altitude (m)	% of power	Altitude (m)	% of power
Sea level	100	1000	89	2000	79	3000	69
100	98	1100	88	2100	78	3100	68
200	97	1200	87	2200	77	3200	67
300	96	1300	86	2300	76	3300	66
400	95	1400	85	2400	75	3400	65
500	94	1500	84	2500	74	3500	64
600	93	1600	83	2600	73	3600	63
700	92	1700	82	2700	72	3700	62
800	91	1800	81	2800	71	3800	61
900	90	1900	80	2900	70	3900	60

II. WITH OUTSIDE TEMPERATURE

Outside temperature (° C)	% of power	
	Gas	Diesel
1b	100	100
20	99	100
25	98	100
30	97	100
35	96,5	99
40	96	97,5
45	95	95,0
50	94	92,5
55	93	90
60	92	87,5

POWER REQUIREMENTS FOR HOISTING

Power spent on the reeving system P_m

where :

$$P_m = \frac{1000\ Pv}{75}$$

P_m = power (ch),
P = load (tf),
v = speed of the hook (m/s),

$$P_m = 10\ Pv$$

P_m = power (kW),
P = load (10^3 daN),
v = speed (m/s).

Power spent at the drawworks P_t

$$P_t = \frac{1000\ FV}{75} = \frac{1000\ nFv}{75}$$

where :

P_t = power (ch),
F = pull load on the fast line (tf),
n = number of lines,

$$P_t = 10\ FV = 10\ nFv$$

P_t = power (kW),
F = load (10^3 daN),
$V = nv$: speed (m/s).

Relation between F and P

$$P = \frac{K\ (1 - K^n)}{1 - K}\ F$$

K = sheave efficiency (0,97 to 0,9625)

	$\frac{K\ (1 - K^n)}{1 - K}$ values			value
n	K = 0.97	K = 0.98	K = 0.9625	API
6	5,40	5,59	5,26	5,244
8	6,99	7,31	6,76	6,728
10	8,49	8,96	8,15	8,100
12	9,90	10,55	9,44	9,240

Relation between P_m and P_t

$$P_m = \frac{K\ (1 - K^n)}{n\ (1 - K)}\ P_t$$

	$\frac{K\ (1 - K^n)}{n\ (1 - K)}$ values			value
n	K = 0.97	K = 0.98	K = 0.9625	API
6	0,900	0,932	0,877	0,874
8	0,874	0,914	0,845	0,841
10	0,849	0,896	0,815	0,810
12	0,825	0,879	0,787	0,770

RELATION BETWEEN HOISTING POWER P_m AND P_t AND INSTALLED POWER

$$\boxed{\text{Power at the drawworks } P_t = \rho \, P_{il}}$$

ρ : mechanical efficiency of the drawworks and compound,
P_{il} : installed power.

$$\boxed{\text{Power at the reeving system } P_m = \rho \frac{K\,(1 - K^n)}{n\,(1 - K)}\, P_{il}}$$

ρ values

About 0,85 for new equipment without torque converter.
About 0,65 for new equipment with torque converter.

Transmission equipment efficiency

Torque converter	0,70 à 0,90
Hydraulic coupling	0,97
Ball bearing crankshaft	0,98
V - belts	0,97
Roller chains in oil bath	0,98
Tooth transmission gear in oil bath	0,98

Powers of 0,97

$0,97^2 = 0,941$	$0,97^5 = 0,859$	$0,97^8 = 0,784$	$0,97^{11} = 0,715$
$0,97^3 = 0,913$	$0,97^6 = 0,833$	$0,97^9 = 0,760$	$0,97^{12} = 0,694$
$0,97^4 = 0,885$	$0,97^7 = 0,808$	$0,97^{10} = 0,737$	

Powers of 0,98

$0,98^2 = 0,960$	$0,98^5 = 0,904$	$0,98^8 = 0,851$	$0,98^{11} = 0,801$
$0,98^3 = 0,941$	$0,98^6 = 0,886$	$0,98^9 = 0,834$	$0,98^{12} = 0,785$
$0,98^4 = 0,922$	$0,98^7 = 0,868$	$0,98^{10} = 0,817$	

APPROXIMATE POWER SPENT FOR ROTATION

from empirical formula $P_r = (10 + \frac{L}{30}) (\frac{N}{100}) (\frac{P}{D})$

where : P_r = power for rotation (ch),
L = depth (m),
N = rotary speed (rpm),
P = weight on bit (10^3daN),
D = bit diameter (in).

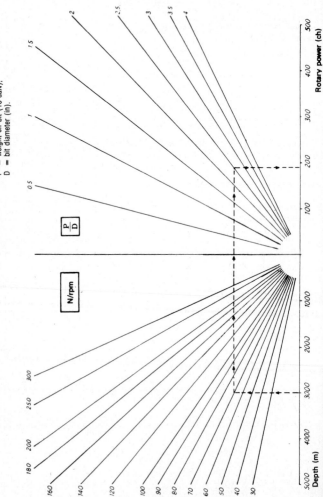

SLUSH PUMPS

I. THEORETICAL OUTPUT Q_t

a. Duplex pumps

$$Q_t = 0,0515 \ nL \ (D^2 - d^2/2)$$

b. Single acting Triplex pumps

$$Q_t = 0,0386 \ nLD^2$$

where :

Q_t = theoretical output (l/min),
n = strokes per minute (str/min),
L = length of stroke (in),
D = diameter of liner (in),
d = diameter of piston rod (in).

II. VOLUMETRIC EFFICIENCY η_v

$$\eta_v = \frac{Q_r}{Q_t}$$

Q_r = true output (l/min).

III. HYDRAULIC POWER P_h

$$P_h \ (ch) = \frac{pQ_r}{440} \qquad \text{or} \qquad P_h \ (kW) = \frac{pQ_r}{600}$$

p : discharge pressure (bar).

IV. ENGINE POWER REQUIRED FOR A GIVEN PUMP OUTPUT Q_r AND A PUMP PRESSURE p

$$P_m \ (ch) = \frac{pQ_r}{440 \eta_m \eta_t} \qquad \text{or} \qquad P_m \ (kW) = \frac{pQ_r}{600 \ \eta_m \eta_t}$$

where :

η_m = mechanical efficiency of the pump (about 0,85),
η_t = compound efficiency (0,65 to 0,90 depending on whether there is a torque converter or not).

V. MAXIMUM SERVICE PRESSURE p_{max} (bar)

$$p_{max} = \frac{F_{max}}{S}$$

F_{max} = maximum load on cross head extension (daN),
S = average area under pressure for a given liner (cm²),
   S equals :
   (a) For Duplex pump : $5,06 \ (D^2 - d^2/2)$,
   (b) For Triplex pump : $5,06 \ D^2$,
where :
D = diameter of liner (in),
d = diameter of piston rod (in).

VI. MAXIMUM HYDRAULIC POWER P_{max}

$$P_{max} \ (ch) = \frac{p_{max}Q_t}{440} \qquad \text{or} \qquad P_{max} \ (kW) = \frac{p_{max}Q_t}{600}$$

OUTPUT IN LITRES PER STROKE
DOUBLE ACTING DUPLEX PUMPS
based on liner size and piston rod diameter

Length of stroke 10″

Ø rod (in) \ Ø piston (in)	3	3 1/4	3 1/2	3 3/4	4	4 1/4	4 1/2	4 3/4	5	5 1/4	5 1/2	5 3/4	6	6 1/4	6 1/2	6 3/4	7	7 1/4	7 1/2	7 3/4	8
1 1/2	4.05	4.86	5.73	6.66	7.66	8.72	9.85	11.04	12.29	13.61	15.02	16.44	17.95	19.53	21.17	22.88	24.65	26.49			
1 5/8	3.94	4.75	5.62	6.55	7.55	8.61	9.74	10.93	12.18	13.50	14.91	16.33	17.84	19.42	21.06	22.77	24.54	26.38			
1 3/4	3.84	4.65	5.52	6.45	7.45	8.51	9.64	10.83	12.08	13.40	14.81	16.23	17.74	19.32	20.96	22.67	24.44	26.27			

Length of stroke 12″

Ø rod (in) \ Ø piston (in)	3	3 1/4	3 1/2	3 3/4	4	4 1/4	4 1/2	4 3/4	5	5 1/4	5 1/2	5 3/4	6	6 1/4	6 1/2	6 3/4	7	7 1/4	7 1/2	7 3/4	8
1 5/8					9.07	10.34	11.69	13.12	14.63	16.21	17.87	19.61	21.42								
1 7/8					8.80	10.07	11.42	12.85	14.36	15.94	17.60	19.34	21.15	23.05	25.02	27.06	29.19	31.39	33.51	35.87	
2					8.65	9.92	11.27	12.70	14.21	15.79	17.45	19.19	21.00	22.90	24.87	26.91	29.04	31.24			
2 1/4					8.32	9.59	10.94	12.37	13.88	15.46	17.12	18.86	20.67	22.57	24.54	26.58	28.71	30.89			

Length of stroke 14″

Ø rod (in) \ Ø piston (in)	3	3 1/4	3 1/2	3 3/4	4	4 1/4	4 1/2	4 3/4	5	5 1/4	5 1/2	5 3/4	6	6 1/4	6 1/2	6 3/4	7	7 1/4	7 1/2	7 3/4	8
1 3/4							13.49	15.16	16.92	18.76	20.70	22.73	24.84	27.05	29.35	31.74	34.21				
2							13.15	14.82	16.58	18.42	20.36	22.39	24.50	26.71	29.01	31.40	33.87	36.44	39.10	41.85	44.68
2 1/8							12.96	14.63	16.39	18.23	20.17	22.20	24.31	26.52	28.82	31.21	33.68	36.25	38.91	41.66	44.49
2 1/4							12.77	14.44	16.20	18.04	19.98	22.01	24.12	26.33	28.63	31.02	33.49	36.06	38.72	41.47	44.30
2 1/2							12.34	14.01	15.77	17.61	19.55	21.58	23.69	25.90	28.20	30.59	33.06	35.63	38.29	41.04	43.87
2 5/8							12.11	13.78	15.54	17.38	19.32	21.35	23.46	25.67	27.97	30.36	32.83	35.40	38.06	40.81	43.64

Length of stroke 15″

Ø rod (in) \ Ø piston (in)	3	3 1/4	3 1/2	3 3/4	4	4 1/4	4 1/2	4 3/4	5	5 1/4	5 1/2	5 3/4	6	6 1/4	6 1/2	6 3/4	7	7 1/4	7 1/2	7 3/4	8
2 1/4									17.35	19.33	21.40	23.58	25.84	28.21	30.67	33.23	35.88	38.63	41.48	44.43	
2 7/8									16.11	18.09	20.16	22.34	24.60	26.97	29.43	31.99	34.64	37.39	40.24	43.19	

OUTPUT IN LITRES PER STROKE (continued)
DOUBLE ACTING DUPLEX PUMPS
based on liner size and piston rod diameter

Length of stroke 16″

Ø piston (in) / Ø rod (in)	4	4 1/4	4 1/2	4 3/4	5	5 1/4	5 1/2	5 3/4	6	6 1/4	6 1/2	6 3/4	7	7 1/4	7 1/2	7 3/4	8	8 1/4	8 1/2	8 3/4	9
2 1/4				16.50	18.51	20.62	22.83	25.15	27.57	30.09	32.71	35.44	38.27	41.21	44.25	47.39	50.63	53.98	57.43		
2 1/2				16.01	18.02	20.13	22.34	24.66	27.08	29.60	32.22	34.95	37.78	40.72	43.76	46.90	50.14	53.49	56.94		
2 3/4				15.48	17.48	19.59	21.80	24.12	26.54	29.06	31.68	34.41	37.24	40.18	43.22	46.36	49.60	52.95	56.40		
3				14.88	16.89	19.00	21.21	23.53	25.95	28.47	31.09	33.82	36.65	39.59	42.63	45.77	49.01	52.36	55.81		
3 1/8				14.56	16.57	18.68	20.89	23.21	25.63	28.15	30.77	33.50	36.33	39.27	42.31	45.45	48.69	52.04	55.49		
3 1/2				13.54	15.55	17.66	19.87	22.19	24.61	27.13	29.75	32.48	35.31	38.25	41.29	44.43	47.67	51.02	54.47		

Length of stroke 18′

Ø piston (in) / Ø rod (in)	4	4 1/4	4 1/2	4 3/4	5	5 1/4	5 1/2	5 3/4	6	6 1/4	6 1/2	6 3/4	7	7 1/4	7 1/2	7 3/4	8	8 1/4	8 1/2	8 3/4	9
2 1/4		14.39	16.41	18.56	20.82	23.19	25.68	28.29	31.01	33.85	36.80	39.87	43.05	46.35	49.77	53.30	56.95	60.72			
2 3/8		14.12	16.14	18.29	20.55	22.92	25.41	28.02	30.74	33.58	36.53	39.60	42.78	46.08	49.50	53.03	56.68	60.45			
2 1/2		13.84	15.86	18.01	20.27	22.64	25.13	27.74	30.46	33.30	36.25	39.32	42.50	45.80	49.22	52.75	56.40	60.17			
2 3/4		13.24	15.26	17.41	19.67	22.04	24.53	27.14	29.86	32.70	35.65	38.72	41.90	45.20	48.62	52.15	55.80	59.57			
3		12.57	14.59	16.74	19.00	21.37	23.86	26.47	29.19	32.03	34.98	38.05	41.23	44.53	47.95	51.48	55.13	58.90			
3 1/8		12.22	14.24	16.39	18.65	21.02	23.51	26.12	28.84	31.68	34.63	37.70	40.88	44.18	47.60	51.13	54.78	58.55			
3 1/4		11.85	13.87	16.02	18.28	20.65	23.14	25.75	28.47	31.31	34.26	37.33	40.51	43.81	47.23	50.76	54.41	58.18			

Length of stroke 20″

Ø piston (in) / Ø rod (in)	4	4 1/4	4 1/2	4 3/4	5	5 1/4	5 1/2	5 3/4	6	6 1/4	6 1/2	6 3/4	7	7 1/4	7 1/2	7 3/4	8	8 1/4	8 1/2	8 3/4	9
2 1/2			17.63	20.01	22.52	25.16	27.93	30.82	33.85	37.00	40.28	43.69	47.23	50.90	54.70	58.62	62.68				
2 7/8			16.59	18.97	21.48	24.12	26.89	29.78	32.81	35.96	39.24	42.65	46.19	49.86	53.66	57.58	61.64				

OUTPUT IN LITRES PER STROKE
SINGLE ACTING TRIPLEX PUMPS
(volumetric efficiency 100 %)

Length of stroke (in)	LINER SIZE (in)															
	7 1/2	7 1/4	7	6 3/4	6 1/2	6 1/4	6	5 3/4	5 1/2	5 1/4	5	4 3/4	4 1/2	4 1/4	4	3 1/2
6 3/4	14.65	13.68	12.77	11.87	11.01	10.17	9.37	8.61	7.88	7.17	6.52	5.87	5.27	4.70	4.17	3.19
7	15.20	14.20	13.24	12.31	11.42	10.55	9.72	8.92	8.18	7.44	6.75	6.09	5.47	4.87	4.32	3.31
7 1/4	15.74	14.70	13.71	12.74	11.82	10.92	10.07	9.24	8.47	7.71	6.99	6.31	5.67	5.05	4.47	3.42
7 1/2	16.28	15.21	14.18	13.18	12.23	11.30	10.42	9.57	8.76	7.97	7.24	6.52	5.87	5.22	4.63	3.55
7 3/4	16.82	15.72	14.66	13.62	12.63	11.67	10.77	9.88	9.05	8.24	7.47	6.74	6.07	5.40	4.78	3.67
8	17.37	16.22	15.13	14.07	13.05	12.06	11.12	10.20	9.34	8.51	7.72	6.96	6.26	5.57	4.94	3.78
8 1/4	17.91	16.72	15.60	14.50	13.45	12.43	11.47	10.52	9.63	8.77	7.96	7.17	6.45	5.75	5.09	3.90
8 1/2	18.46	17.24	16.07	14.94	13.86	12.81	11.82	10.84	9.93	9.04	8.20	7.40	6.65	5.92	5.25	4.02
8 3/4	19.00	17.74	16.55	15.37	14.27	13.18	12.16	11.16	10.22	9.30	8.44	7.62	6.84	6.09	5.40	4.13
9	19.54	18.25	17.02	15.82	14.68	13.57	12.51	11.48	10.51	9.57	8.68	7.83	7.04	6.27	5.56	4.26
9 1/4	20.08	18.76	17.49	16.26	15.09	13.93	12.85	11.80	10.80	9.83	8.92	8.05	7.23	6.44	5.71	4.37
9 1/2	20.63	19.27	17.97	16.70	15.49	14.32	13.20	12.12	11.10	10.10	9.17	8.27	7.42	6.62	5.87	4.49
10	21.72	20.28	18.92	17.58	16.31	15.07	13.90	12.76	11.68	10.63	9.65	8.70	7.82	6.97	6.17	4.72
10 1/2	22.80	21.29	19.86	18.46	17.12	15.82	14.59	13.39	12.26	11.17	10.13	9.14	8.21	7.32	6.48	4.97
11	23.88	22.31	20.81	19.33	17.94	16.57	15.29	14.03	12.85	11.70	10.62	9.57	8.60	7.67	6.79	5.20
11 1/2	24.97	23.32	21.75	20.22	18.75	17.33	15.98	14.67	13.43	12.22	11.10	10.01	8.99	8.01	7.10	5.44
12	26.06	24.33	22.70	21.09	19.57	18.08	16.67	15.31	14.02	12.76	11.58	10.44	9.38	8.36	7.41	5.67

ENDLESS V-BELTS
(API - Std IB)

Standard V-belt dimensions are international. Their designation consists of a letter representing the nominal cross-section with a number affixed representing the pitch length.

Cross sectional dimensions are the top width and the thickness expressed in inches or millimetres. Table I gives the standardised cross-sections.

TABLE I

STANDARDS	CROSS-SECTIONS				
	A	B	C	D	E
French (mm)	13 × 8	17 × 11	22 × 14	32 × 19	38 × 25
API (in)	1/2 × 5/16	21/32 × 13/32	7/8 × 17/32	1/4 × 3/4	1 1/2 × 29/32
(mm)	13 × 8	17 × 10	22 × 14	32 × 19	38 × 23

Note. Due to tolerance on the thickness API and French standards are identical.

V-belts in a set of matched belts shall be marked with the same letter from G to U and shall not have pitch length variation in excess of the limits given below :

Nominal pitch length		Matched set pitch length tolerance									
		V-belt cross-section symbol									
		A		B		C		D		E	
(in)	(mm)	in	mm	in	mm	in	mm	in	mm	in	mm
26 à 75	660 à 1905	0,10	2,5	0,10	2,5	0,10	2,5	—	—	—	—
80 à 144	2032 à 3650	0,20	5,1	0,20	5,1	0,20	5,1	0,20	5,1	—	—
158 à 270	4016 à 6858	0,30	7,6	0,30	7,6	0,30	7,6	0,30	7,6	0,30	7,6
300 à 390	7620 à 9906	0,40	10,2	0,40	10,2	0,40	10,2	0,40	10,2	0,40	10,2
420 à 660	10668 à 16700	0,50	12,7	0,50	12,7	0,50	12,7	0,50	12,7	0,50	12,7

The pitch length shall be determined under tension :

		V-belt cross section symbol				
		A	B	C	D	E
Tension	lb	50	65	165	300	400
	daN	22,4	29,1	74	134,5	179,3

Pitch length can be determined knowing the outside or the inside length as shown in Table II.

TABLE II

To obtain pitch length	CROSS SECTION SYMBOL									
	A		B				D		E	
	mm	in	mm	in	mm	in	mm	in	mm	in
substract from the outside length	17	0,67	23	0,90	29	1,15	40	1,6	53	2,1
add to the inside length	33	1,30	47	1,85	59	2,32	80	3,2	108	4,25

ENDLESS V-BELTS (continued)

Calculation of a V-belt length

$$L = \frac{D + d}{2} \times \Pi + 2A + \frac{(D - d)^2}{4A} \qquad \text{(approximate formula)}$$

where : L = outside length (mm or in),
 D = outside diameter of the big sheave (mm or in),
 d = outside diameter of the small sheave (mm or in),
 A = distance between axes (mm or in).

Recommended practice for measuring tension in V-belt drives

(1) Measure length of span (distance between axes).

(2) At center of span apply a force P with a spring scale in a direction perpendicular to span until belt is deflected (with reference to an adjacent belt) an amount equal to 1/64 in (or 0,156 mm) for every inch (or cm) of span length. For example the deflexion of 100 in (254 cm) span would be 100/64 (1 9/16'') or 39,7 mm.

(3) Note force P and compare it with the values given in Table III. If force P is between the values given for normal tension and 1 1/2 times normal tension, the tension should be satisfactory. A new drive may be tightened initially to two times the normal tension since the tension drops rapidly during the run-in period.

TABLE III

Cross-section symbol	Required force P at centre of span					
	normal tension		1 1/2 normal tension		2 normal tension	
	daN	lb	daN	lb	daN	lb
A	0,67	1,5	1,0	2,25	1,3	3
B	1,60	3,5	2,4	5,25	3,2	7
C	3,1	7	4,7	10,50	6,2	14
D	5,7	12,75	8,6	19,25	11,4	25,50
E	10	22	15	·33	20	44

Ordering a set of matched V-belts

When ordering a set of 6 matched V-belts, cross-section symbol C, the pitch length must be calculated. Assuming that the measured outside length is 8970 mm, Table II gives the pitch length : 8970 — 29 = 8941 mm. The manufacturer must deliver a set of 6 V-belts each belt marked with the symbol :

$$C - 8941 - K \text{ (K = letter for matched V-belt)}$$

Sheave speed

$$N_1 D_1 = N_2 D_2$$

where :

N_1 = rpm of the sheave, the diameter of which is D_1,
N_2 = rpm of the sheave, the diameter of which is is D_2.

Peripheral speed of a belt in m/s

$$V = \frac{N_1 D_1}{1900} \text{ or } \frac{N_2 D_2}{1900} \qquad \begin{array}{l} \text{D in mm} \\ N_2 \text{ rpm} \end{array}$$

It is recommended that the figure of V = 25 m/s should not be exceeded.

CHAINS
(API - Std 7F)

I. SINGLE AND MULTIPLE CHAIN ASSEMBLIES

The standardised designation of a chain is made up of two numbers separated by a hyphen or by the letter H for heavy chains.

For example : chain 160-6 or 160-H-6

The right digit of the first number can be :

0 for standard chain,

1 for light weight chain,

5 for chain without roller.

The one or two digits to the left of the first number is the pitch of the chain expressed in 1/8 of an inch (here 16/8 or 2'').

The number at the right is the number of strands (here 6).

In the H (heavy) series only the thickness of the flanges are different.

CHAIN DIMENSIONS

Chain N°	Pitch		Roller diameter Dr		Inner link width W		Pin diameter Dp		Tension for measuring length		Flange thickness			
											Standard		Heavy	
	in	mm	in	mm	in	mm	in	mm	lb	kg	in	mm	in	mm
25	1/4	6,4	0,130	3,3	1/8	3,2	0,0905	2,3	18	7,8	0,030	0,8	—	—
35	3/8	9,5	0,200	5,1	3/16	4,8	0,141	3,6	18	7,8	0,050	1,3	—	—
41	1/2	12,7	0,306	7,8	1/4	6,4	0,141	3,6	18	7,8	0,050	1,3	—	—
40	1/2	12,7	0,312	7,9	5/16	7,9	0156	4,0	31	14,1	0,060	1,5	—	—
50	5/8	15,9	0,400	10,2	3/8	9,5	0,200	5,1	49	22,2	0,080	2	—	—
60	3/4	19,1	0,469	11,9	1/2	12,7	0,234	5,9	70	31,8	0,094	2,4	0,125	3,2
80	1	25,4	0,625	15,9	5/8	15,9	0,312	7,9	125	56,7	0,125	3,2	0,156	4,0
100	1 1/4	31,8	0,750	19,1	3/4	19,1	0,375	9,5	195	88,5	0,156	4,0	0,187	4,7
120	1 1/2	38,1	0,875	22,2	1	25,4	0,437	11,1	281	127,5	0,187	4,7	0,219	5,6
140	1 3/4	44,5	1,000	25,4	1	25,4	0,500	12,7	383	173,7	0,219	5,6	0,250	6,4
160	2	50,8	1,125	28,6	1 1/4	31,8	0,562	14,3	500	226,8	0,250	6,4	0,281	7,1
180	2 1/4	57,2	1,406	35,7	1 13/32	35,7	0,687	17,4	633	287,1	0,281	7,1	0,312	7,9
200	2 1/2	63,5	1,562	39,7	1 1/2	38,1	0,781	19,8	781	354,3	0,312	7,9	0,375	9,5
240	3	76,2	1,875	47,6	1 7/8	47,6	0,937	23,8	1125	510,3	0,375	9,5	0,500	12,7

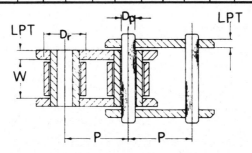

CHAINS (continued)
(API - Std 7F)

II. ROTARY CHAINS

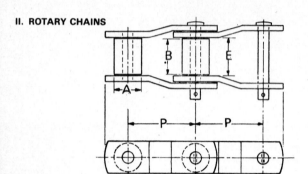

Standard rotary chains are given in the Table below.

	NOMINAL SIZE (in) (CHAIN NUMBER)					
	3		3 1/8		4	
	in	mm	in	mm	in	mm
Pitch P	3,075	78,1	3,125	79,4	4,063	103,2
Roller diameter A	1 1/4	31,7	1 5/8	41,3	1 3/4	44,4
Roller length B	1 7/16	36,5	1 19/32	40,5	1 7/8	47,6
Distance between flanges E	1 1/2	38,1	1 5/8	41,3	1 15/16	49,2
Distance between center lines for Duplex chains	—	—	3 3/16	81,0	—	—
Number of links in 10 ft (3,05 m)		39		39		30

Note. For the purpose of measuring standard length, the chains should be under a tensile load of 500 lb (225 daN)

III. LENGTH OF A CHAIN

$$L = 2C + \frac{N + n}{2} + 39,5 \frac{(N - n)^2}{C}$$

where :

L = chain length in pitches,
C = distance between sprocket centres in pitches,
N = number of teeth on the large sprocket,
n = number of teeth on the small sprocket.

IV. PROPER CHAIN TENSION

For a check of a chain tension, turn one sprocket to tighten the lower strand of chain ; then measure the sag of upper strand. This sag measured at midpoint should be approximately two to three per cent of the length of the tangent to the sprockets.

Example. If the length of the tangent between the sprockets is 200 cm, the sag shall be between 4 and 6 cm.

NOTES

H DRILLING MUD PRESSURE LOSSES

Summary

RELATION BETWEEN MUD WEIGHT AND
PRESSURE HEAD OF MUD (psi/ft)

MUD WEIGHT			Fluid head (psi/ft)				
kg/dm³	lb/gal	lb/cu.ft	(psi/ft)				
0,90	7,50	56,0	0,3901		CONVERSION FACTORS		
0,96	8,00	59,8	0,4162				
1,00	8,35	62,4	0,4335				
1,02	8,51	63,7	0,4422		TO CONVERT	INTO	MULTIPLY BY
1,04	8,68	64,9	0,4508				
1,06	8,85	66,2	0,4595		↘	↘	↘
1,08	9,01	67,4	0,4682	0,11983	kgf/l	lb/gal	8,3452
1,10	9,18	68,7	0,4769				
1,12	9,35	69,9	0,4855	0,016019	kgf/l	lb/cu.ft	62,427
1,14	9,51	71,2	0,4942				
1,16	9,68	72,4	0,5029	0,13368	lb/gal	lb/cu.ft	7,48052
1,18	9,85	73,7	0,5115				
1,20	10,01	74,9	0,5202	2,3067	kgf/l	psi/ft	0,433518
1,22	10,18	76,2	0,5289				
1,24	10,35	77,4	0,5375	↖	↖	↖	
1,26	10,52	78,7	0,5462	MULTIPLY BY	INTO	TO CONVERT	
1,28	10,68	79,9	0,5549				
1,30	10,85	81,2	0,5636				
1,32	11,02	82,4	0,5722				
1,36	11,35	84,9	0,5896				
1,40	11,68	87,4	0,6069		HYDROSTATIC PRESSURE		
1,44	12,02	89,9	0,6242				
1,48	12,35	92,4	0,6416				
1,52	12,69	94,9	0,6589		$P = \dfrac{Hd}{10}$		
1,56	13,02	97,4	0,6763				
1,60	13,35	99,9	0,6936				
1,64	13,69	102,4	0,7109				
1,68	14,02	104,9	0,7283			H (m)	
1,72	14,35	107,4	0,7456		P (kgl/cm²) if	d (kgf/l)	
1,76	14,69	109,9	0,7630				
1,80	15,02	112,4	0,7803				
1,84	15,36	114,9	0,7976			H (m)	
1,88	15,69	117,4	0,8150		P (bar) if	d (daN/l)	
1,92	16,02	119,9	0,8324				
1,96	16,36	122,4	0,8497				
2,00	16,69	124,9	0,8670				
2,04	17,02	127,4	0,8844		$P = 0,981\ \dfrac{Hd}{10}$		
2,08	17,36	129,9	0,9017				
2,12	17,69	132,4	0,9191				
2,16	18,02	134,9	0,9364			H (m)	
2,20	18,36	137,4	0,9537		P (bar) if	d (kgf/l)	
2,24	18,70	139,9	0,9710				
2,28	19,03	142,4	0,9884				
2,30	19,21	143,7	0,9970				

INCREASE OF MUD WEIGHT

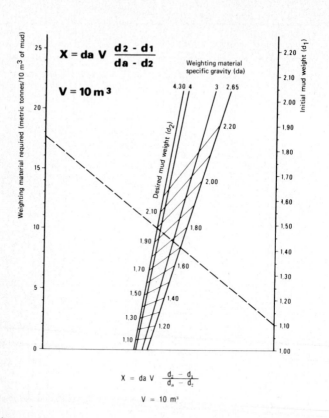

$$X = da \ V \ \frac{d_2 - d_1}{d_a - d_2}$$

$$V = 10 \ m^3$$

Example.

To increase the mud weight from 1,10 to 1,70, 17,8 metric tonnes of limestone (da = 2,65) must be added to each 10³ m³ of mud.

INCREASE OF MUD WEIGHT WITH BARITE (d = 4,2)

Barite required (kilogramme per 1 m³ of mud) — upper-right of table
Water required (litres per 1 m³ of mud) — lower-left of table

DESIRED MUD WEIGHT

INITIAL MUD WEIGHT	1.05	1.10	1.15	1.20	1.25	1.30	1.35	1.40	1.45	1.50	1.55	1.60	1.65	1.70	1.75	1.80	1.85	1.90	1.95	2.00	2.05	2.10	2.15	2.20	2.25
1.00	67	135	207	280	356	434	516	600	687	778	872	969	1071	1176	1286	1400	1519	1643	1773	1909	2051	2200	2356	2520	2692
1.05	—	68	138	210	285	362	442	525	611	700	793	888	988	1092	1200	1313	1430	1552	1680	1814	1953	2100	2254	2415	2585
1.10	1000	—	69	140	214	290	368	450	535	622	713	808	906	1008	1114	1225	1340	1461	1587	1718	1856	2000	2151	2310	2477
1.15	2000	500	—	70	142	217	295	375	458	544	634	727	824	924	1029	1138	1251	1370	1493	1623	1758	1900	2049	2205	2369
1.20	3000	1000	333	—	71	145	221	300	382	467	555	646	741	840	943	1050	1162	1278	1400	1527	1660	1800	1946	2100	2261
1.25	4000	1500	667	250	—	72	147	225	305	389	476	565	659	756	857	963	1072	1187	1307	1432	1563	1700	1844	1995	2154
1.30	5000	2000	1000	500	200	—	74	150	229	311	396	485	576	672	771	875	983	1096	1213	1336	1465	1600	1741	1890	2046
1.35	6000	2500	1333	750	400	167	—	75	153	233	317	404	494	588	686	788	894	1004	1120	1241	1367	1500	1639	1785	1938
1.40	7000	3000	1667	1000	600	333	143	—	76	156	238	323	412	504	600	700	804	913	1027	1145	1270	1400	1537	1680	1831
1.45	8000	3500	2000	1250	800	500	286	125	—	78	159	242	329	420	514	613	715	822	933	1050	1172	1300	1434	1575	1723
1.50	9000	4000	2333	1500	1000	667	429	250	111	—	79	162	247	336	429	525	626	730	840	955	1074	1200	1332	1470	1615
1.55	10000	4500	2667	1750	1200	833	571	375	222	100	—	81	165	252	343	438	536	639	747	859	977	1100	1229	1365	1508
1.60	11000	5000	3000	2000	1400	1000	714	500	333	200	91	—	82	168	257	350	447	548	653	764	879	1000	1127	1260	1400
1.65	12000	5500	3333	2250	1600	1167	857	625	444	300	182	83	—	84	171	263	357	457	560	668	781	900	1024	1155	1292
1.70	13000	6000	3667	2500	1800	1333	1000	750	556	400	273	167	77	—	86	175	268	365	467	573	684	800	922	1050	1185
1.75	14000	6500	4000	2750	2000	1500	1143	875	667	500	364	250	154	71	—	88	179	274	373	477	586	700	820	945	1077
1.80	15000	7000	4333	3000	2200	1667	1286	1000	778	600	455	333	231	143	67	—	89	183	280	382	488	600	717	840	969
1.85	16000	7500	4667	3250	2400	1833	1429	1125	889	700	545	417	308	214	133	63	—	91	187	286	391	500	615	735	862
1.90	17000	8000	5000	3500	2600	2000	1571	1250	1000	800	636	500	385	286	200	125	59	—	93	191	293	400	512	630	754
1.95	18000	8500	5333	3750	2800	2167	1714	1375	1111	900	727	583	462	357	267	188	118	56	—	95	195	300	410	525	646
2.00	19000	9000	5667	4000	3000	2333	1857	1500	1222	1000	818	667	538	429	333	250	176	111	53	—	98	200	307	420	538
2.05	20000	9500	6000	4250	3200	2500	2000	1625	1333	1100	909	750	615	500	400	313	235	167	105	50	—	100	205	315	431
2.10	21000	10000	6333	4500	3400	2667	2143	1750	1444	1200	1000	833	692	571	467	375	294	222	158	100	48	—	102	210	323
2.15	22000	10500	6667	4750	3600	2833	2286	1875	1556	1300	1091	917	769	643	533	438	353	278	211	150	95	43	—	105	215
2.20	23000	11000	7000	5000	3800	3000	2429	2000	1667	1400	1182	1000	846	714	600	500	412	333	263	200	143	87	42	—	108
2.25	24000	11500	7333	5250	4000	3167	2571	2125	1778	1500	1273	1083	923	786	667	563	471	389	316	250	190	136	87	42	—
2.30	25000	12000	7667	5500	4200	3333	2714	2250	1889	1600	1364	1167	1000	857	733	625	529	444	368	300	238	182	130	83	40

DESIRED MUD WEIGHT

MUD WEIGHT REDUCTION WITH WATER

VOLUME OF MUD AFTER ADDING WEIGHTING MATERIAL TO ONE CUBIC METRE OF MUD

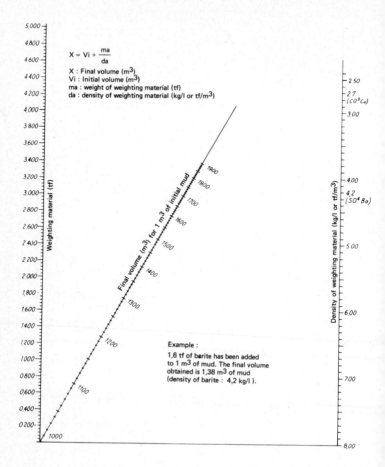

$$X = Vi + \frac{ma}{da}$$

X : Final volume (m^3)
Vi : Initial volume (m^3)
ma : weight of weighting material (tf)
da : density of weighting material (kg/l or tf/m^3)

Weighting material (tf)

Final volume (m^3) for 1 m^3 of initial mud

Density of weighting material (kg/l or tf/m^3)

2 50
2 7 ($CO^3 Ca$)
3.00
4.00
4,2 ($SO^4 Ba$)
5.00
6.00
7.00
8,00

Example :

1,6 tf of barite has been added to 1 m^3 of mud. The final volume obtained is 1,38 m^3 of mud (density of barite : 4,2 kg/l).

MUD WEIGHT REDUCTION WITH WATER

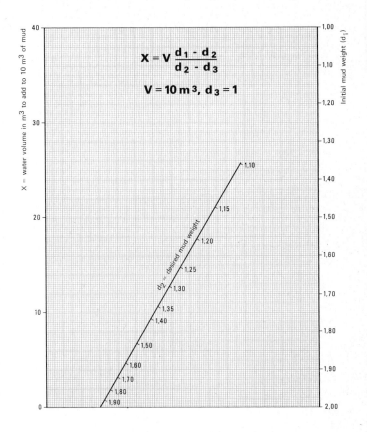

X = water volume in m³ to add to 10 m³ of mud

Initial mud weight (d_1)

$$X = V \frac{d_1 - d_2}{d_2 - d_3}$$

$$V = 10 \text{ m}^3, \ d_3 = 1$$

d_2 = desired mud weight

MUD WEIGHT REDUCTION WITH OIL
(d_3 = 0,85)

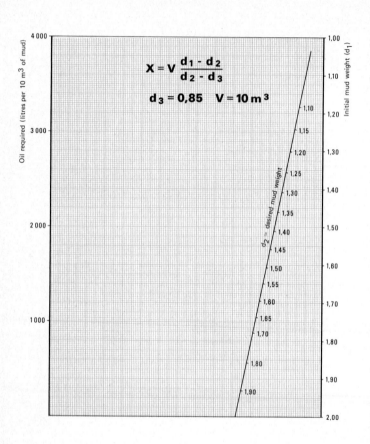

TERNARY DIAGRAM FOR DETERMINING SOLID CONTENT OF MUD
(water base muds)

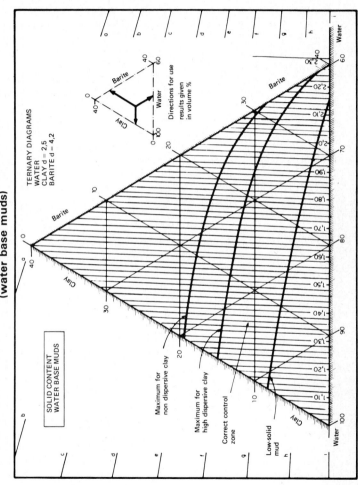

TERNARY DIAGRAMS
WATER
CLAY d = 2.5
BARITE d = 4.2

Directions for use
results given
in volume %

SOLID CONTENT
WATER BASE MUDS

Maximum for non dispersive clay

Maximum for high dispersive clay

Correct control zone

Low-solid mud

TERNARY DIAGRAM FOR DETERMINING SOLID CONTENT OF MUD
(saturated salt water mud)

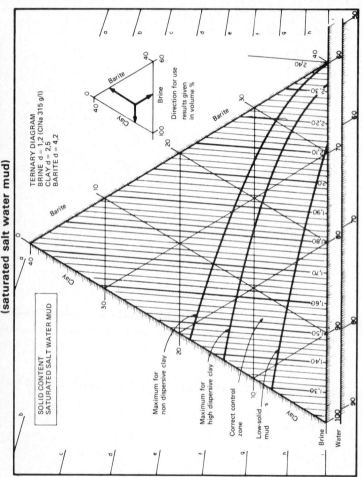

TERNARY DIAGRAM
BRINE d = 1,2 (ClNa 315 g/l)
CLAY d = 2,5
BARITE d = 4,2

Direction for use
results given
in volume %

SOLID CONTENT
SATURATED SALT WATER MUD

Maximum for
non dispersive clay

Maximum for
high dispersive clay

Correct control
zone

Low-solid
mud

CALCULATION BASIS FOR OIL BASE MUD

Specific gravity of oil : 0,85,
Specific gravity of ClNa : 2,72,
Specific gravity of $CaCl_2$: 3,70.

Amount of oil and NaCl brine for various H/E (oil/water) ratios

Water : 884 l
NaCl : 116 l } 1 m³ of brine d = 1,20

H/E ratio	50/50	60/40	65/35	70/30	75/25	80/20
Specific gravity	1,030	0,996	0,976	0,957	0,938	0,920
Oil	469,5	570,3	622	674	726,75	779,7
Brine	530,5	429,7	378	326	273,25	220,3

Note. 1 kg NaCl plus 2,806 l of water give 3,173 of brine at 315 g/l.

Amount of oil and $CaCl_2$ brine (at 320 g $CaCl_2$ pure per litre) for various H/E ratios

Water : 909,73 l
$CaCl_2$ at 96 % : 90,27 l } 1 m³ of brine d = 1,243

H/E ratio	50/50	60/40	65/35	70/30	75/25	80/20
Specific gravity	1,051	1,010	0,990	0,969	0,948	0,927
Oil	476,4	577	628	680	731,9	784,5
Brine	523,6	423	372	320	268,1	215,5

Note. 1 kg $CaCl_2$ plus 2,724 l of water give 2,994 l of brine at 320 g/l.

Amount of oil and brine (at 700 g of pure $CaCl_2$ per litre) for various H/E ratios

Water : 811 l
$CaCl_2$ at 96 % : 189 l } 1 m³ of brine d = 1,51

H/E ratio	50/50	60/40	65/35	70/30	75/25	80/20
Specific gravity	1,210	1,142	1,109	1,071	1,035	0,998
Oil	447,8	549	601	655	708,7	764,4
Brine	552,2	451	399	345	291,3	235,6

Note. 1 kg $CaCl_2$ plus 1,158 l of water give 1,426 l of brine at 700 g/l.

MUD CYCLE TIME

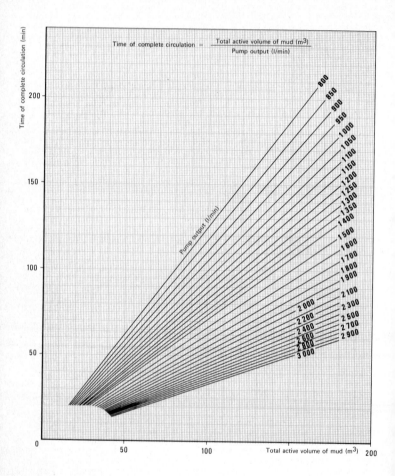

$$\text{Time of complete circulation} = \frac{\text{Total active volume of mud (m}^3)}{\text{Pump output (l/min)}}$$

Note. Total active volume of mud (m³) = mud in hole (m³) + active surface volume (m³)

ANNULAR OR RISING VELOCITY OF MUD

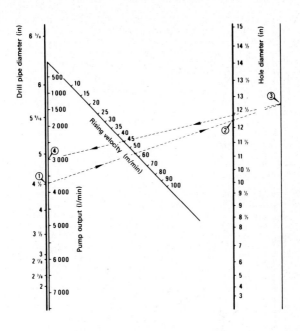

Example :   1 = drill pipe diameter 4 1/2'',
2 = hole diameter 12 1/4'',
3 = base line,
4 = pump output 3000 l/min,
5 = final result 45 m/min.

AMOUNT OF DRILLED CUTTINGS IN MUD

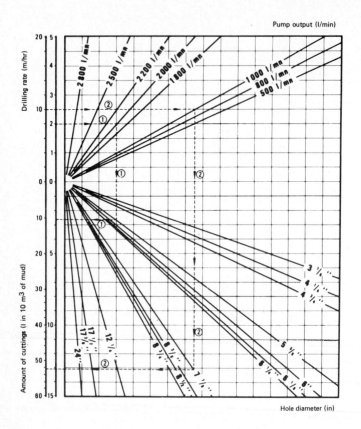

Examples :

(1) Drilling rate   = 2 m/h,
    Pump output   = 2000 l/min,
    Hole diameter   = 5 5/8'',
    Amount of cuttings = 2,65 l in 10 m³ of mud.

(2) Drilling rate   = 10 m/h,
    Pump output   = 1000 l/min,
    Hole diameter   = 7 7/8'',
    Amount of cuttings = 52,2 l in 10 m³ of mud.

SHALE SHAKER SCREENS

Mesh per inch	Wire diameter		Width of opening		Approximate open area (%)
	in	μ	in	μ	
8 × 8	.028	711	.097	2463	60.2 %
10 × 10	.025	635	.083	2108	56.3 %
12 × 12	.023	584	.060	1524	51.8 %
14 × 14	.020	508	.054	1371	51.0 %
16 × 16	.018	457	.045	1143	50.7 %
18 × 18	.018	457	.038	965	45.8 %
20 × 20	.016	406	.034	863	46.2 %
8 × 20	.032/.020	813/508	.094/.030	2380/760	45.7 %
20 × 30	.015	381	.035/.018	890/457	39.6 %
30 × 30	.013	330	.021	533	40.8 %
40 × 40	.010	254	.015	380	36.0 %
40 × 40	.013	330	.012	330	27.0 %
50 × 50	.009	229	.011	280	30.3 %
60 × 60	.0075	191	.009	230	30.5 %
70 × 30	.0075	191	.026/007	610/178	40.3 %
80 × 80	.0055	140	.007	178	31.4 %

SCREEN STANDARDS

FRANCE		GERMANY	GREAT BRITAIN		ITALY	USSR		USA				Wentworth and J. Boucart scale
AFNOR		Deutsche normen	British Standard Institution		Unificazione italiano			TYLER		ASTM		
NF X 11·501 1970		DIN 4100 1957	BS-410 1943		UNI-2332 1943	COST-3584-53 1953		TYLER STANDARD SCREEN SCALE SIEVES		ASTM-E 11·61 1961		
Opening (mm)	module	Opening (mm)	Opening (mm)	Designation No	Opening (mm)	Opening (mm)	Designation	Opening	Designation mesh	Opening (mm)	Designation No	
0.040	17	0.04			0.04	0.040	004	0.038	400 ·	0.037	400	Silt loess 62,5 to 4 μ
0.050	18	0.045 0.05	0.044	350		0.045 0.050	0045 005	0.043	325	0.044	325	
		0.056	0.053	300	0.05	0.056	0056	0.053 0.061	270 250	0.053	270 ·	
0.063	19	0.063	0.064	240	0.063	0.063	0063			0.063	230	
0.080	20	0.071 0.08	0.076	200	0.075 0.08	0.071 0.080	0071 008	0.074	200 ·	0.074	200 ·	Very fine sand 125 to 62,5 μ
		0.09	0.089	170	0.09	0.090	009	0.088	170	0.088	170	
0.100	21	0.1			0.1	0.100	01					
			0.104	150	0.106	0.112	0112	0.104	150 ·	0.105	140 ·	
0.125	22	0.125	0.124	120	0.125	0.125 0.140	0125 014	0.124	115	0.125	120	
0.160	23	0.16	0.152	100	0.15 0.16	0.160	016	0.147	100 ·	0.149	100 ·	Fine sand 250 to 125 μ
0.200	24	0.2	0.178	85	0.18 0.2	0.180 0.200	018 02	0.175	80	0.177	80	
			0.211	72	0.212	0.224	0224	0.208	65 ·	0.210	70 ·	
0.250	25	0.25	0.251	60	0.25	0.250 0.280	025 028	0.246	60	0.250	60	
0.315	26	0.315	0.295	52	0.3 0.315	0.315		0.295	48 ·	0.297	50 ·	Medium sand 0,5 to 0,25 mm
0.40	27	0.4	0.353	44	0.355 0.4	0.355 0.400	0355 04	0.351	42	0.354	45	
			0.422	36	0.425	0.450	045	0.417	35 ·	0.420	40 ·	
0.50	28	0.5	0.500	30	0.5	0.500	05	0.495	32	0.500	35	
0.63	29	0.63	0.599	25	0.6 0.63	0.560 0.630	056 063	0.589	28 ·	0.595	30 ·	Coarse sand 1 to 0,5 mm
			0.699	22	0.71 0.75	0.700	07	0.701	24	0.707	25	
0.80	30	0.8			0.8	0.800	08					
			0.853	18	0.85	0.900	09	0.833	20 ·	0.841	20 ·	
1.00	31	1	1.003	16	1	1.00	1	0.991	16	1.00	18	
1.25	32	1.25	1.204	14	1.18 1.25	1.25	1.25	1.168	14 ·	1.19	16	Very coarse sand 2 to 1 mm
1.60	33	1.6	1.405	12	1.4 1.6	1.60	1.6	1.397	12	1.41	14	
2.00	34	2	1.676 2.057	10 8	1.7 2	2.00	2	1.651 1.981	10 · 9	1.68 2.00	12 · 10	
2.50	35	2.5	2.411	7	2.36 2.5 2.8	2.50	2.5	2.362 2.794	8 · 7	2.38 2.83	8 7	Granulated material 2 to 4 mm
3.15	36	3.15	2.812	6	3.15							
4.00	37	4	3.353	5	3.35 4			3.327 3.962 4.699	6 · 5 4 ·	3.36 4.00 4.76	6 · 5 4	
5.00	38	5						5.613	3 1 2	5.66	3 1 2	

PRESSURE LOSSES - GENERAL

Flow of liquids through pipes

When a liquid is flowing through a pipe, a part of its energy is lost, absorbed by the work of friction forces :
— internal friction due to its viscosity,
— external friction due to the roughness of the pipe.

This loss of energy is called pressure loss and it is the difference in pressure between two points in the pipe. For example, the discharge pump pressure can be considered as the total energy of the circulating fluid. This entire energy is lost through the mud circuit since, when back at the mud tank, the pressure of the mud is zero. In this case, the discharge pump pressure represents the total amount of pressure losses in the mud circuit.

Pressure losses through the mud circuit can be differentiated :
(1) Through surface equipment,
(2) Through drill pipe and drill-collar bore,
(3) Through the rock bit,
(4) Between drill pipe OD, drill collar OD and wall of hole.

For turbulent flow, pressure losses are given by the formula :

$$P = k\rho \frac{LQ^2}{D^5} \text{ (approximate formula)}$$

where :
P = pressure losses,
k = proportionality factor,
ρ = specific gravity of mud,
L = length of pipe,
Q = pump output,
D = inside diameter of pipe.

Pressure losses are directly proportional to :
the specific gravity of the mud (really ρ at a power between 0,75 and 1),
the length of the pipe,
the square of the pump output (really at a power between 1,75 and 2).
and inversely proportional to :
the fifth power of the pipe diameter (really at a power beween 4,75 and 5).

PRESSURE LOSSES - GENERAL (continued)

Pressure losses as a function of pump output

Pressure losses are proportional to the square of the pump output. Then if p_1 are the pressure losses for Q_1 pump output, pressure losses p_2 for Q_2 pump output will be :

$$p_2 = p_1 \times \left[\frac{Q_2}{Q_1}\right]^2$$

Example. Given : pump output Q_1 2000l/min,
pressure losses P_1 50 bars.
Find : pressure losses P_2 for a pump output Q_2 2600 l/min

$$p_2 = p_1 \left[\frac{Q_2}{Q_1}\right]^2 = 50 \times \left[\frac{2,6}{2}\right]^2 = 50 \times 1,69 = 84,5 \text{ bars}$$

Multiplying the pump output by 1,3 the pressure losses are multiplied by 1,69.

Pressure losses as a function of pipe diameter

Pressure losses are inversely proportional to the fifth power of the diameter. Then, if the diameter is slightly increased or reduced the influence on the pressure losses will be very great :

$$p_2 = p_1 \left[\frac{D_1}{D_2}\right]^5$$

Example. Given : pressure losses 10 bars P_1,
diameter of pipe 3 1/2'' D_1.
Find : pressure losses P_2 through a 2 1/2'' pipe D_2

$$p_2 = p_1 \left[\frac{D_1}{D_2}\right]^5 = 10 \times \left[\frac{3,5}{2,5}\right]^5 = 10 \times (1,4)^5 = 10 \times 5,38 = 53,8 \text{ bars}$$

Here, pressure losses are multiplied more than 5 times. Before changing the size of a pipe it will be advisable to check the influence of this change on the pressure losses which can overload the pump.

Note. The figures on the following pages have been established to easily give the pressure losses through the different portions of the hydraulic circuit.

ANNULAR VELOCITY OF THE MUD (m/min)
(Circulation rate 0 to 2500 l/min)

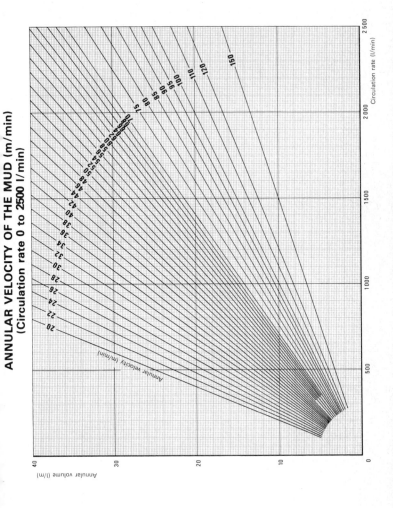

ANNULAR VELOCITY OF THE MUD (m/min)
(Circulation rate 0 to 5000 l/min)

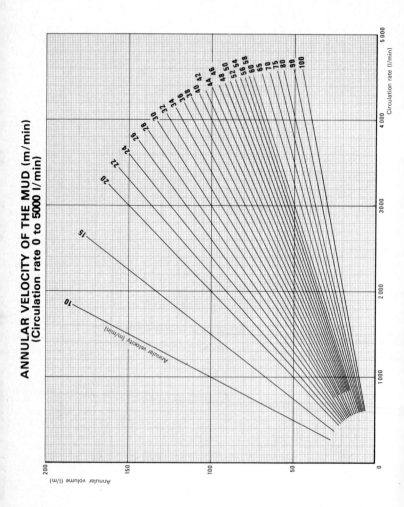

Annular velocity (m/min)

Annular volume (l/m)

Circulation rate (l/min)

PRESSURE LOSSES THROUGH SURFACE EQUIPMENT

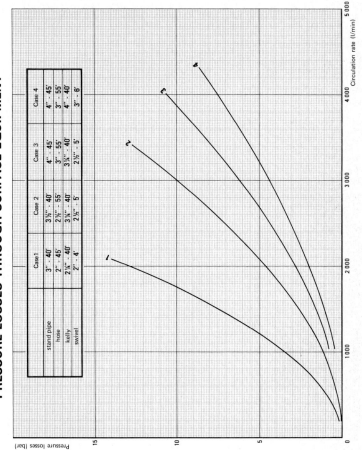

	Case1	Case 2	Case 3	Case 4
stand pipe	3" - 40'	3½" - 40'	4" - 45'	4" - 45'
hose	2" - 45'	2" - 55'	3" - 55'	3" - 55'
kelly	2¼" - 40'	3¼" - 40'	3¼" - 40'	4" - 40'
swivel	2" - 4'	2½" - 5'	2½" - 5'	3" - 6'

Pressure losses (bar)

Circulation rate (l/min)

PRESSURE LOSSES THROUGH 2 3/8" AND 2 7/8" DRILL PIPE (SG = 1)

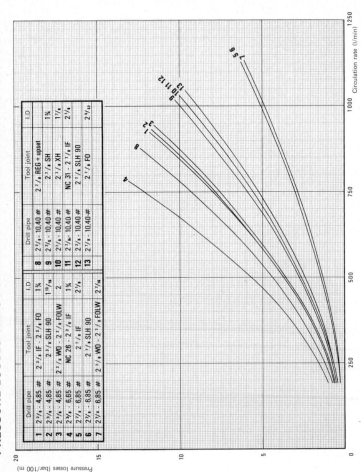

	Drill pipe	Tool joint	I.D
1	2 ³/₈ - 4.85 #	2 ³/₈ IF - 2 ³/₈ FO	1 ¾
2	2 ³/₈ - 4.85 #	2 ³/₈ SLH 90	1 ¹⁵/₁₆
3	2 ³/₈ - 4.85 #	2 ³/₈ WO - 2 ³/₈ FOLW	2
4	2 ³/₈ - 6.65 #	NC 26 - 2 ³/₈ IF	1 ¾
5	2 ³/₈ - 6.65 #	2 ³/₈ IF	2 ¹/₈
6	2 ³/₈ - 6.85 #	2 ³/₈ SLH 90	2 ¹/₈
7	2 ³/₈ - 6.85 #	2 ³/₈ WO - 2 ³/₈ FOLW	2 ¹/₁₆

	Drill pipe	Tool joint	I.D
8	2 ⁷/₈ - 10.40 #	2 ⁷/₈ REG + upset	2 ⁷/₈ REG + upset
9	2 ⁷/₈ - 10.40 #	2 ⁷/₈ SH	1 ¾
10	2 ⁷/₈ - 10.40 #	2 ⁷/₈ XH	1 ¹¹/₁₆
11	2 ⁷/₈ - 10.40 #	NC 31 - 2 ⁷/₈ IF	2 ¹/₈
12	2 ⁷/₈ - 10.40 #	2 ⁷/₈ SLH 90	
13	2 ⁷/₈ - 10.40 #	2 ⁷/₈ FO	2 ⁹/₃₂

Pressure losses (bar/100 m)

Circulation rate (l/min)

PRESSURE LOSSES THROUGH 3 1/2" DRILL PIPE (SG = 1)

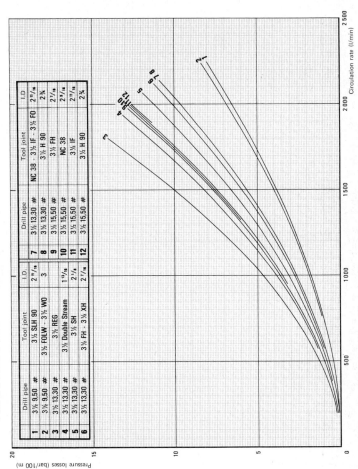

	Drill pipe	Tool joint	I.D.
1	3½ 9.50 #	3½ SLH 90	2 ⁹/₁₆
2	3½ 9.50 #	3½ FOLW - 3½ WO	3
3	3½ 13.30 #	3½ REG	—
4	3½ 13.30 #	3½ Double Stream	1¹³/₁₆
5	3½ 13.30 #	3½ SH	2¹/₄
6	3½ 13.30 #	3½ FH - 3½ XH	2⁷/₁₆

	Drill pipe	Tool joint	I.D.
7	3½ 13.30 #	NC 38 - 3½ IF - 3½ FO	2¹¹/₁₆
8	3½ 13.30 #	3½ H 90	2¾
9	3½ 15.50 #	3½ FH	2¹/₁₆
10	3½ 15.50 #	NC 38	2³/₁₆
11	3½ 15.50 #	3½ IF	2¹¹/₁₆
12	3½ 15.50 #	3½ H 90	2¾

PRESSURE LOSSES THROUGH 4" DRILL PIPE (SG = 1)

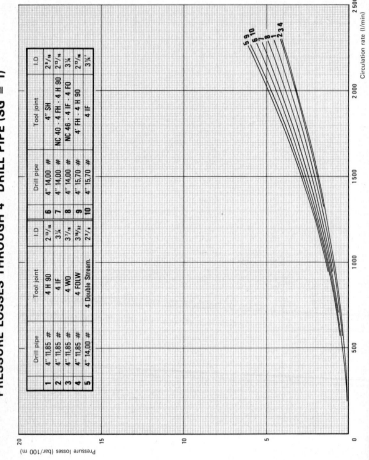

	Drill pipe	Tool joint	I.D
1	4" 11.85 #	4 H 90	2 ¹³/₁₆
2	4" 11.85 #	4 IF	3 ¼
3	4" 11.85 #	4 WO	3 ⁷/₁₆
4	4" 11.85 #	4 FOLW	3 ⁹/₃₂
5	4" 14.00 #	4 Double Stream.	2 ⁷/₈

	Drill pipe	Tool joint	I.D
6	4" 14.00 #	4" SH	2 ⁷/₁₆
7	4" 14.00 #	NC 40 - 4 FH - 4 H 90	2 ¹³/₁₆
8	4" 14.00 #	NC 46 - 4 IF - 4 FO	3 ¼
9	4" 15.70 #	4" FH - 4 H 90	2 ¹³/₁₆
10	4" 15.70 #	4 IF	3 ¼

Pressure losses (bar/100 m)

Circulation rate (l/min)

PRESSURE LOSSES THROUGH 4 1/2" DRILL PIPE (SG = 1)

	Drill pipe	Tool joint	I.D.		Drill pipe	Tool joint	I.D
1	4½" 13,75 #	4½ H 90	3¼	5	4½" 16,60 #	4½ REG	
2	4½" 13,75 #	4½ IF	3¾	6	4½" 16,60 #	4½ SH - 4½ Double St.	2 11/16
3	4½" 13,75 #	4½ WO	3 7/8	7		4½ FH	3
4	4½" 13,75 #	4½ FOLW	3 31/32	8		NC 46 - 4½ H 90 - 4½ XH	3¼
				9		NC 50 - 4½ IF - 4½ FO	3¾
				10	4½" 20 #	4½ REG	
				11		NC 46 - 4½ FH	3
				12		4½ H 90 - 4½ XH	3¼
				13		4½ IF	3¾

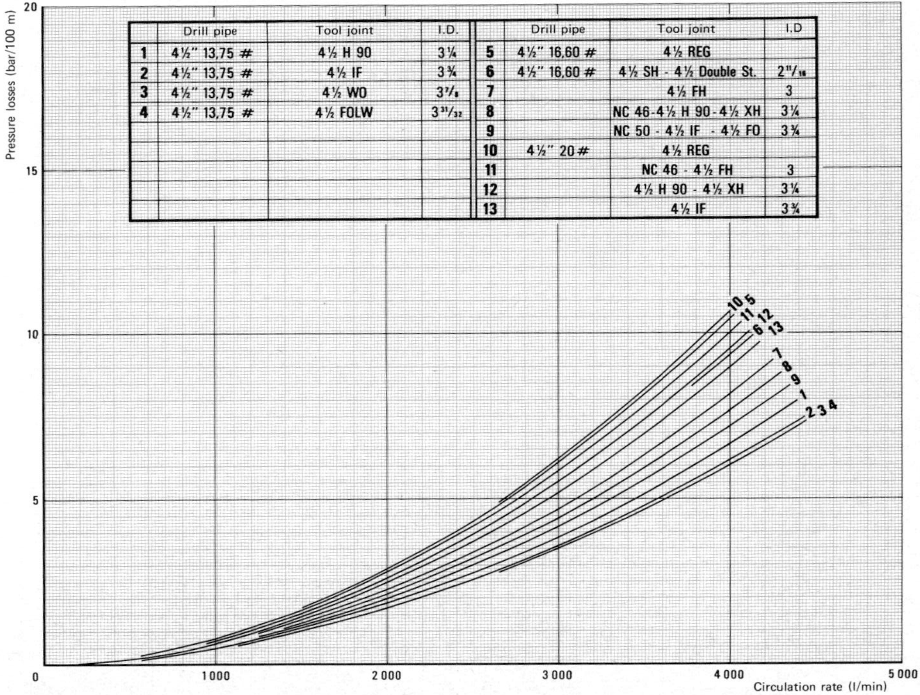
Pressure losses (bar/100 m)
20
15
10
5
0
1000
2 000
3 000
4 000
5 000
Circulation rate (l/min)
10 5
11 12
6 13
7
8
9
1
2 3 4

PRESSURE LOSSES THROUGH 5″, 5 1/2″, 6 5/8″ DRILL PIPE (SG = 1)

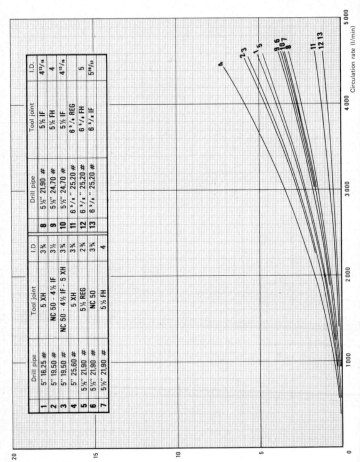

	Drill pipe	Tool joint	I.D
1	5″ 16,25 #	5 XH	3¾
2	5″ 19,50 #	NC 50 - 4½ IF	3½
3	5″ 19,50 #	NC 50 - 4½ IF - 5 XH	3¾
4	5″ 25,60 #	5 XH	3¾
5	5½″ 21,90 #	5½ REG	2¾
6	5½″ 21,90 #	NC 50	3¾
7	5½″ 21,90 #	5½ FH	4

	Drill pipe	Tool joint	I.D.
8	5½″ 21,90 #	5½ IF	4¹³/₁₆
9	5½″ 24,70 #	5½ FH	4
10	5½″ 24,70 #	5½ IF	4¹³/₁₆
11	6⅝″ 25,20 #	6⅝ REG	5
12	6⅝″ 25,20 #	6⅝ FH	5
13	6⅝″ 25,20 #	6⅝ IF	5³⁵/₃₂

Pressure losses (bar/100 m)

Circulation rate (l/min)

PRESSURE LOSSES THROUGH DRILL COLLAR BORE (SG = 1)

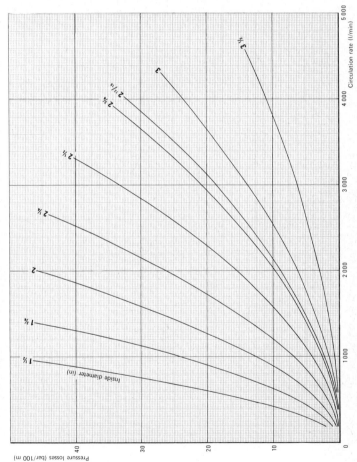

PRESSURE LOSSES AND VELOCITY ACCROSS BIT NOZZLE (SG = 1) (turbulent flow)

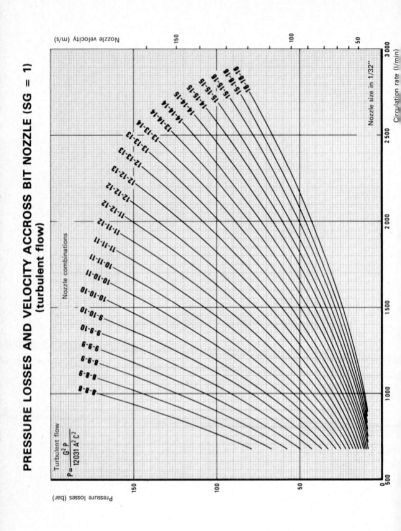

PRESSURE LOSSES AND VELOCITY ACCROSS BIT NOZZLES (SG = 1)

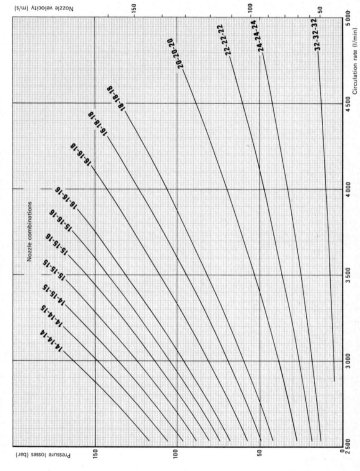

PRESSURE LOSSES BETWEEN DRILL COLLAR OD AND WALL OF HOLE (SG = 1)
(For 5 5/8", 5 3/4", 5 7/8", 6", 6 1/8" and 6 3/4" hole size)

Curve No.	1	2	3	4	5	6	7	8	9	2	10	11
Hole diameter	6 1/4	6 1/4	6 1/8	6 1/8	6	6	5 7/8	5 7/8	5 3/4	5 3/4	5 5/8	5 5/8
Drill collar O.D.	4 1/8	4 3/4	4 1/8	4 3/4	4 1/8	4 3/4	4 1/8	4 3/4	4 3/4	4 1/8	4 1/8	4 3/4

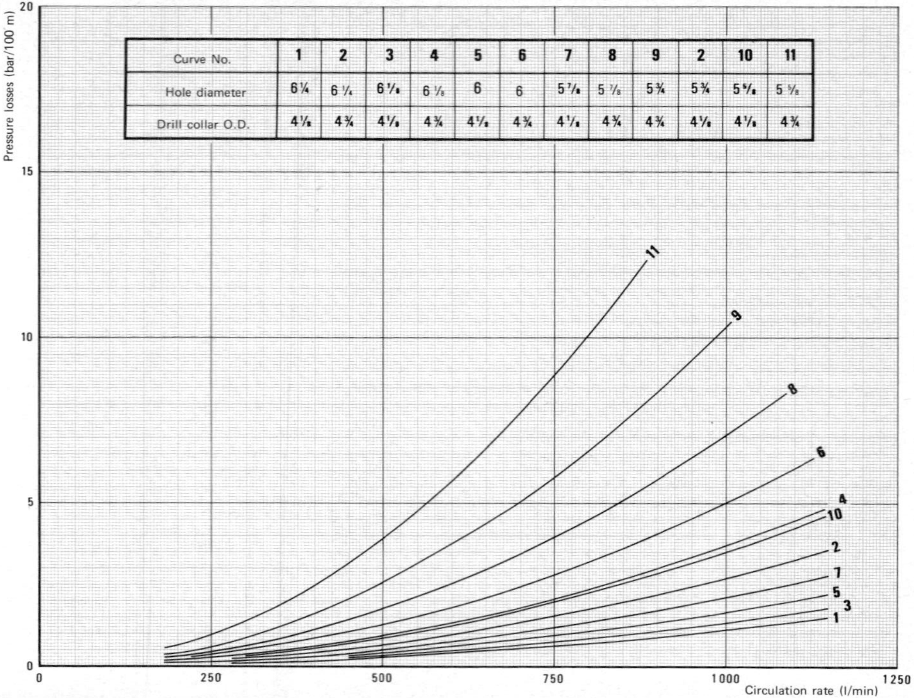
Pressure losses (bar/100 m)
20
15
10
5
0
250
500
750
1000
1250
Circulation rate (l/min)
11
9
8
6
4
10
2
7
5
3
1

PRESSURE LOSSES BETWEEN DRILL COLLAR OD AND WALL OF HOLE (SG = 1)
(For 6 3/4" and 7 7/8" hole size)

Curve No.	1	2	3	4	5	6	7	8	9
Hole diameter	6¾	6¾	6¾	6¾	7⅞	7⅞	7⅞	7⅞	7⅞
Drill collar O.D.	4¾	5¼	5½	5¾	6	6¼	6½	6¾	7

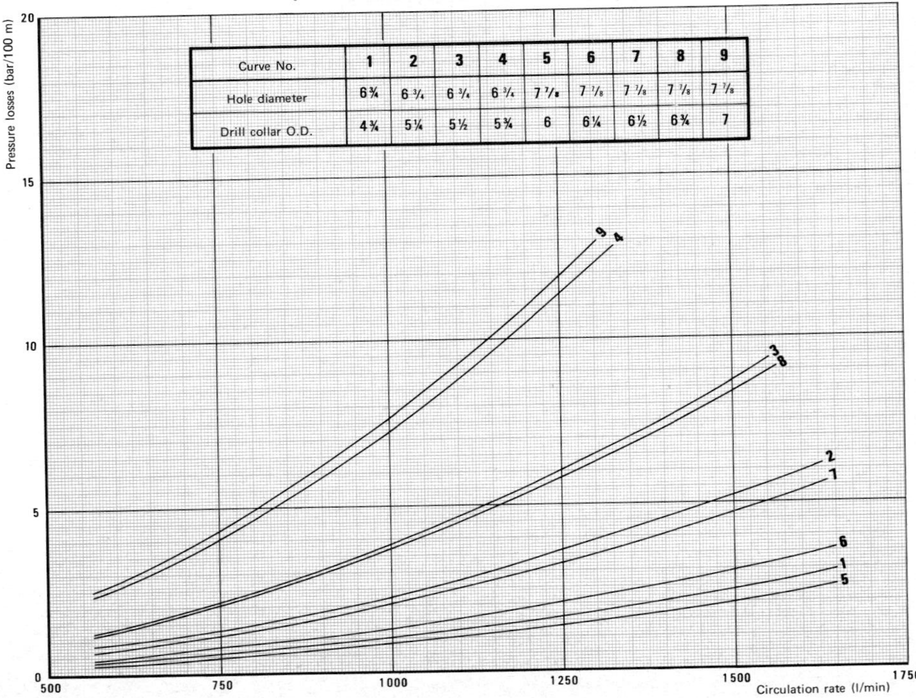

Pressure losses (bar/100 m)

Circulation rate (l/min)

PRESSURE LOSSES BETWEEN DRILL COLLAR OD AND WALL OF HOLE (SG = 1)
(For 8 3/8'', 8 1/2'', 8 5/8'' and 8 3/4'' hole size)

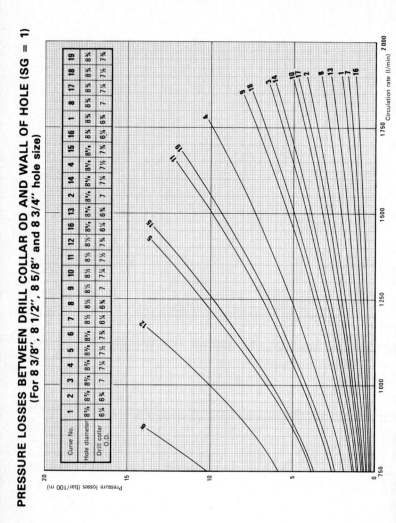

PRESSURE LOSSES BETWEEN DRILL COLLAR OD AND WALL OF HOLE (SG = 1)
(For 9 7/8" and 12 1/4" hole size)

Curve No.	1	2	3	4	5	6	7	8	1
Hole diameter	12¼	12¼	12¼	12¼	12¼	9⅞	9⅞	9⅞	12¼
Drill collar O.D.	7¾	9½	10	11	11¼	7¾	8	8¾	8

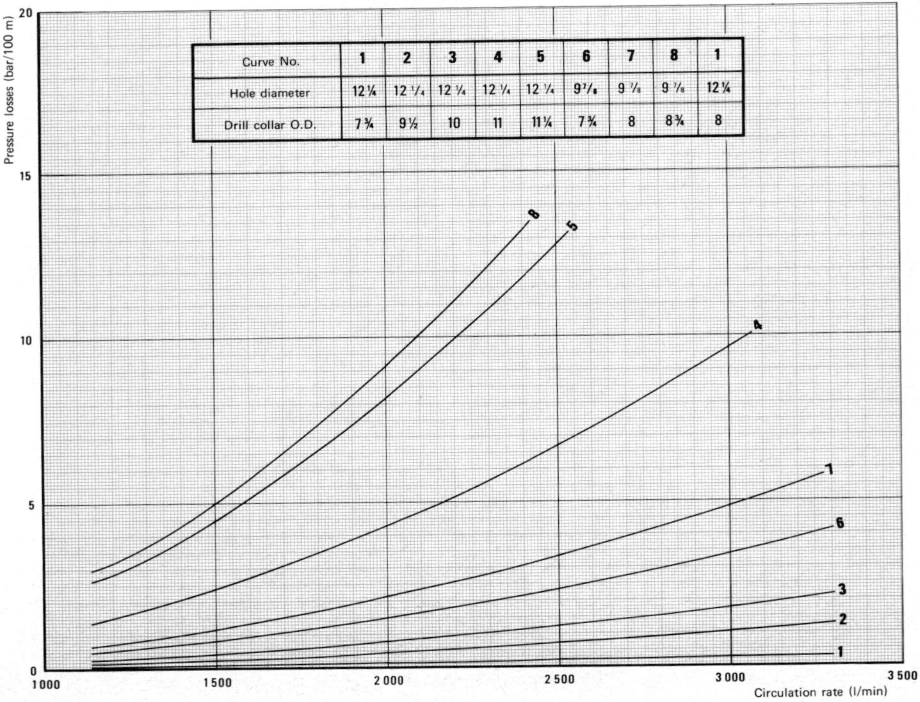
Pressure losses (bar/100 m)
20
15
10
5
0
1000
1500
2000
2500
3000
3500
Circulation rate (l/min)
8
5
4
1
6
3
2
1

PRESSURE LOSSES BETWEEN DRILL COLLAR OD AND WALL OF HOLE (SG = 1)
(For 15" and 17 1/2" hole size)

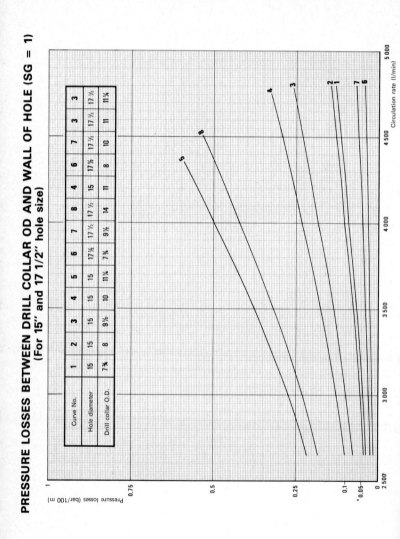

Curve No.	1	2	3	4	5	6	7	8	4	6	7	3	3
Hole diameter	15	15	15	15	15	17½	17½	17½	15	17½	17½	17½	17½
Drill collar O.D.	7¾	8	9½	10	11¼	7¾	9½	14	11	8	10	11	11¼

Pressure losses (bar/100 m)

Circulation rate (l/min)

PRESSURE LOSSES BETWEEN DRILL PIPE OD AND WALL OF HOLE (SG = 1)
(For 5 5/8" and 5 3/4" hole size)

Curve No.	1	2	3	4	5	6	7	8	9	10	10	11	12	9	3	10	13
Hole diameter	5 5/8	5 5/8	5 5/8	5 5/8	5 5/8	5 5/8	5 5/8	5 5/8	5 3/4	5 3/4	5 3/4	5 3/4	5 7/8	5 7/8	5 7/8	5 7/8	5 7/8
Drill pipe O.D.	2 7/8	2 7/8	2 7/8	3 1/2	3 1/2	3 1/2	3 1/2	3 1/2	2 7/8	2 7/8	3 1/2	3 1/2	2 7/8	2 7/8	3 1/2	3 1/2	3 1/2
Tool joint O.D.	3 5/8 to 4	4 1/8 to 4 1/4	4 3/8	4 to 4 1/4	4 5/8	4 3/4	4 7/8	5	3 5/8 to 3 3/4	4 to 4 3/8	4	5	3 5/8 to 4 1/8	4 1/4 to 4 3/8	4 and 4 1/8	4 1/4 to 4 3/4	4 7/8 and 5

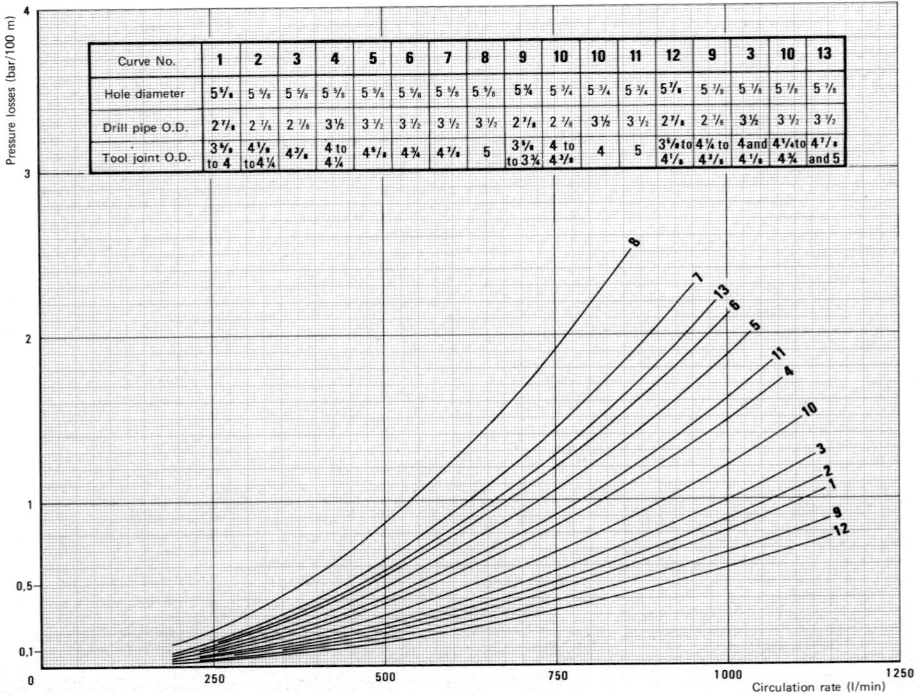
Pressure losses (bar/100 m)
4
3
2
1
0,5
0,1
0
250
500
750
1000
1250
Circulation rate (l/min)
8
7
13
6
5
11
4
10
3
2
1
9
12

PRESSURE LOSSES BETWEEN DRILL PIPE OD AND WALL OF HOLE (SG = 1)
(For 6", 6 1/8" and 6 1/4" hole size)

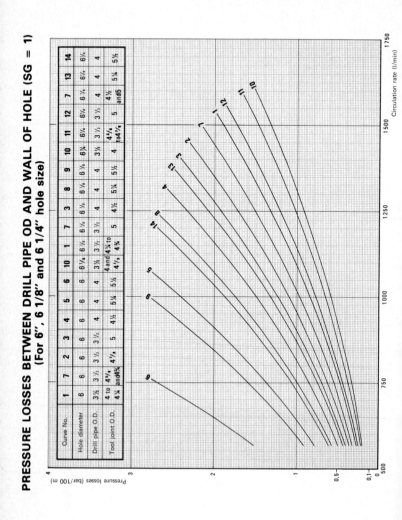

Curve No.	1	7	2	3	4	5	6	10	1	7	3	8	9	10	11	12	7	13	14
Hole diameter	6	6	6	6	6	6	6	6 1/8	6 1/8	6 1/8	6 1/8	6 1/8	6 1/8	6 1/4	6 1/4	6 1/4	6 1/4	6 1/4	6 1/4
Drill pipe O.D.	3 1/2	3 1/2	3 1/2	3 1/2	4	4	4	3 1/2	3 1/2	3 1/2	4	4	4	3 1/2	3 1/2	3 1/2	4	4	4
Tool joint O.D.	4 to 4 1/4	4 5/8 and 4 3/4	4 7/8	5	4 1/2	5 1/4	5 1/2	4 and 4 1/8	4 1/4 to 4 3/4	5	4 1/2	5 1/4	5 1/2	4	4 5/8 to 4 7/8	5	4 1/2 and 5	5 1/4	5 1/2

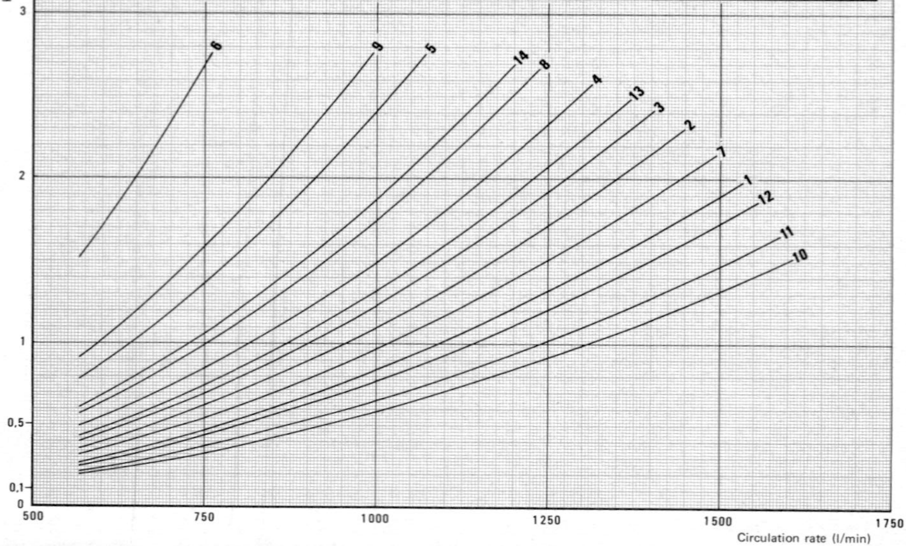
Pressure losses (bar/100 m)
4
3
2
1
0,5
0,1
0
500
750
1000
1250
1500
1750
Circulation rate (l/min)
6
9
5
14
8
4
13
3
2
7
1
12
11
10

PRESSURE LOSSES BETWEEN DRILL PIPE OD AND WALL OF HOLE (SG = 1)
(For 6 3/4'' hole size)

Curve No.	1	2	3	4	5	6	7	8
Hole Diameter	6¾	6 3/4	6 3/4	6 3/4	6 3/4	6 3/4	6 3/4	6 3/4
Drill pipe O.D.	3½	3 1/2	3 1/2	3 1/2	4	4	4	4
Tool joint O.D.	4 and 4 1/8	4¼ and 4 5/8	4¾ and 4 7/8	5	5¼	5½	5¾	6

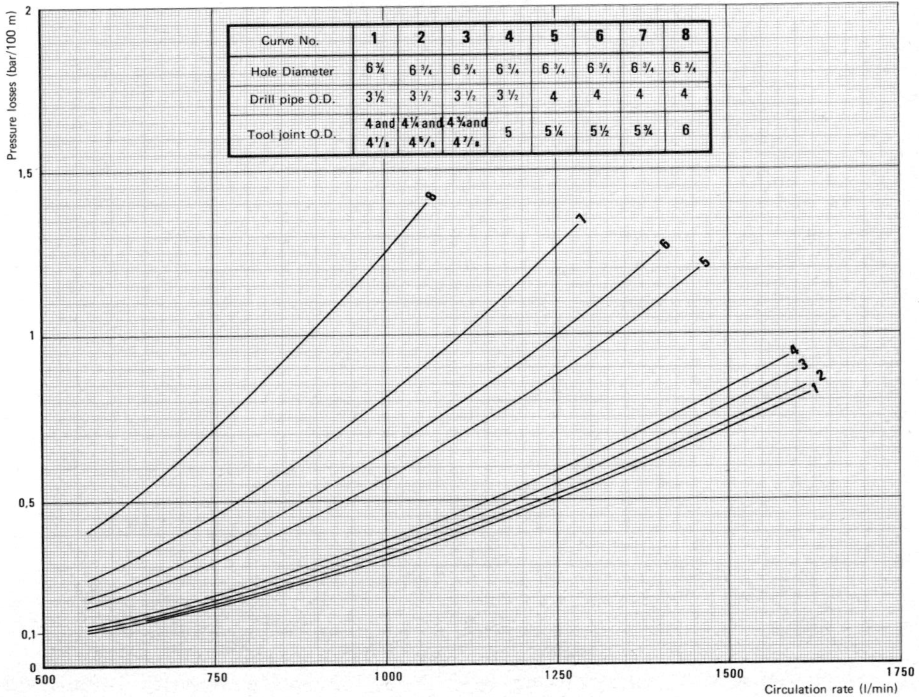
Pressure losses (bar/100 m)
2
1,5
1
0,5
0,1
0
500
750
1000
1250
1500
1750
Circulation rate (l/min)
8
7
6
5
4
3
2
1

PRESSURE LOSSES BETWEEN DRILL PIPE OD AND WALL OF HOLE (SG = 1)
(For 7 7/8" hole size)

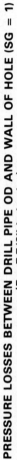

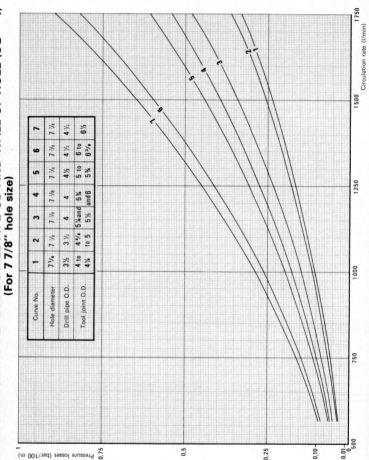

Curve No.	1	2	3	4	5	6	7
Hole diameter	7 7/8	7 7/8	7 7/8	7 7/8	7 7/8	7 7/8	7 7/8
Drill pipe O.D.	3 ½	3 ½	4	4	4 ½	4 ½	4 ½
Tool joint O.D.	4 to 4 ¼	4 ½ to 5	5 ¼ and 5 ½	5 ¾ and 6	5 to 5 ¾	6 to 6 ¾	6 ½

Pressure losses (bar/100 m)

Circulation rate (l/min)

PRESSURE LOSSES BETWEEN DRILL PIPE OD AND WALL OF HOLE (SG = 1)
(For 8 3/8", 8 1/2", 8 5/8" and 8 3/4" hole size)

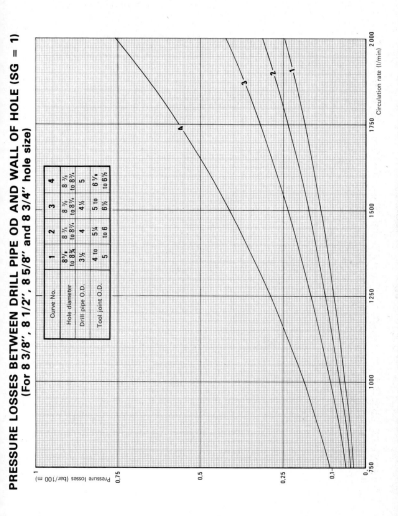

Curve No.	1	2	3	4
Hole diameter	8 ³⁄₈ to 8 ³⁄₄	8 ³⁄₈ to 8 ³⁄₄	8 ³⁄₈ to 8 ³⁄₄	8 ³⁄₈ to 8 ³⁄₄
Drill pipe O.D.	3½	4	4½	5
Tool joint O.D.	4 to 5	5¼ to 6	5 to 6½	6¼ to 6½

PRESSURE LOSSES BETWEEN DRILL PIPE OD AND WALL OF HOLE (SG = 1)
(For 9 7/8" and 12 1/4" hole size)

Curve No.	1	2	3	4	5	6
Hole diameter	12 1/4	12 1/4	12 1/4	9 7/8	9 7/8	9 7/8
Drill pipe O.D.	4 1/2	5	5 1/2	4 1/2	5	5 1/2
Tool joint O.D.	5 to 6 1/2	6 1/8 to 6 1/2	6 to 7 1/4	5 to 6 1/2	6 1/8 to 6 1/2	6 to 7 1/4

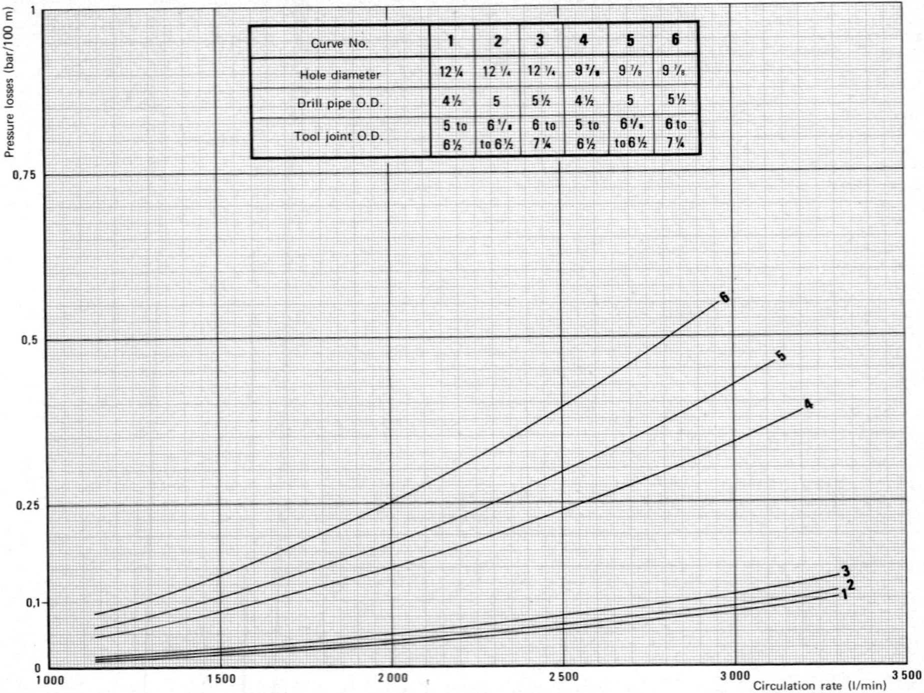

PRESSURE LOSSES BETWEEN DRILL PIPE OD AND WALL OF HOLE (SG = 1)
(For 15″ and 17 1/2″ hole size)

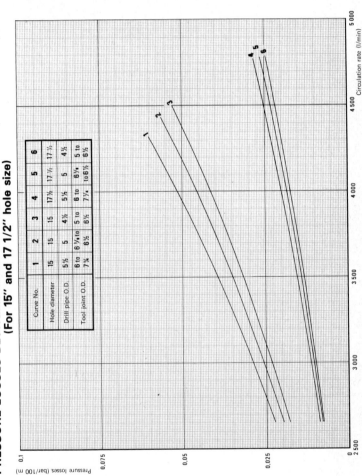

Curve No.	1	2	3	4	5	6
Hole diameter	15	15	15	17½	17½	17½
Drill pipe O.D.	5½	5	4½	5½	5	4½
Tool joint O.D.	6 to 7¼	6¼ to 6½	5 to 6½	6 to 7¼	6¼ to 6½	5 to 6½

Pressure losses (bar/100 m)

Circulation rate (l/min)

VISCOSITY CORRECTION FOR WEIGHTED DRILLING FLUIDS
(Mud weights > 1,7 and light muds with abnormal viscosity)
(This correction should not be applied to pressure losses through the nozzles)

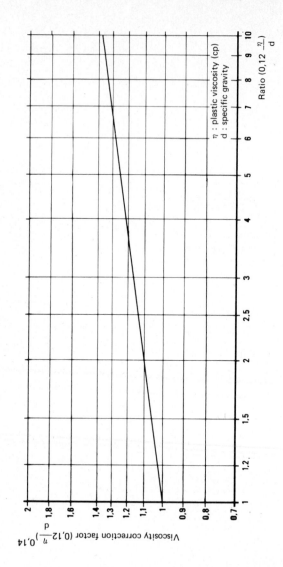

INPUT POWER REQUIRED FOR MUD PUMP

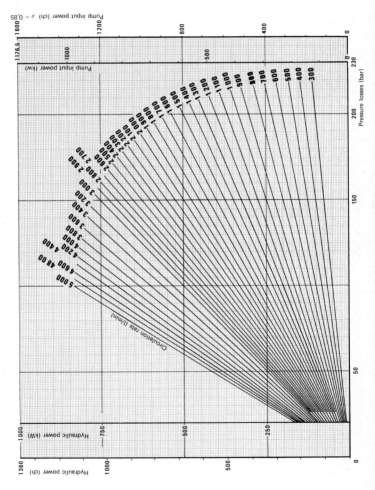

PRESSURE LOSSES FOR WATER THROUGH 1 km OF LINE PIPE

from Tool pusher's Manual

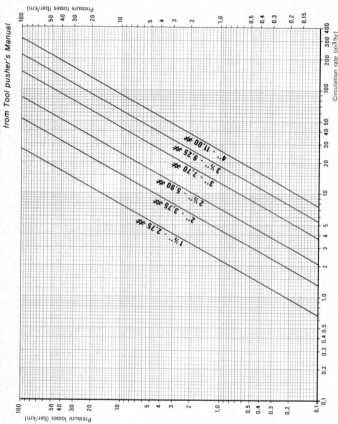

Note. Where inlet and outlet are at different elevations, pressure should be added or subtracted (approximately 1 bar per 10 metres of elevation difference) : added if outlet is at a higher level than inlet, and subtracted if inlet is at a higher level than outlet.

DETERMINING THE REYNOLDS NUMBER
Re between 1000 and 100 000

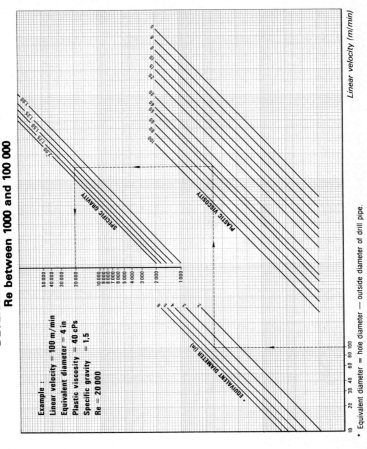

Example :

Linear velocity = 100 m/min
Equivalent diameter = 4 in
Plastic viscosity = 40 cPs
Specific gravity = 1,5
Re = 20 000

SPECIFIC GRAVITY

PLASTIC VISCOSITY

EQUIVALENT DIAMETER (in)

Linear velocity (m/min)

* Equivalent diameter = hole diameter — outside diameter of drill pipe.

EQUIVALENT LENGTHS FOR SPECIAL LINE CONNECTIONS (in metres)

Values given in the table are approximate averages.
However the table gives a quick way to find pressure losses through low pressure circuit.
When there is a change in pipe diameter, calculation is always done in relation to the smallest diameter (d).

Left group

Fitting	3"	4"	6"	8"	10"
mudtank outlet	0.8	1.1	1.8	2.5	3.3
suction by a pipe sunk in a tank	1.9	2.7	4.5	6.4	8.3
mud tank inlet	0.4	0.54	0.9	1.3	1.7
	3.4	4.8	8.1	11.5	14.8
$\phi/D = 0.5$ A-10°	0.8	1.1	1.8	2.5	3.3
$\phi/D = 0.5$ A-20°	1.9	2.7	4.5	6.4	8.3
	3.8	5.4	9.0	12.7	16.5
elbows R > 5D	1.2	1.6	2.7	3.8	5.0
R - 2D	1.5	2.1	3.6	5.1	6.6
R - D	2.7	3.8	6.3	8.9	11.6
	4.6	2.7	10.8	15.2	19.8
A - 22°	0.6	0.8	1.4	1.9	2.5
A - 45°	1.5	2.1	3.6	5.1	6.6
A - 60°	2.3	3.2	5.4	7.6	10.0

Right group

Fitting	3"	4"	6"	8"	10"
TES	1.0	1.3	2.2	3.2	4.1
	5.4	7.5	12.6	17.8	23.0
	4.6	6.4	10.8	15.2	19.8
sudden change in diameter (narrowing) $d/D = .08$	0.6	0.8	1.3	1.9	2.5
$d/D = .06$	1.2	1.6	2.7	3.8	5.0
$d/D = .04$	1.5	2.1	3.6	5.1	6.6
$d/D = .02$	1.7	2.4	4.0	5.7	7.4
sudden change in diameter (widening) $d/D = .08$	0.6	0.8	1.4	1.9	2.5
$d/D = .06$	1.5	2.1	3.6	5.1	6.6
$d/D = .04$	2.7	3.8	6.3	8.9	11.5
$d/D = .02$	3.6	5.1	8.5	12.1	15.7
divergent A-10° $d/D = .06$	0.4	0.6	0.9	1.3	1.7
A-10° $d/D = .04$	0.6	0.8	1.4	1.9	2.5
A-20° $d/D = .06$	0.8	1.1	1.8	2.5	3.3
A-20° $d/D = .04$	1.3	1.7	3.2	4.4	5.8
convergent A - 20°	0.5	0.7	1.2	1.6	2.1
A - 30°	0.6	0.9	1.5	2.0	2.6
A - 45°	1.2	1.7	2.9	4.0	5.3
butterfly valve	0.5	0.7	1.1	1.5	2.0
direct flow valve	1.2	1.6	2.7	3.8	5.0

I CEMENTING

Summary

GENERAL DATA
UNITS COMMONLY USED

1 sack of cement (USA) =	$\begin{cases} 94 \text{ lb} \\ 42,64 \text{ kg} \end{cases}$
1 sack of cement (France) =	$\begin{cases} 49,5 \text{ kg} \\ 109 \text{ lb} \end{cases}$
Volume in one sack of 94 lb cement =	$\begin{cases} 1 \text{ cu ft} \\ 28,32 \text{ l} \end{cases}$
Useful consequence (an x cu ft silo holds x sacks)	
Volume in one sack of 50 kg cement =	$\begin{cases} 32,89 \text{ l} \\ 1,16 \text{ cu ft} \end{cases}$
True specific gravity of powdered cement =	3,15
Apparent specific gravity of powdered cement =	1,5
True volume occupied by 1 kg of powdered cement =	0.3175 l

$$\text{Slurry density} = \frac{\text{Mass of cement} + \text{Mass of water}}{\text{Volume of cement} + \text{Volume of water}} = \frac{\text{Total mass}}{\text{Volume of slurry}}$$

Example. 100 kg of cement + 46 l of water give :

$$\text{slurry density} = \frac{100 + 46}{\dfrac{100}{3,15} + 46} = \frac{146}{31,75 + 46} = \frac{146}{77,75} = 1,88 \text{ kg/l.}$$

and slurry volume = 77,75 l.

Densities :

1 kg/l ou dm³	=	8,345 lb/gal
1 kg/l ou dm³	=	62,428 lb/cu ft
1 lb/gal	=	0,1198 kg/dm³
1 lb/gal	=	7,48 lb/cu ft
1 lb/cu ft	=	0,01602 kg/dm³
1 lb/cu ft	=	0,1337 lb/gal

Pressure head in psi/ft = density in kg/l × 0,4333

1 short ton =	$\begin{cases} 2000 \text{ lb} \\ 907 \text{ kg} \\ 21,28 \text{ 94 lb sacks of cement} \\ 18,33 \text{ 50 kg sacks of cement} \end{cases}$
1 metric tonne =	$\begin{cases} 2205 \text{ lb} \\ 23,45 \text{ 94 lb sacks of cement} \\ 1,10 \text{ short ton} \\ 0,984 \text{ long ton} \end{cases}$
1 long ton =	$\begin{cases} 2240 \text{ lb} \\ 1016 \text{ kg} \\ 23,83 \text{ 94 lb sacks of cement} \\ 20,53 \text{ 50 kg sacks of cement} \end{cases}$
1 cu ft	$\begin{cases} 7,48 \text{ gal (US)} \\ 28,32 \text{ l} \end{cases}$
1 gallon (US) =	3,785 l
1 lb/94 lb sack =	1,06 kg/100 kg

CORRELATION BETWEEN SACKS AND TONS OF CEMENT

94 lb sacks

Sacks	Metric tonnes 2205 lb 1000 kg	Short tons 2000 lb 907 kg	Long tons 2240 lb 1016 kg
100	4,26	4,70	4,20
120	5,11	5,64	5,04
140	5,96	6,58	5,88
160	6,82	7,52	6,72
180	7,67	8,46	7,56
200	8,53	9,40	8,39
220	9,37	10,34	9,24
240	10,22	11,28	10,08
260	11,07	12,22	10,92
280	11,93	13,16	11,76
300	12,79	14,10	12,59
320	13,63	15,04	13,44
340	14,48	15,98	14,28
360	15,34	16,92	15,12
380	16,19	17,86	15,96
400	17,05	18,80	16,79
420	17,89	19,74	17,64
440	18,74	20,68	18,48
460	19,60	21,62	19,32
480	20,45	22,56	20,16
500	21,32	23,50	20,98
520	22,15	24,44	21,84
540	23,00	25,38	22,68
560	23,86	26,32	23,52
580	24,71	27,26	24,36
600	25,58	28,21	25,18
620	26,41	29,14	26,04
640	27,26	30,08	26,88
660	28,12	31,02	27,72
680	28,97	31,96	28,56
700	29,85	32,90	29,38
720	30,67	33,84	30,24
740	31,52	34,78	31,08
760	32,38	35,72	31,92
780	33,23	36,66	32,76
800	34,11	37,61	33,57
820	34,93	38,54	34,44
840	35,78	39,48	35,28
860	36,64	40,42	36,12
880	37,49	41,36	36,96
900	38,37	42,31	37,77
920	39,19	43,24	38,64
940	40,04	44,18	39,48
960	40,90	45,12	40,32
980	41,75	46,06	41,16

50 kg sacks

Sacks	Metric tonnes 2205 lb 1000 kg	Short tons 2000 lb 907 kg	Long tons 2240 lb 1016 kg
100	4,95	5,45	4,87
120	5,94	6,53	5,84
140	6,93	7,62	6,82
160	7,92	8,71	7,79
180	8,91	9,80	8,76
200	9,90	10,89	9,74
220	10,89	11,98	10,72
240	11,88	13,07	11,69
260	12,87	14,16	12,66
280	13,86	15,25	13,64
300	14,85	16,34	14,62
320	15,84	17,43	15,59
340	16,83	18,52	16,56
360	17,82	19,61	17,54
380	18,81	20,69	18,52
400	19,80	21,78	19,48
420	20,79	22,87	20,46
440	21,78	23,96	21,44
460	22,77	25,05	22,42
480	23,76	26,14	23,38
500	24,75	27,23	24,36
520	25,74	28,32	25,34
540	26,73	29,41	26,30
560	27,72	30,49	27,28
580	28,71	31,58	28,25
600	29,70	32,67	29,22
620	30,69	33,76	30,20
640	31,68	34,85	31,17
660	32,67	35,94	32,15
680	33,66	37,03	33,12
700	34,65	38,12	34,10
720	35,64	39,21	35,07
740	36,63	40,29	36,05
760	37,62	41,38	37,02
780	38,61	42,47	37,99
800	39,60	43,56	38,97
820	40,59	44,65	39,95
840	41,58	45,74	40,91
860	42,57	46,83	41,89
880	43,56	47,92	42,86
900	44,55	49,01	43,84
920	45,54	50,09	44,82
940	46,53	51,18	45,79
960	47,52	52,27	46,76
980	48,51	53,36	47,73

API SPECIFICATIONS FOR OIL-WELL CEMENTS
(API - Spec 10 A and RP 10B)

CLASS	Maximum depth		Minimum thickening time (min)	Minimum compressive strength						% by weight of cement	Mixing water			
				Curing temperature and pressure		After 8 h curing time		After 24 h curing time			per 94 lb sack		per 50 kg sack	
	m	ft		T (°C)	P (bar)	bar	psi	bar	psi		litres	gal.	litres	gal.
A	0-1800	0-6000	at 310 m: 90 at 1800 m: 90	38	atm	18	260	127	1800	46	19,6	5,19	23	6,07
B	0-1800	0-6000	at 310 m: 90 at 1800 m: 90	38	atm	14	200	106	1500	46	19,6	5,19	23	6,07
C	0-1800	0-6000	at 310 m: 90 at 1800 m: 90	38	atm	21	300	141	2000	56	23,9	6,32	28	7,39
D	1800-3000	6000-10000	at 1800 m: 90 at 3000 m: 100	77 110	211 211	— 35	— 500	70 141	1000 2000	38	16,2	4,29	19	5,02
E	3000-4300	10000-14000	at 3000 m: 100 at 4300 m: 154	77 143	211 211	— 35	— 500	70 141	1000 2000	38	16,2	4,29	19	5,02
F	3000-4900	10000-16000	at 3000 m: 100 at 4900 m: 190	110 160	211 211	— 35	— 500	70 70	1000 1000	38	16,2	4,29	19	5,02
G	0-2450	0-8000	at 2450 m: 90	35 60	56 211	21 106	300 1500	— —	— —	44	18,8	4,97	22	5,81
H	0-2450	0-8000	at 2450 m: 90	35 60	56 211	21 106	300 1500	— —	— —	38	16,2	4,29	19	5,02
J (1)	3660-4900	12000-16000	at 3050 m: 180 at 4880 m: 180	143 177	211 211	35 —	500 —	— 70	— 1000	(2)	(2)	(2)	(2)	(2)

Note 1. The addition of bentonite to cement requires that the amount of water be increased. It is recommended, for testing purposes, that 5.3 % water be added for each 1 % bentonite in all API classes of cement. For example for a class A cement slurry having a water-cement ratio of 46 %, to which is added 4 % bentonite, will require an increase in water-cement ratio to 67.2 % (46 + 4 × 5.3 = 46 + 21.2 = 67.2).

Note 2. The addition of barite to cement requires that the amount of water be increased. It is recommended, for testing purposes, that 0.2 % water be added for each 1 % barite. For example, for cement slurry having a water-cement ratio of 38 %, to which is added 60 % barite, will require an increase in water-cement ratio to 50 % (38 + 60 × 0.2 = 38 + 12 = 50 %).

(1) API tentative. (2) As recommended by the manufacturer.

OIL-WELL CEMENT CLASSES AND TYPES

CLASS	TYPE
A	Intended for use from surface to 1 800 m (6 000 ft) depth when special properties are not required. Available only in ordinary (O) (1).
B	Intended for use from surface to 1 800 m (6 000 ft) depth, when conditions require moderate to high sulfate-resistance. Available in both moderate (MSR) (1) and high sulfate resistance type (HSR) (1).
C	Intended for use from surface to 1 800 m (6 000 ft) depth, when conditions require high early strength. Available in (o), (MSR) and (HSR) types.
D	Intended for use from 1 800 m (6 000 ft) to 3 000 m (10 000 ft) depth under conditions of moderately high temperatures and pressures. Available in both (MSR) an (HSR) types.
E	Intended for use from 3 000 m (10 000 ft) to 4 300 m (14 000 ft) depth under conditions of high temperatures and pressures. Available in both (MSR) and (HSR) types.
F	Intended for use from 3 000 m (10 000 ft) to 4 300 m (14 000 ft) depth under conditions of extremely high temperatures and pressures. Available in both (MSR) and (HSR) types.
G	Intended for use as basic cement from surface to 2 450 m (8 000 ft) depth as manufactured or can be used with accelerators and retarders to cover a wide range of well depths and temperatures. Available in both (MSR) and (HSR) types.
H	Intended for use as basic cement from surface to 2 450 m (8 000 ft) depth as manufactured, and can be used with accelarators and retarders to cover a wide range of well depths and temperatures. Available only in (MSR) type.
J	Intended for use as manufactured from 3 660 m (12 000 ft) to 4 900m (16 000 ft) depth under conditions of extremely high temperatures and pressures or can be used with accelerators and retarders to cover a range of well depths and temperatures.

(1) (O) Ordinary type.
   (MSR) Moderate Sulfate - Resistant type,.
   (HSR) High Sulfate - resistant type.
Note. For all information concerning chemical requirements for all the classes of cement see API Spec 10 A.

CEMENT SLURRY (soft water)

Slurry density		Volume of water								Volume of slurry					Hydrostatic head (psi/ft)
kg/l	lb/gal	l/100 kg	lb/cu ft	gal/sack 94 lb	cu ft/sack 94 lb	l/sack 94 lb	l/sh tn	gal/sh tn	Cu ft/ sh tn	l/100 kg	l/sack 94 lb	cu ft/sack 94 lb	l/sh tn	Cu ft/ sh tn	
1.74	14.5	60.6	108.1	6.83	0.92	25.9	550	145.3	19.58	92.4	39.4	1.39	838	29.58	0.754
1.75	14.6	59.4	109.2	6.69	0.90	25.3	539	142.4	19.15	91.0	38.8	1.37	826	29.15	0.759
1.76	14.7	58.2	110.0	6.56	0.88	24.8	528	139.6	18.73	89.6	38.2	1.35	813	28.73	0.763
1.77	14.8	57.0	110.7	6.42	0.86	24.3	517	136.6	18.30	88.9	37.9	1.34	807	28.52	0.767
1.78	14.85	55.9	111.1	6.29	0.84	23.8	507	133.9	17.88	87.7	37.4	1.32	796	28.09	0.772
1.79	14.9	54.8	111.5	6.17	0.83	23.4	497	131.3	17.66	86.3	36.8	1.30	783	27.66	0.776
1.80	15.0	53.7	112.2	6.05	0.81	22.9	487	128.7	17.24	85.6	36.5	1.29	777	27.45	0.780
1.81	15.1	52.6	113.0	5.92	0.79	22.4	477	126.0	16.81	84.4	36.0	1.27	766	27.03	0.785
1.82	15.2	51.6	113.7	5.81	0.78	22.0	468	123.6	16.60	83.0	35.4	1.25	753	26.60	0.789
1.83	15.3	50.6	114.5	5.70	0.76	21.6	459	121.3	16.17	82.3	35.1	1.24	747	26.39	0.794
1.84	15.35	49.7	114.9	5.59	0.75	21.2	451	119.0	15.96	81.1	34.6	1.22	736	25.96	0.798
1.85	15.4	48.7	115.2	5.49	0.74	20.8	442	116.8	15.74	80.4	34.3	1.21	730	25.75	0.802
1.86	15.5	47.7	116.0	5.37	0.72	20.3	433	114.3	15.32	79.7	34.0	1.20	724	25.54	0.807
1.87	15.6	46.8	116.7	5.27	0.71	19.9	424	112.1	15.11	78.3	33.4	1.18	711	25.11	0.811
1.88	15.7	45.9	117.5	5.17	0.69	19.6	416	110.0	14.68	77.6	33.1	1.17	704	24.90	0.815
1.89	15.8	45.0 (1)	118.2	5.07	0.68	19.2	408	107.9	14.47	76.5	32.6	1.15	694	24.47	0.820
1.90	15.85	44.2	118.6	4.98	0.67	18.8	401	106.0	14.26	75.8	32.3	1.14	687	24.26	0.824
1.91	15.9	43.4	118.9	4.89	0.66	18.5	394	104.1	14.04	75.0	32.0	1.13	681	24.05	0.828
1.92	16.0	42.5	119.7	4.79	0.64	18.1	385	101.9	13.62	74.3	31.7	1.12	675	23.83	0.832
1.93	16.1	41.7	120.4	4.70	0.63	17.8	378	100.0	13.41	73.6	31.4	1.11	668	23.62	0.837
1.94	16.2	41.0	121.2	4.62	0.62	17.5	372	98.3	13.19	72.5	30.9	1.09	658	23.20	0.841
1.95	16.3	40.2	121.9	4.53	0.61	17.1	365	96.4	12.98	71.8	30.6	1.08	651	22.98	0.845
1.97	16.4	38.7	122.7	4.36	0.58	16.5	351	92.8	12.34	70.4	30.0	1.06	638	22.56	0.854
2.00	16.7	36.3	124.9	4.09	0.55	15.5	329	87.0	11.70	68.5	29.2	1.03	621	21.92	0.867
2.02	16.85	35.6	126.0	3.96	0.53	15.0	323	84.3	11.28	67.1	28.6	1.01	609	21.49	0.876
2.04	17.0	34.0	127.2	3.83	0.51	14.5	308	81.5	10.85	65.7	28.0	0.99	596	21.07	0.884
2.10	17.5	30.4	130.9	3.42	0.46	12.8	276	72.8	9.79	62.4	26.6	0.94	566	20.00	0.910
2.16	18.0	27.2	134.7	3.06	0.41	11.6	247	65.1	8.72	59.1	25.2	0.89	536	18.94	0.936

Note. When using sea water instead of soft water the density of the slurry will be increased by one point. Example : a cement slurry having a water-cement ratio of 45 l/100 kg, with sea water the density will be 1.90 kg/l instead of 1.89 kg/l as shown in the table (see (1) in the table).

CEMENT SLURRY (salt saturated water)
(315 g/l d = 1,20)

Slurry density		lb/Cu ft	Volume of salt saturated water				Volume of slurry			Hydrostatic head (psi/ft)
kg/l	lb/gal	lb/Cu ft	l/100 kg	l/sack 94 lb	gal/sack 94 lb	Cu ft/sack 94 lb	l/100 kg	l/sack 94 lb	Cu ft/sack 94 lb	
1.80	15.0	112.2	71.3	30.4	8.03	1.07	103.0	43.9	1.55	0.780
1.81	15.1	113.0	69.7	29.7	7.85	1.05	101.5	43.3	1.53	0.785
1.82	15.2	113.7	68.0	29.0	7.66	1.02	100.0	42.6	1.50	0.789
1.83	15.3	114.5	66.5	28.4	7.50	1.00	98.4	42.0	1.48	0.794
1.84	15.35	114.9	65.0	27.7	7.32	0.98	96.8	41.3	1.46	0.798
1.85	15.4	115.2	63.4	27.0	7.13	0.95	95.1	40.6	1.43	0.802
1.86	15.5	116.0	62.1	26.5	7.00	0.94	93.8	40.0	1.41	0.807
1.87	15.6	116.7	60.7	25.9	6.84	0.91	92.3	39.4	1.39	0.811
1.88	15.7	117.5	59.3	25.3	6.68	0.89	90.9	38.6	1.36	0.815
1.89	15.8	118.2	58.0	24.7	6.53	0.87	89.5	38.2	1.35	0.820
1.90	15.85	118.6	56.5	24.1	6.37	0.85	88.2	37.6	1.33	0.824
1.91	15.9	118.9	55.3	23.6	6.24	0.83	86.9	37.1	1.31	0.828
1.92	16.0	119.7	53.9	23.0	6.08	0.81	85.6	36.5	1.29	0.832
1.93	16.1	120.4	52.9	22.6	5.97	0.80	84.4	36.0	1.27	0.837
1.94	16.2	121.2	51.7	22.0	5.81	0.78	83.3	35.5	1.25	0.841
1.95	16.3	121.9	50.7	21.6	5.71	0.76	82.4	35.1	1.24	0.845
1.97	16.4	122.7	48.6	20.7	5.47	0.73	81.2	34.6	1.22	0.854
2.00	16.7	124.9	45.6	19.4	5.13	0.69	77.3	33.0	1.17	0.867
2.02	16.85	126.0	43.7	18.6	4.91	0.66	75.4	32.2	1.14	0.876
2.04	17.0	127.2	41.9	17.9	4.73	0.63	73.6	31.4	1.11	0.884
2.06	17.2	128.6	40.2	17.1	4.52	0.60	71.9	30.7	1.08	0.893
2.08	17.4	129.9	38.5	16.4	4.33	0.58	70.2	29.9	1.06	0.901
2.10	17.5	130.2	37.0	15.8	4.17	0.56	68.7	29.3	1.03	0.910
2.12	17.7	132.4	35.4	15.1	3.99	0.53	67.1	28.6	1.01	0.919
2.14	17.85	133.7	34.1	14.5	3.83	0.51	65.7	28.0	0.99	0.927
2.16	18.0	134.9	32.7	13.9	3.67	0.49	64.5	27.5	0.97	0.936

CORRELATION BETWEEN WEIGHT OF DRY CEMENT, VOLUME OF MIXING WATER, VOLUME AND DENSITY OF SLURRY

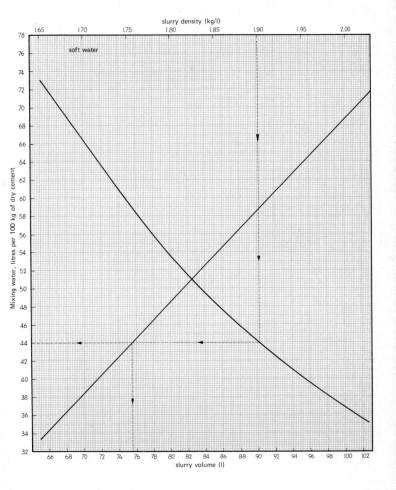

CORRELATION BETWEEN WEIGHT OF DRY CEMENT, VOLUME OF SALT SATURATED WATER, VOLUME AND DENSITY OF SLURRY

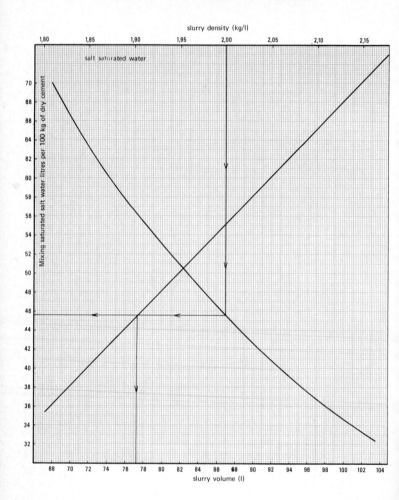

CORRELATION BETWEEN WEIGHT OF DRY CEMENT VOLUME AND DENSITY OF SLURRY

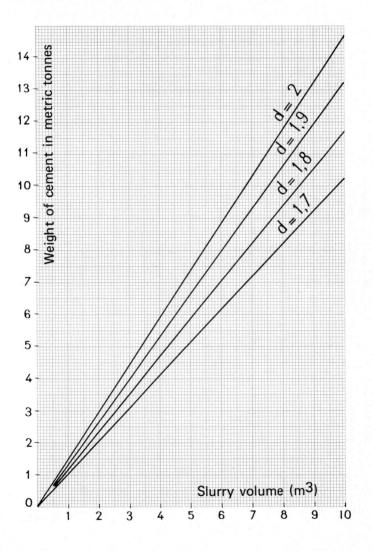

CORRELATION BETWEEN PER CENT OF BENTONITE, VOLUME OF WATER, DENSITY AND VOLUME OF SLURRY

Percent of weight of cement	Density	Mixing water			Volume of slurry		Percent of weight of cement + bentonite
		l/100 kg cement	gal/94 lb sack	cu ft/94 lb sack	l/100 kg cement	cu ft/94 lb sack	
2	1,73	62	6,98	0,94	94,5	1,43	1,96
	1,77	58	6,53	0,88	90,5	1,37	
	1,80	54	6,08	0,82	86,5	1,31	
	1,84	50	5,63	0,76	82,5	1,25	
	1,88	46	5,18	0,69	78,5	1,19	
4	1,63	78	8,79	1,18	111,5	1,68	3,85
	1,66	74	8,34	1,12	107,5	1,62	
	1,68	70	7,89	1,06	103,5	1,56	
	1,71	66	7,44	1,00	99,5	1,50	
	1,74	62	6,98	0,94	95,5	1,44	
6	1,59	88	9,91	1,33	122	1,84	5,66
	1,61	84	9,46	1,27	118	1,78	
	1,63	80	9,01	1,21	114	1,72	
	1,65	76	8,56	1,15	110	1,66	
	1,68	72	8,11	1,09	106	1,60	
8 (1)	1,53	104	11,72	1,57	139	2,10	7,41
	1,54	100	11,27	1,51	135	2,04	
	1,56	96	10,81	1,45	131	1,98	
	1,57	92	10,36	1,39	127	1,92	
	1,59	88	9,91	1,33	123	1,86	
10	1,52	108	12,17	1,63	144	2,17	9,09
	1,53	104	11,72	1,57	140	2,11	
	1,55	100	11,27	1,51	136	2,05	
12	1,50	114	12,84	1,72	150,5	2,27	10,71
	1,51	110	12,39	1,66	146,5	2,21	
	1,53	106	11,94	1,60	142,5	2,15	
14	1,46	128	14,42	1,93	165,5	2,50	12,28
	1,48	122	13,74	1,84	159,5	2,41	
	1,50	116	13,07	1,75	153,5	2,32	
16	1,44	140	15,77	2,11	178	2,69	13,79
	1,46	132	14,87	1,99	170	2,57	
	1,47	126	14,19	1,90	164	2,48	
18	1,40	160	18,02	2,42	199	3,00	15,25
	1,42	150	16,90	2,27	189	2,85	
	1,44	140	15,77	2,11	179	2,70	
20	1,38	170	19,15	2,57	210	3,17	16,67
	1,40	160	18,02	2,42	200	3,02	
	1,42	150	16,90	2,27	190	2,87	
25	1,36	190	21,40	2,87	232	3,50	20,00
	1,38	180	20,28	2,72	222	3,35	
	1,39	170	19,15	2,57	212	3,20	

(1) For 8 % and higher it is advisable to add a thinner.
Note. All calculations for 3,15 specific gravity of cement and 2,5 specific gravity of bentonite.

CEMENT-BENTONITE MIXTURES

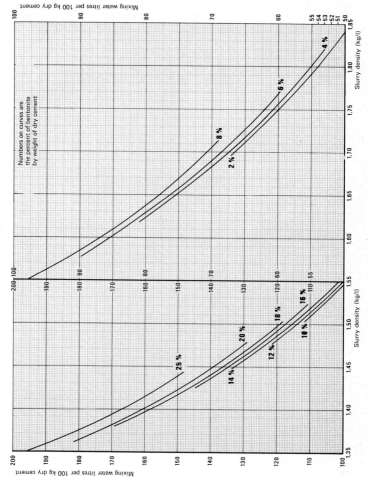

Mixing water litres per 100 kg dry cement

Numbers on curves are
the percent of bentonite
by weight of dry cement

Slurry density (kg/l)

Slurry density (kg/l)

Mixing water litres per 100 kg dry cement

DURATION OF CEMENT JOB D$_c$

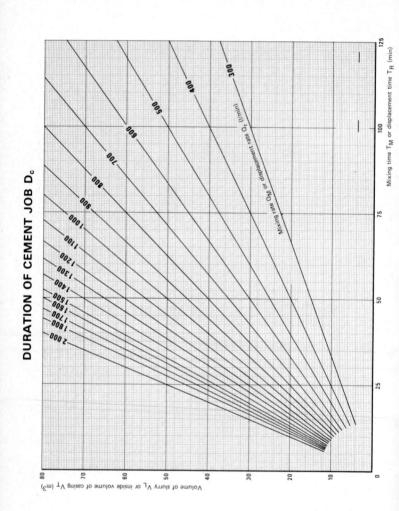

EFFECTS OF SOME ADDITIVES ON CEMENT PROPERTIES

		Bentonite	Perlite	Diatomaceous earth	Pozzolan	Sand	Barite	Hematite	Calcium chloride	Sodium chloride	Lignosulfonate	CMHEC (1)	Diesel oil	Water loss additive	Lost circulation material
Density	decreased	●	●	●	●										
	increased					●	●	●	x	x	x				
Water required	less										●				
	more	●	x	●	x	x	x	x							x
Viscosity	decreased								x		●				
	increased	x	x	x	x	x	x	x							
Thickening time	accelerated	x					x	x	●	●					
	retarded			x							x	●	●	x	x
Setting time	accelerated						x	x	●	●					
	retarded	x	x	x	x						●	●		x	
Early strength	decreased	x	x	x	x		x	x			●	●		x	x
	increased								●	●					
Final strength	decreased	x	x	●	x		x					x		x	x
	increased														
Duration	decreased	x	x	x									x		x
	increased				●										
Water loss	decreased	●									x	●	x	●	x
	increased		x	x											

x Denotes minor effect,
● Denotes major effect,
(1) Caboxymethyl hydroxyethyl cellulose.

DIFFERENTIAL HYDROSTATIC PRESSURE
between a column of cement and a column of mud of same height

$$\Delta P = 0,981 \frac{H}{10} (D - D_1)$$

where : D = slurry density (kg/l),
$\quad\quad$ D_1 = mud density (kg/l),
$\quad\quad$ H = height of the column (m),
$\quad\quad$ ΔP = differential pressure (bar).

Plug balance formula

The following formula is used to find the height of cement for plug balancing purposes

$$h = \frac{N}{a+b}$$

where : h = height of balanced cement column (m),
$\quad\quad$ N = volume of cement slurry used (l),
$\quad\quad$ a = capacity of casing, tubing or drill pipe (l/m),
$\quad\quad$ b = annular volume (l/m).

QUALITATIVE INFLUENCE OF VARIOUS OPERATIONS
with the stresses of a partially cemented casing string

OPERATIONS	Tension	Collapse	Bursting	Buckling tendancy
Decrease of average temperature	increases			decreases
Increase of average temperature	decreases			increases
Increase of internal pressure	increases		increases	increases
Decrease of internal pressure	decreases		decreases	decreases
Increase of external pressure	decreases	increases		decreases
Decrease of external pressure	increases	decreases		increases
Substitution of internal fluid by a heavier fluid	increases		increases	increases
Substitution of internal fluid by a lighter fluid	decreases		decreases	decreases
Substitution of external fluid by a heavier fluid	decreases	increases		decreases
Substitution of external fluid by a lighter fluid	increases	decreases		increases
Swabbing	decreases		decreases	decreases

QUANTITATIVE INFLUENCE OF TEMPERATURE
AND PRESSURE VARIATIONS
with the stresses of a partially cemented
casing string

Influence of temperature changes

The increase or decrease of the tension at the top of a casing string due to a decrease or increase of the average temperature is given by :

$$T = 25,5 \, S \, \Delta t \quad \text{or} \quad T = 32,7 \, W \, \Delta t$$

where :

T = tension variation (daN),
S = cross-section of casing (cm²),
W = linear weight of casing (kg/m),
Δt = average temperature variation of casing (°C),

The average temperature of the free part of the casing is given by the formula :

$$t = t_0 + \frac{(t_1 - t_0) \, L_2}{2 \, L_1}$$

where :

t = average temperature of the free part of the casing (°C),
t_0 = surface temperature (°C),
t_1 = bottom hole temperature (°C),
L_1 = depth of hole (m),
L_2 = depth of top of cement (m).

Influence of internal pressure changes

The increase or decrease of the tension at the top of a casing string due to an increase or decrease of the internal pressure is given by the formula :

$$T = 0,6 A_i \, \Delta p_i$$

where :

T = tension variation (daN),
A_i = internal section area of casing (cm²),
Δp_i = variation of internal average pressure (bar).

Influence of external pressure changes

The increase or decrease of the tension at the top of a casing string due to a decrease or increase of the external pressure is given by the formula :

$$T = 0,6 A_e \, \Delta p_e$$

where :

T = tension variation (daN),
A_e = external section area of casing (cm²),
Δp_e = variation of external pressure (bar).

QUANTITATIVE INFLUENCE OF TEMPERATURE
AND PRESSURE VARIATIONS
with the stresses of a partially cemented casing string (continued)

If internal or external average pressure change is due to a change in the mud weight, the average pressure change is given by the formula :

$$\Delta p = 0,1 \, (d_2 - d_1) \, \frac{L_2}{2}$$

where :

d_1 = initial mud weight,
d_2 = new mud weight,
L_2 = depth of the top of cement (m).

Critical force of buckling

The critical force of buckling is given by the formula :

$$F_c = P_e \, A_e - P_i \, A_i$$

where :

F_c = critical force of buckling (daN),
P_e = annulus pressure at top of cement level (bar),
P_i = internal pressure at top of cement level (bar),
A_e, A_i = external and internal section area of casing (cm²).

If F_c is positive the casing string will support a compressive strain at the top of cement level equal to F_c without buckling.

If F_c is negative the casing string will buckle for a tensile strength lower than F_c.

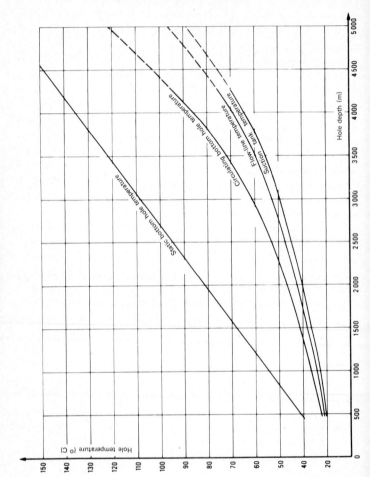

HOLE TEMPERATURES (by Gulf)

J DIRECTIONAL DRILLING

Summary

DIRECTIONAL DRILLING

I. REFERENCE COORDINATES

The azimuths measured in holes refer either :
— to geographic North,
— or to magnetic North,
— or to Lambert or UTM North.

Definitions :

Declination : angle between geographic and magnetic North.

Convergence : angle between geographic and Lambert North.

The values of declination and convergence are usually given on ordnance maps.

The geographic azimuth is equal to the magnetic azimuth read on the survey instruments (single shot as an example) plus or minus the declination following two criteria :
— declination East or West,
— quarter in which is located the magnetic azimuth measured.

The following figure shows us if we must add (+ sign) or subtract (— sign) the declination to the magnetic azimuth to obtain the geographic azimuth.

Example. Declination 8° West,
Magnetic azimuth N30°E,
Geographic azimuth is : from the above figure we are in the shaded quarter,

from which : geographic azimuth = N (30 — 8)°E = N22°E

The Lambert azimuth is equal to the geographical azimuth found as above plus or minus the convergence following two criteria :
— convergence East or West,
— quarter in which is located the geographic azimuth found.

The following figure shows us if we must add (+ sign) or substract (— sign) the convergence to the geographic azimuth to obtain Lambert azimuth.

Example. Convergence 2° East,
Geographic azimuth N22°E,
Lambert azimuth : from the above figure we are in the shaded quarter - from which :

Lambert azimuth = N (22 — 2)°E = N20°E

DIRECTIONAL DRILLING
(continued)

II. VERTICAL PROFILE

The figures on the following pages serve to find rapidly the vertical profile of the hole when we know :

— the displacement which is the horizontal distance between a later position and the original position,

— the kickoff point (KOP),

— the rate of build-up.

Example : Displacement 1200 m.

    KOP at 350 m,

    Vertical depth to be reached = 2750 m,

    Rate of build-up 1°/10 m.

We plot on the figure (page J 4) :

— the displacement : 1200 m,

— the vertical depth from the KOP : 2750 — 350 = 2400. We find the maximum inclination angle (or drift angle) : 28°.

III. RAGLAND DIAGRAM

The Ragland diagram serves to determine the parameters entering into the calculation of the orientation of the deflecting tool.

It is a diagram which includes four lines representing the characteristics of the hole and the deflecting tools.

These four lines are :

(1) Original direction and inclination of a part of the hole.

Example : 5° and N25E

(2) Dog-legs circle, *i.e.* the variation of inclination we are looking for.

Example : 1 1/2° dog-leg

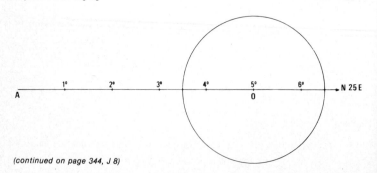

(continued on page 344, J 8)

DIRECTIONAL DRILLING
(continued)

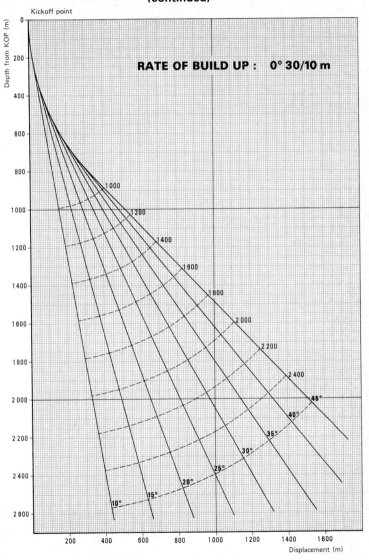

DIRECTIONAL DRILLING
(continued)

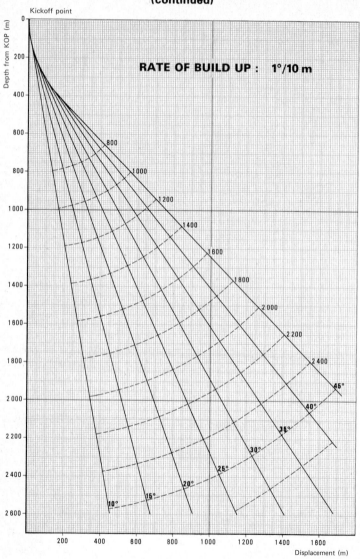

RATE OF BUILD UP : 1°/10 m

DIRECTIONAL DRILLING
(continued)

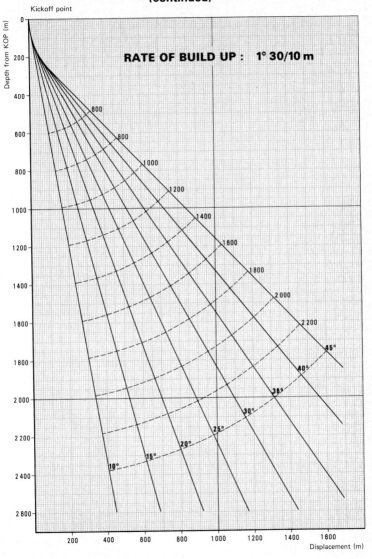

Kickoff point

RATE OF BUILD UP : 1° 30/10 m

Depth from KOP (m)

Displacement (m)

DIRECTIONAL DRILLING
(continued)

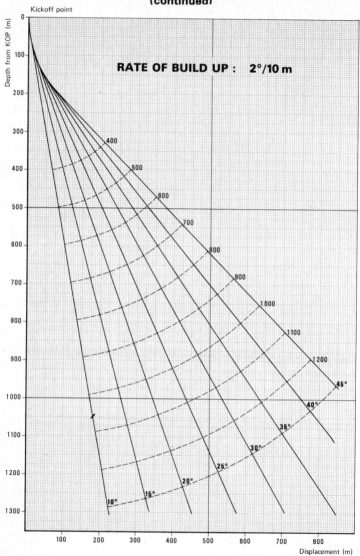

Kickoff point

RATE OF BUILD UP : 2°/10 m

Depth from KOP (m)

Displacement (m)

DIRECTIONAL DRILLING
(continued)

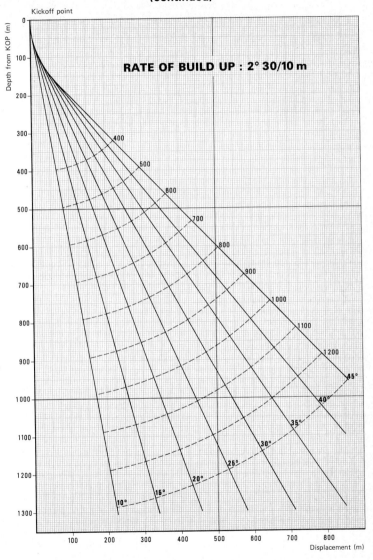

Kickoff point

RATE OF BUILD UP : 2° 30/10 m

DIRECTIONAL DRILLING
(continued)

(3) Orientation of the deflecting tool with respect to the original direction taking into account the roll-off.

Example. Deflecting tool 45° to the right of the original direction, *i.e.* N70°E.

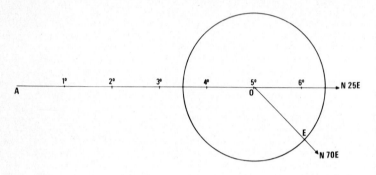

(4) Direction obtained with respect to the original direction.

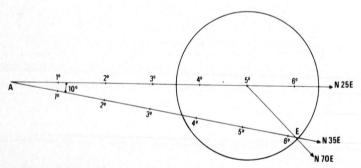

The angle between the two directions is 10°.
  The new direction is N35E.
  Point E is common to three lines :
  — orientation of the deflecting tool,
  — dog-leg circle,
  — direction obtained.
  It is necessary to know only two of them to obtain the third one.

DIRECTIONAL DRILLING
(continued)

IV. CONTROL OF THE SHAPE OF THE HOLE

Calculation of the elements

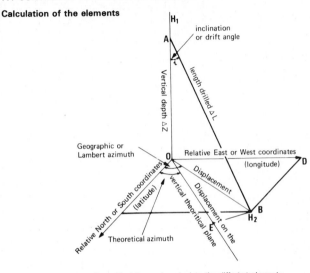

The table below gives the method to use to calculate the different elements.

Element	Basic data	Reference angle	Std field table		Trigonometric table	
			Latitude	Displacement	Sine	Cosine
Vertical depth ΔZ	Length drilled between two consecutive surveys ΔL	Drift angle : i	+ (ΔL – i)			+ – (ΔL cos i)
Horizontal displacement D	Length drilled between two consecutive surveys ΔL	Drift angle : i		+ (ΔL – i)	+ (ΔLsin i)	
Relative North or South coordinates	Horizontal displacement D	Geographic azimuth Ag	+ (D – Ag)			+ (D cos Ag)
Relative East or West coordinates	Horizontal displacement D	Geographic azimuth Ag		+ (D – Ag)	+ (D sin Ag)	
Projection on theoretical plane	Horizontal displacement D	Angular difference between geographic azimuths : survey and target (Ag — Agth)	D- (Ag – Agth)			+ D cos (Ag – Agth)

DIRECTIONAL DRILLING
(continued)

V. DISPLACEMENT AND VERTICAL DEPTH AS A FUNCTION OF DRIFT ANGLE FOR A FIXED DEPTH DRILLED OF 100 m

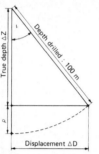

	ΔD (m)	ΔZ (m)	p (m)	
8°30'	14,78	98,90	1,10	81°30'
9°	15,64	98,77	1,23	81°
9°30'	16,50	98,63	1,37	80°30'
10°	17,36	98,48	1,52	80°
10°30'	18,22	98,32	1,68	79°30'
11°	19,08	98,16	1,84	79°
11°30'	19,94	97,99	2,01	78°30'
12°	20,79	97,81	2,19	78°
12°30'	21,64	97,63	2,37	77°30'
13°	22,49	97,44	2,56	77°
13°30'	23,34	97,24	2,76	76°30'
14°	24,19	97,03	2,97	76°
14°30'	25,04	96,81	3,19	75°30'
15°	25,88	96,59	3,41	75°
16°	27,56	96,13	3,87	74°
17°	29,24	95,63	4,37	73°
18°	30,90	95,11	4,89	72°
19°	32,56	94,55	5,45	71°
20°	34,20	93,97	6,03	70°
21°	35,84	93,36	6,64	69°
22°	37,46	92,72	7,28	68°
23°	39,07	92,05	7,95	67°
24°	40,67	91,35	8,65	66°
25°	42,26	90,63	9,37	65°
26°	43,84	89,88	10,12	64°
27°	45,40	89,10	10,90	63°
28°	46,95	88,29	11,71	62°
29°	48,48	87,46	12,54	61°
30°	50,00	86,60	13,40	60°
31°	51,50	85,72	14,28	59°
32°	52,99	84,80	15,20	58°
33°	54,46	83,87	16,13	57°
34°	55,92	82,90	17,10	56°
35°	57,36	81,91	18,09	55°
36°	58,78	80,90	19,10	54°
37°	60,18	79,86	20,14	53°
38°	61,57	78,80	21,20	52°
39°	62,93	77,71	22,29	51°
40°	64,28	76,60	23,40	50°
41°	65,61	75,47	24,53	49°
42°	66,91	74,31	25,69	48°
43°	68,20	73,13	26,87	47°
44°	69,47	71,93	28,07	46°
45°	70,71	70,71	29,29	45°

	ΔD (m)	ΔZ (m)	p (m)
0°15'	0,44		
0°30'	0,87		
0°45'	1,31	99,99	0,01
1°	1,74	99,98	0,02
1°15'	2,18	99,98	0,02
1°30'	2,62	99,97	0,03
1°45'	3,05	99,95	0,05
2°	3,49	99,94	0,06
2°15'	3,93	99,92	0,08
2°30'	4,36	99,90	0,10
2°45'	4,80	99,88	0,12
3°	5,23	99,86	0,14
3°15'	5,67	99,84	0,16
3°30'	6,10	99,81	0,19
3°45'	6,54	99,79	0,21
4°	6,98	99,76	0,24
4°15'	7,41	99,72	0,28
4°30'	7,85	99,69	0,31
4°45'	8,28	99,66	0,34
5°	8,72	99,62	0,38
5°15'	9,15	99,58	0,42
5°30'	9,58	99,54	0,46
5°45'	10,02	99,50	0,50
6°	10,45	99,45	0,55
6°30'	11,32	99,36	0,64
7°	12,19	99,25	0,75
7°30'	13,05	99,14	0,86
8°	13,92	99,03	0,97

Example. If i = 25° and ΔL = 36 m

ΔZ	ΔD	p = 100 − ΔZ	

$$\Delta Z = \frac{90,63 \times 36}{100} = 32.63 \text{ m and } \Delta D = \frac{42,26 \times 36}{100} = 15,21 \text{ m}$$

DIRECTIONAL DRILLING
(continued)

VI. DOG-LEG AND RADIUS OF CURVATURE - PRACTICAL RULES

API Dog-leg

It is the change of the inclination angle per 100 ft (Or 30,5 m).

A dog-leg per 10 m is sometimes used.

To calculate the dog-leg both changes in inclination and direction must be taken into account.

Graphic calculation of a dog-leg

In the NS - EW - plane draw :

(1) The direction of the first survey from D.

(2) Plot on this direction the value of the inclination angle A.

(3) The direction of the second survey.

(4) Plot on this direction the value of the inclination angle B.

The dog-leg is the length of the segment AB measured with the same scale as the one used to plot A and B, and related to 100 ft (or 10 m).

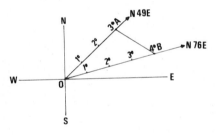

Example : Drilled from A to B. *i.e.* L metres. At A the survey gives an azimuth N49°E and inclination 3°.

At B the survey gives an azimuth N76E and inclination 4°. Dog-leg AB = 2° for L.

dog-leg per 100 ft = $\dfrac{2°}{L} \times 30,5$ m

dog-leg per 10 m = $\dfrac{2°}{L} \times 10$

Radius of curvature

$$R_c = \frac{180 \Delta L}{\Pi \Delta i}$$

ΔL = length drilled,

Δi = change of inclination for ΔL.

Possible azimuth variation with a deflecting tool

The maximum angle which it is possible to turn to the right or the left with a deflecting tool is given by the convenient formula :

$$\Delta A = \frac{180}{i}$$

ΔA = maximum azimuth variation when the inclination of the hole is i.

Orientation of a deflecting tool to turn at a maximum towards the right or the left

The hole is deviated at a maximum towards the right or the left when the deflecting tool is oriented at 90° or 100° to the right or the left related to the direction of the hole.

DIRECTIONAL DRILLING
(continued)

VII. COMPUTATION METHODS

a. Backward angle method

b. Average angle method

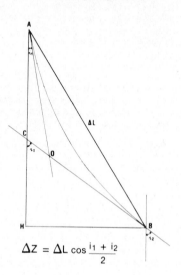

$$\Delta Z = \Delta L \cos i_2$$

$$\Delta Z = \Delta L \cos \frac{i_1 + i_2}{2}$$

c. Radius of curvature method

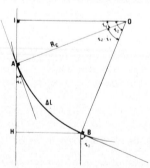

$$\Delta Z = AH = R_c [\sin i_2 - \sin i_1] \quad \text{with} \quad R_c = \frac{180 \Delta L}{\Pi \times i}$$

K KICK CONTROL FISHING

Summary

PRINCIPAL LETTER SYMBOLS USED

H = bit depth (m),

Z = vertical bit depth (m),

Q = drilling mud flow rate (l/min),

Q_r = reduced mud flow rate to kill the well (l/min),

d_1 = drilling mud weight,

d_r = required mud weight to kill the well,

d_2, d_3, etc. = intermediate mud weights,

V_{int} = internal volume of drill string (m³),

V_{an} = annular volume (m³),

V_{tot} = total mud volume to be weighted (including tank volume) (m³),

G = volume of gain (m³),

P_H = hydrostatic pressure (bar),

P_G = formation pressure (bar),

P_F = bottom hole pressure (bar),

P_{adm} = maximum allowable casing pressure (well closed) (bar),

P_M = maximum allowable casing pressure when circulating (bar),

P_{t_1} = initial shut-in drill pipe pressure (bar),

P_{a_1} = initial shut-in casing pressure (bar),

P = circulating pressure when drilling (bar),

P_{R_1} = initial circulating pressure at reduced mud flow rate with d_1 mud weight (bar),

P_{R_r} = final circulating pressure at reduced mud flow rate with d_r mud weight (bar),

P_{C_1} = pressure losses at reduced mud flow rate with d_1 mud weight (bar),

P_{C_2}, P_{C_3}, etc. = pressure losses at reduced mud flow rate with d_2, d_3 mud weights (bar),

P_{C_r} = pressure losses at reduced mud flow rate with d_r mud weight (bar),

P_A = circulating casing pressure (bar),

S = safety margin (bar).

PRELIMINARY DATA REQUIRED

So as not to waste time when a kick occurs it is necessary to keep a certain amount of information up to date as drilling progresses :

I. VOLUMES OF CIRCULATING FLUIDS

V_{int} = inside volume of drill string (m³),
V_{an} = annular volume (m³),
V_{bac} = volume of mud in tanks (m³).

II. REDUCED MUD FLOW RATE TO CONTROL THE KICK

This reduced mud flow rate (Q_r) is generally taken between 1/4 and 1/2 of the drilling mud flow rate Q (very often $Q_r = 1/2$ Q).

III. DISPLACEMENT TIMES AT REDUCED MUD FLOW RATE (min or pump strokes)

(a) Drill pipe displacement time (surface to bit) :

$$t_1 \text{ (min)} = \frac{V_{int}}{Q_r} \times 1000$$

or pump strokes N = t_1 × strokes per minute.

(b) Annulus displacement time (bit to surface) :

$$t_2 \text{ (min)} = \frac{V_{an}}{Q_r} \times 1000$$

or pump strokes N = t_2 × strokes per minute.

Note. Figures 1 and 2 and 2bis give these displacement times.

IV. PRESSURE LOSSES AT REDUCED MUD FLOW RATE WITH d_1 MUD (P_{C_1})

These pressure losses can be determined as drilling progresses :
— either every day at the beginning of morning shift,
— or after running in a new bit before resuming drilling,
— or using Fig. 3 knowing pressure losses when drilling at rate Q.

V. MAXIMUM ALLOWABLE CASING PRESSURE (P_{adm})

Calculation of P_{adm} is based on :
— drilling mud weight,
— casing head working pressure (137 bars for 2000 series, 206 bars for 3000 series, 343 bars for 5000 series, etc.),
— internal yield strength of casing :

$$P_{adm} = \text{int. yield strength} - \frac{H_s d_1}{10} \quad (H_s = \text{depth of casing shoe})$$

— formation fracture gradient at casing shoe :

$$P_{adm} = H_s \left(g_r - \frac{d_1}{10}\right)$$

(g_r = formation fracture gradient may vary from 0,12 to 0,23 as a function of the type of formation and the depth).

SEQUENCES FOLLOWING DETECTION OF A KICK

The main objective when controlling a kick is to circulate the total amount of the influx out of the hole. The best way to prevent further entry of formation fluid into the hole is to permanently slightly overbalance the formation pressure.

I. INITIAL ACTION FOLLOWING KICK DETECTION :
— pull the kelly out of the preventers,
— shut down the pump,
— close the pipe rams preventer and flow line,
— record the volume of mud gained in the pits,
— after 10 to 15 min read and record the stabilized pressures :
  shut-in drill pipe pressure (P_{t_1}),
  shut-in casing pressure (P_{a_1}) (check that P_{a_i} never reaches P_{adm}).

II. CALCULATIONS
Required mud weight to slightly overbalance formation pressure

$$d_r = d_1 + \frac{10\,(P_{t1} + S)}{H}$$

$$\frac{10\,(P_{t1} + S)}{H} = \Delta d \quad \text{(given in Figure 4)}$$

(for a deviated hole take Z instead of H)
Initial circulating pressure at the reduced mud flow rate Q_r with mud d_1 (P_{R_1}) :

$$P_{R_1} = P_{C_1} + P_{t1} + S$$

Final circulating pressure at the reduced mud flow rate Q_r
(a) With required mud weight d_r :

$$P_{R_r} = P_{C_r} = P_{C_1}\frac{d_r}{d_1} \quad \text{(Figure 3 bis)}$$

(b) With intermediate mud d_2 :

$$P_{R_2} = P_{R_1} - \frac{P_{R_1} - P_{R_r}}{d_r - d_1}\,(d_2 - d_1) \quad \text{(Figure 5)}$$

(c) With intermediate mud d_3 :

$$P_{R_3} = P_{R_2} - \frac{P_{R_2} - P_{Rr}}{d_r - d_2}\,(d_3 - d_2) \quad \text{(Figure 6)}$$

III. KICK KILLING PROCEDURES
There are two basic methods for killing flowing wells using drill-pipe pressure control when $P_{a_1} < P_{adm}$.
The driller's method : accomplished in two steps (Fig. 8)
(a) First step : the kick is circulated out with the drilling mud (d_1) at the reduced flow rate Q_r and the choke is adjusted to obtain a casing pressure $P_A = P_{a_1} + S$. In a short while the drill pipe pressure will stabilize at P_{R_1}, then hold this drill pipe pressure constant by adjusting the choke during the displacement time $t_1 + t_2$ or longer if the well is caved in.

Stop the pump and shut in the well. P_t must be equal to P_a.
(b) Second step : the well is circulated, with the mud d_r which has been prepared during the first step at the reduced flow rate Q_r. Adjust the choke to have $P_A = P_{t_1} + S$ and hold this pressure during the displacement time t_1. The drill pipe pressure will fall from P_{R_1} to $P_{R_r} = P_{C_1}\,d_r/d_1$. During the displacement time t_2 adjust choke to hold the drill-pipe pressure at this value P_{R_r}. If the hole is caved in circulate until mud d_r reaches the surface. Stop the pump, the well is killed if $P_a = P_t = 0$.

SEQUENCES FOLLOWING DETECTION OF A KICK
(continued)

The balanced bottom hole pressure method (Fig. 9)

Accomplished in one or several steps (generally no more than two steps) depending upon formation pressure, amount of formation fluid entry into the hole, density of drilling fluid, availability of mud weighting material, etc.

In the one step method the well is circulated with mud d_r the same way as with mud d_2 in the first step of the two step method described hereunder :

(a) First step : the kick is circulated out of the hole with an intermediate mud weight d_2 such as :

$$d_2 = \frac{d_1 + d_r}{2}$$

Start circulating at rate Q_r with mud d_2 and adjust the choke to obtain the casing pressure $P_A = P_{a_1} + S$. At start the drill-pipe pressure will be $P_{R_1} = P_{C_1} + P_{t_1} + S$. Keep on circulating at constant flow rate Q_r during displacement time t_1, adjust the casing choke to lower the drill pipe pressure in graduated steps from P_{R_1} to :

$$P_{R_2} = P_{R_1} - \frac{P_{R_1} - P_{R_r}}{d_r - d_1} (d_2 - d_1) \quad \text{(see Fig 5 and Fig 9)}$$

During displacement time t_2 adjust the choke to keep the drill pipe pressure P_{R_2} constant. If the hole is caved keep on circulating until the mud d_2 reaches the surface. Then stop the pump and shut-in the well. The pressures will be :

$$P_{t2} = P_{a2} = P_{t_1} - \frac{H (d_2 - d_1)}{10} \quad \text{(see Fig 7)}$$

(b) Second step : replacement of mud d_2 by mud d_r.

Start circulating at rate Q_r with mud d_r and maintain the casing head pressure at $P_A = P_{t_2} + S$ during displacement time t_1. The drill-pipe pressure will fall from P_{R_2} to P_{R_r}.

Keep on circulating at rate Q_r during displacement time t_2 maintaining the drill pipe pressure at P_{R_r} by adjusting the casing choke. If the hole is caved in keep on circulating until mud d_r reaches the surface. Stop the pump, shut-in the well. The well is killed if $P_a = P_t = 0$.

SEQUENCES FOLLOWING DETECTION OF A KICK
(continued)

Notes

I. To raise mud weights (from d_1 to d_2, from d_2 to d_r) refer to section H_3 to find the amount of weighting material needed.

II. Fig 8 and 9 : give the behaviour of drill pipe and casing head pressures during the control of the well with the two methods described. The continuous lines show the pressures which have to be followed in adjusting the casing choke. Note that t_2 is a minimum displacement time valid only if the hole is not caved.

III. ANALYSIS OF THE FIGURES

Figure 1 : is used to find the displacement times (from surface to bit through the drill string and from bit to surface through the annulus) from the inside volume of the drill string or from the volume of the annulus.

Figure 2 and 2 bis : are used to find the displacement times either from the volume per metre of drill pipe and volume per metre of annulus or from the total volumes.

Figure 3 : is used to find from the pressure losses for a given fluid flow rate, the pressure losses for an other fluid flow rate (with mud characteristics remaining unchanged).

Figure 3bis : is used to find from the pressure losses for a given mud weight, the pressure losses for an other mud weight (with fluid flow rate remaining unchanged).

Figure 4 : is used to find the amount of mud weight increase necessary to slightly overbalance the formation pressure.

Figure 5 : is used to find, from the initial drill pipe circulating pressure (P_{R_1}), the final circulating pressure with mud d_2 (P_{R_2}) when the two step balanced bottom hole pressure method is used with intermediate mud weight d_2 such as :

$$d_2 = \frac{d_1 + d_r}{2}$$

Figure 6 : is used to find, from the initial drill pipe circulating pressure (P_{R_1}), the final circulating pressures P_{R_2} and P_{R_3} when using the three step balanced bottom hole pressure method with two intermediate mud weights d_2 and d_3 such as :

$$d_2 = d_1 + \frac{d_r - d_1}{3} \text{ and } d_3 = d_1 + \frac{2 (d_r - d_1)}{3}$$

Figure 7 : is used to find the static drill pipe pressures after a certain amount of mud weight increase Δd. For example P_{t_2} when mud d_3 is replaced by mud d_2 (mud weight increase $\Delta d = d_2 - d_1$ and static drill pipe pressure P_{t_1} with mud d_1).

Figure 1
DISPLACEMENT TIME : SURFACE TO BIT, BIT TO SURFACE

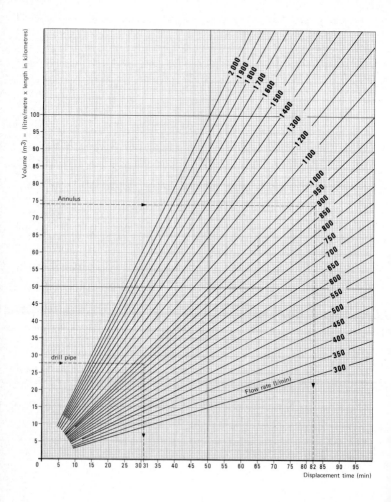

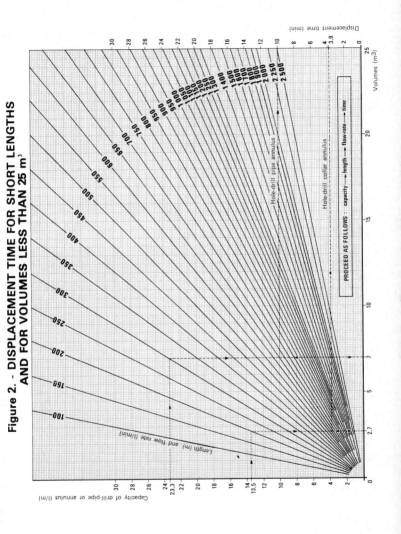

Figure 2. - DISPLACEMENT TIME FOR SHORT LENGTHS AND FOR VOLUMES LESS THAN 25 m³

Figure 2 bis
DISPLACEMENT TIME FOR LONG LENGTHS
AND VOLUMES HIGHER THAN 25 m³

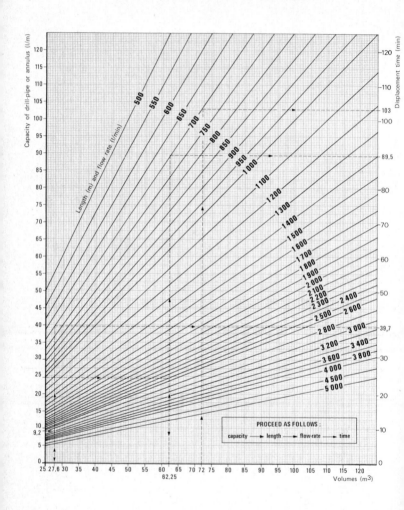

359

Figure 3
PRESSURE LOSSES RELATED TO FLOW-RATE

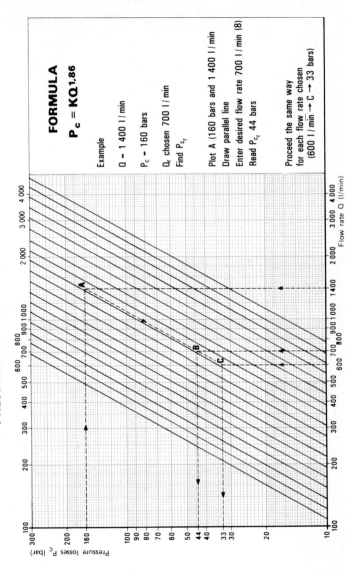

FORMULA
$$P_c = KQ^{1,86}$$

Example

Q — 1 400 l/min

P_c — 160 bars

Q_r chosen 700 l/min

Find P_{c_r}

Plot A (160 bars and 1 400 l/min)

Draw parallel line

Enter desired flow rate 700 l/min (8)

Read P_{c_r} 44 bars

Proceed the same way
for each flow rate chosen
(600 l/min → C → 33 bars)

Flow rate Q (l/min)

Pressure losses P_c (bar)

Figure 3 bis
PRESSURE LOSSES RELATED TO MUD WEIGHTS

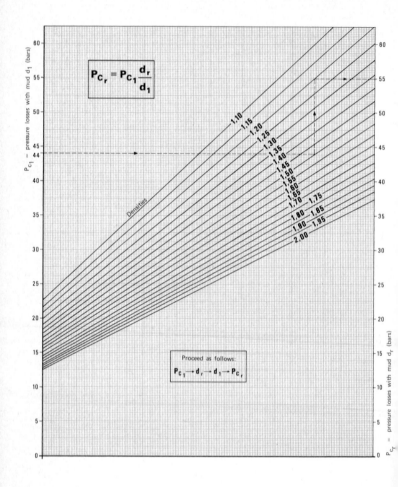

Figure 4
CALCULATION OF REQUIRED MUD WEIGHT d_r

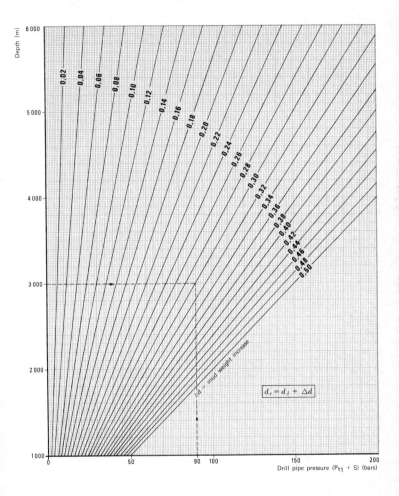

Depth (m)

Drill pipe pressure (P_{t1} + S) (bars)

Δd = mud weight increase

$$d_r = d_i + \Delta d$$

362 K 12

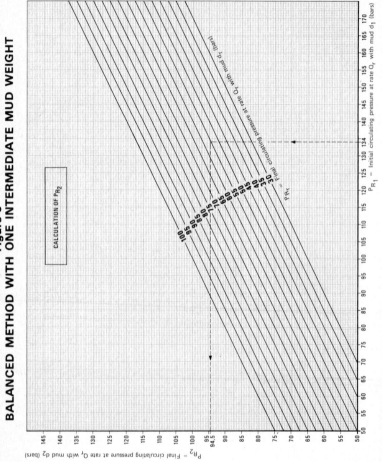

Figure 5
BALANCED METHOD WITH ONE INTERMEDIATE MUD WEIGHT

CALCULATION OF P_{R_2}

P_{R_1} = Initial circulating pressure at rate Q_r with mud d_1 (bars)

P_{R_2} = Final circulating pressure at rate Q_r with mud d_2 (bars)

Final circulating pressure at rate Q_r with mud d_r (bars)

Figure 6
BALANCED METHOD WITH TWO INTERMEDIATE MUD WEIGHTS

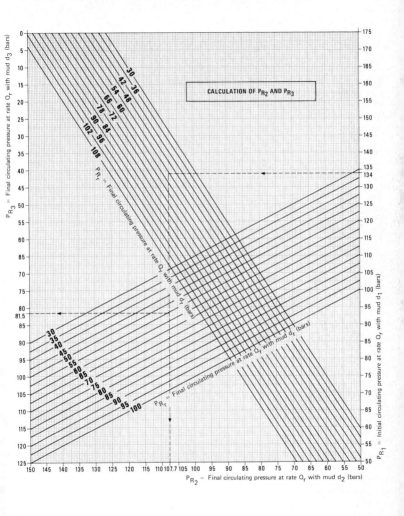

Figure 7
CALCULATION OF STATIC DRILL PIPE PRESSURES
WITH INTERMEDIATE MUD WEIGHTS

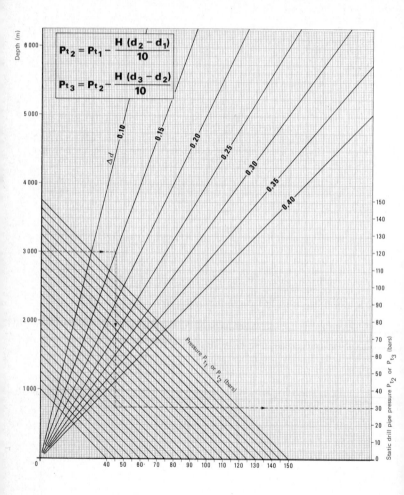

$$Pt_2 = Pt_1 - \frac{H(d_2 - d_1)}{10}$$

$$Pt_3 = Pt_2 - \frac{H(d_3 - d_2)}{10}$$

365

Figure 8. - DRILLER'S METHOD

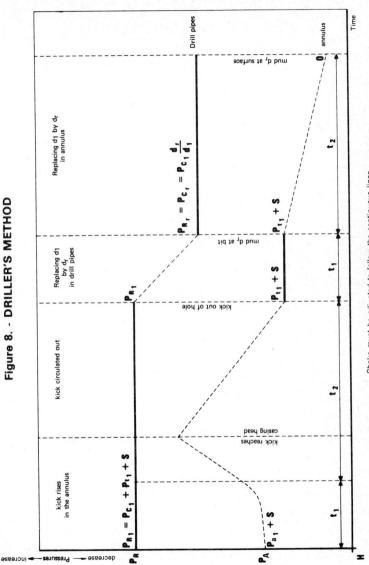

Choke must be adjusted to follow the continuous lines

Figure 9. - BALANCED BOTTOM HOLE PRESSURE METHOD

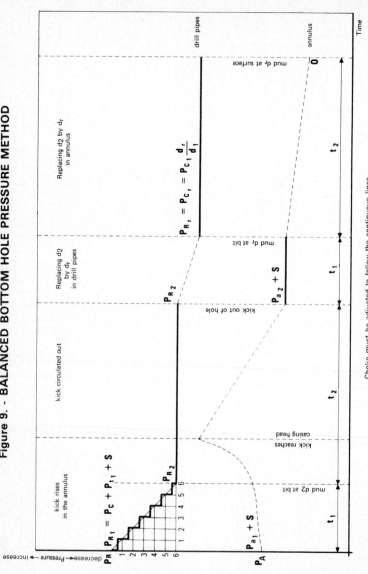

Choke must be adjusted to follow the continuous lines

EXAMPLE OF KICK CONTROL

I. RECORDED KNOWN INFORMATION AS KICK OCCURS :

— Well depth	$H = 3000$ m
— Casing size 9 5/8'' (N80-47 #) - Shoe at	$H_s = 2500$ m
— Bit size 8 1/2''	

— Drill pipe 5''-19,50 # - length = 2800 m
— Drill collar 6 3/4 × 2 1/4 - length = 200 m
— Drilling mud weight $\qquad$ $d_1 = 1,20$
— Drilling mud flow rate (Triplex pump 10'' × 6'') $\quad Q = 1400$ l/min (100 strokes/min)
— Reduced mud flow rate chosen $\qquad Q_r = 700$ l/min (50 strokes/min)
— Pressure losses with mud d_1 : when drilling $\qquad P_{C_1} = 160$ bars
   at flow rate Q_r : (1) measured on rig $\qquad P_{C_r} = 45$ bars
      (2) from Fig 3 $\qquad P_{C_r} = 44$ bars

— Maximum allowable casing head pressure :
  — 5000 series casing head $\qquad P_{adm} = 350$ bars
  — Casing bursting pressure ($P_e = 474$ bars)
    Hydrostatic pressure at 2500 m $P_H = 300$ bars
    $P_{adm} = P_e - P_H = 474 - 300$ $\qquad = 174$ bars
  — Formation breaking pressure at casing shoe
    $P_{adm} = H_S (0,18 - 0,12)$ $\qquad\qquad = 150$ bars
    0,18 : formation fracture gradient chosen
    0,12 : mud gradient (d = 1,20)
— Inside volume of drill string (9,2 l/m) 3000 × 9,2 $\qquad = 27,6$ m³
— Volumes of annular spaces
    Annulus hole-drill collars 13,5 l/m × 200 m $\qquad = 2700$ l
    Annulus hole-drill pipes 23,3 l/m × 300 m $\qquad = 6990$ l
    Annulus casing-drill pipe 24,9 l/m × 2500 m $\qquad \underline{= 62250}$ l
    Total V_{an} $\qquad\qquad\qquad = 71940$ l # 72 m³

— Displacement times at reduced flow rate Q_r (Fig 1,2 or 2bis)

$$\text{Surface to bit } \frac{27,6}{700} \times 1000 \# 40 \text{ min} = t_1$$

$$\text{Bit to surface } \frac{72}{700} \times 1000 \# 103 \text{ min} = t_2 \text{ (minimum)}$$

— Volume of gain in mud tanks $G = 3$ m³

II. KICK CONTROL

Shut-in the well and note down :
— Stabilized drill pipe pressure $\qquad\qquad\qquad P_{t1} = 75$ bars
— Stabilized casing head pressure $\qquad\qquad\qquad P_{a1} = 90$ bars
    Then $P_{a_1} < P_{adm}$

Calculation of initial circulating pressure at flow rate Q_r with drilling mud d_1 :

$$P_{R_1} = P_{t_1} + P_{C_1} + S = 75 + 44 + 15 \qquad\qquad P_{R_1} = 134 \text{ bars}$$

safety margin S has been chosen equal to 15 bars.

Calculation of the required mud weight d_r to slightly overbalance the formation pressure :

$$d_r = d_1 + \frac{10 (P_{t1} + S)}{H} = 1,2 + \frac{10 (75 + 15)}{3000} = 1,5$$

(Fig 4 gives $\Delta d = 0,3$ $D_r = d_1 + 0,3 = 1,5$)

EXAMPLE OF KICK CONTROL
(continued)

Calculation of an intermediate mud weight d_2 :

$$d_2 = \frac{d_r + d_1}{2} = \frac{1,5 + 1,2}{2} = 1,35$$

Calculation of the final circulating pressure at reduced flow rate Q_r
(a) With the required mud weight d_r :

$$P_{R_r} = P_{C_r} = P_{C_1} \frac{d_r}{d_1} = 44 \times \frac{1,5}{1,2} = 55 \text{ bars (see Fig 3 bis)}$$

(b) With the intermediate mud weight d_2 :

$$P_{R_2} = P_{R_1} - \frac{R_{R_1} - P_{R_r}}{d_r - d_1}(d_2 - d_1) = 94,5 \text{ bars (see Fig 5)}$$

A. Driller's Method

First step : Circulation with drilling mud d_1.

Start circulating at reduced flow rate (700 l/min or 50 strokes/min) and open the casing choke to have a casing pressure $P_A = P_{a_1} + S = 90 + 15 = 105$ bars. Start a stop-watch. Read the circulating drill pipe pressure which must be $P_{R_1} = 134$ bars. Maintain this P_{R_1} pressure by adjusting the casing choke during $t_1 + t_2 = 143$ min total displacement time. By this time the kick should be circulated out, but if the hole is caved in keep on circulating until the casing pressure remains steady at $P_{t_1} + S = 90$ bars.

Stop the pump and shut-in the well, shut-in pressures must be $P_{t_1} = P_{a_1} = 75$ bars.

Second step : Circulation mud d_r.

For this second step we need a minimum volume of 140 m³ ($V_{int} + V_{an} + 40$) of mud at the required weight $d_r = 1,50$. Part or totality of these 140 m³ can be prepared during the first step depending on the possibilities of the mud circuit of the rig.

Fig 3 section H gives the amount of barite needed : 467 kg/m³. If we weight 130 m³ we need $130 \times 467 = 60710$ kg of barite and we obtain 144,3 m³ of mud (Fig 4 section H gives a volume factor of 1,11).

Start circulating with this weighted mud d_r at rate Q_r (700 l/mn or 50 strokes/min) and open choke to have a casing pressure $P_A = P_{t_1} + S = 75 + 15 = 90$ bars and maintain this casing pressure by adjusting the choke during displacement time $t_1 = 40$ min. The drill pipe pressure should drop from $P_{R_1} = 134$ bars to $P_{R_r} = 55$ bars.

Keep on circulating and hold the drill pipe pressure at $P_{R_r} = 55$ bars during $t_2 = 103$ min or if the hole is caved in, until mud d_r reaches the surface. The well is under control.

B. Balanced Bottom Hole Pressure Method

The two step method will be used.

First step : Circulation with intermediate mud $d_2 = 1,35$ (see above).

Mud d_2 must be prepared either before starting circulation or in proportion as it is pumped into the well.

To weight the mud, from 1,2 to 1,35 221 kg/m³ of barite are necessary (Fig 3, section H) *i.e.* a total of $140 \times 221 = 30940$ kg.

To circulate the mud in proportion as it is prepared, the mud rate through the hopper must be the same as the circulation rate (700 l/min) and the mixing rate of the barite must be $\frac{700 \times 221}{1000} = 155$ kg/min.

EXAMPLE OF KICK CONTROL
(continued)

Calculation of the steps to lower the drill pipe pressure from $P_{R_1} = 134$ bars to $P_{R_2} = 94,5$ bars during displacement time $t_1 = 40$ min. Let's take 5 steps of $\frac{40}{5} = 8$ min each for a pressure drop of $\frac{134 - 94,5}{5} \neq 8$ bars per step.

Steps	Pressure drop (bars)		Average pressure for each step (bars)	Duration (min)
	from	to		
1	134	126	130	8
2	126	118	122	8
3	118	110	114	8
4	110	102	106	8
5	102	94,5	98	8

Circulation procedure :
Start circulating at rate Q_r (700 l/min or 50 strokes/min) and adjust choke to have casing head pressure :

$$P_A = P_{a_1} + S = 90 + 15 = 105 \text{ bars}$$

The drill pipe pressure must be $P_{R_1} = 134$ bars.

Adjust choke to lower P_{R_1} to 130 bars (first step of above table) and keep that pressure constant during 8 min. Then lower the pressure to the value of the second step, 122 bars, for 8 min, and so on to the last step at 98 bars.

Lower the pressure to 94,5 bars and keep that pressure constant during the displacement time $t_2 = 103$ min or until mud d_2 arrives at the surface, if the hole is caved in.

$$P_{a2} = P_{t2} = P_{t1} - \frac{H(d_2 - d_1)}{10} = 30 \text{ bars} \quad \text{(see Fig 7)}$$

Second step : Circulation with mud $d_r = 1,50$.

Mud $d_2 = 1,35$ must be weighted to $d_r = 1,50$. This will need 233 kg/m³ of barite (see Fig 3, section H) *i.e.* a total of 32600 kg for 140 m³ of mud d_2. If mud d_r is pumped in proportion as it is prepared the mud rate through the hopper must be the same as the circulation rate (700 l/min) and the mixing rate of the barite must be $\frac{700 \times 233}{1000} = 163$ kg/min.

Circulation procedure : Start circulating at rate Q_r and adjust choke to have

$$P_A = P_{a_2} + S = 30 + 15 = 45 \text{ bars}$$

The drill pipe pressure will be the same $P_{R_2} = 94,5$ bars as when circulating with mud d_2. Maintain 45 bars at the casing head during $t_1 = 40$ min, drill pipe pressure will drop to $P_{R_f} = 55$ bars. Then maintain the drill pipe pressure at 55 bars during $t_2 = 103$ min or until the mud d_r arrives at the surface if the hole is caved in.

Stop the pump and shut-in the well $P_a = P_t = 0$. The well is under control.

DETERMINATION OF THE LENGTH
OF FREE PIPE IN A FROZEN STRING

The relation between the differential stretch and the length of free pipe in a frozen string due to a differential pull is :

$$L = \frac{2,675 \times W \times I}{P_2 - P_1}$$

where :
L   = length of free pipe (m),
W = weight per metre of pipe (kg/m),
I   = differential stretch (mm),
P_2-P_1 = differential pull (10^3daN).
Note. This method does not give great accuracy for L.

EXAMPLE ON HOW TO USE THE METHOD :

Bit at 2247 min in a 8 1/2'' hole with mud density 1,40.
Drilling string :
2000 m 5''-19,50 # (NC 50) grade 95, premium class (S) drill pipes, weight per metre 31,4 kg/m (see page B 23)
247 m 6 3/4 × 2 13/16'' drill collars, weight per metre 149,4 kg/m (see page B 35).

First step. Calculation of maximum pull on drill pipe :
Tension load at minimum yield strength : 176 10^3daN (see page B 6).
Maximum allowable pull : 176 × 0,9 = 158 10^3daN (0,9 = a constant relating proportional limit to yield strength).

Second step. Calculation of the weight of drill string in mud :
Drill collars : 247 m × 149,4 kg/m = 38900 kg = 38,2 10^3daN (see page A 4).
Drill pipes : 2000 m × 31,4 kg/m = 62800 = 61,6 10^3daN.
Total # 100 10^3daN
In mud d = 1,4 the buoyancy factor is 0,822 (see page E 14).
Weight of drill string in mud 100,6 ×0,822 = 82,2 10^3daN.
The allowable pull margin is 158 — 82,2 = 75,8 10^3daN.

Third step. Apply P_1 = 105 10^3daN to the drill string. Draw a chalk-mark at the kelly-drive bushing level. Apply 110 10^3daN and comme back to P_1 = 105 10^3daN, draw a second chalk mark at the kelly-drive bushing level. This second mark must be distinct from the first one (the difference comes from the friction of drill pipe in the hole). Draw a datum line in the middle of the two marks.

Fourth step. As above draw two chalk-marks with a pull of P_2 = 135 10^3daN, then 140 10^3daN and back to 135 10^3daN. Draw a datum-line in the middle of the two marks. Measure the distance I in millimetres between the two datum-lines. Assume I = 700 mm.

Fifth step. Apply the formula with :

$$W = 31,4 \text{ kg/m},$$
$$I = 700 \text{ mm},$$
$$P_1 = 105 \ 10^3 \text{daN},$$
$$P_2 = 135 \ 10^3 \text{daN},$$
$$L = \frac{2,675 \times 31,4 \times 700}{30} = 1960 \text{ m}$$

1960 m is the length of free pipe, *i.e.* that the stuck point is nearly at the top of the drill collars.

MAXIMUM ALLOWABLE NUMBER OF TURNS WHICH CAN BE GIVEN TO 1000 m OF DRILL PIPE UNDER A GIVEN AXIAL TENSION
(Grade E)

Nominal weight and diameter of drill pipe	NUMBER OF TURNS FOR 1000 m OF DRILL PIPE UNDER A TENSION OF (10³ daN)																					
	0	10	20	30	40	50	60	70	80	90	100	110	120	130	140	150	160	170	180	190	200	210
2 3/8" - 6.65 #	18 3/4	18 1/4	17 3/4	16 1/2	14																	
2 7/8" - 10.40 #	15 1/2	15 1/4	14 1/2	14	14	13	11 3/4															
3 1/2" - 9.50 #	12 3/4	12 1/2	12 1/4	11 3/4	11 1/4	10 1/4	9	7 1/4	4 3/4													
3 1/2" - 13.30 #	12 3/4	12 1/2	12 1/4	12	11 3/4	11 3/4	11 1/2	10 1/4	9 1/4	8 1/4	7	5 1/2										
3 1/2" - 15.50 #	12 3/4	12 3/4	12 1/2	12 1/2	12	11 3/4	11 1/2	11	10 1/2	10	9	8	6 1/2	5 1/2	2 3/4							
4" - 11.85 #	11	11	11	10 1/2	10	9 1/2	9	8	7	5 1/2	2 1/2											
4" - 14.00 #	11	11	10 1/2	10 1/2	10 1/2	10	9 1/2	9	8 1/2	7 3/4	6 1/2	5 1/2	3 1/2									
4" - 15.70 #	11	11	10 1/2	10 1/2	10 1/2	10 1/4	9 1/2	9	8 1/2	8 1/4	7	6 1/4	5 1/4	3								
4 1/2" - 13.75 #	9 3/4	9 3/4	9 3/4	9 1/2	9 1/4	9 1/4	8 1/2	8 1/2	8 1/4	7 3/4	7	6 1/4										
4 1/2" - 16.60 #	9 3/4	9 3/4	9 1/2	9 1/2	9 1/4	9 1/4	8 3/4	8 1/2	8 1/4	8	8 1/2	8 1/4	7 3/4	7 1/4	6 1/2	5 3/4	5	3 3/4	1 1/2			
4 1/2" - 20.00 #	9 3/4	9 3/4	9 1/2	9 1/2	9 1/4	9 1/2	9	9 1/2	9	8 1/2	8 1/4	8 1/4	7 3/4	7 1/4	6 1/2			2				
5" - 16.25 #	8 3/4	8 3/4	8 3/4	8 1/2	8 1/4	8 1/4	8	8	7 1/2	7 1/2	7 1/4	5 3/4	5	4	2 1/4	4 1/2	3 1/2					
5" - 19.50 #	8 3/4	8 3/4	8 1/2	8 1/2	8 1/2	8 1/4	8 1/2	8	7 3/4	7	7 1/4	8	6 1/2	6	5 1/4	6 1/2	6 1/2	6 1/4				
5" - 25.60 #	8 3/4	8 3/4	8 1/2	8 1/2	8 1/2	8 1/2	8 1/2	8	8 1/4	8 1/4	7 1/4	8	7 3/4	6	5 1/4	6 3/4	6 1/2	6 1/4	5 3/4	5 1/4	4 3/4	4
5 1/2" - 19.20 #	8	8	8	8	7 3/4	7 3/4	8	7 1/2	7 1/4	7	6 1/4	6 1/4	5 3/4	5	4 1/4	3 1/2	2	3 3/4				
5 1/2" - 21.90 #	8	8	8	8	7 3/4	7 3/4	7 1/2	7 1/2	7 1/4	6 3/4	6 3/4	6 1/2	6 1/4	6 1/4	5 1/4	6	5 1/4	4 3/4	3			
5 1/2" - 24.70 #	8	8	8	8	7 3/4	7 3/4	7 1/2	7 1/2	7	7 1/2	7	7	6 3/4	6 3/4	6 1/4	6	5 1/4	5	4 1/2	4 1/4	3 1/2	2 1/2
6 5/8" - 25.20 #	6 3/4	6 3/4	6 1/2	6 1/2	6 1/2	6 1/4	6 1/4	6 1/4	6 1/4	6	6	5 3/4	5 1/4	5 1/4	5	4 3/4	4 1/2	4	3 3/4	3 1/4	2 1/2	1 1/2

Note. Tabulated values are based on the following formula :

$$N = \frac{100 \, L \, S_s}{\Pi D G} \sqrt{1 - \frac{P^2}{4A^2 S_s^2}}$$

where :

N = number of turns which can be given for a tension P,
L = length of drill pipe string (m).
S_s = maximum unit shear strength ($S_s = 0.577$ minimum unit yield strength),
D = outside diameter (cm),
G = modulus of elasticity in shear : 8400 hbars.
P = total load in tension (daN),
A = cross section area (mm²).

MAXIMUM ALLOWABLE NUMBER OF TURNS WHICH CAN BE GIVEN TO 1000 m OF DRILL PIPE UNDER A GIVEN AXIAL TENSION
(Grade 95)

NUMBER OF TURNS FOR 1000 M OF DRILL PIPE UNDER A TENSION OF (10^3 daN)

Nominal weight and diameter of drill pipe	0	10	20	30	40	50	60	70	80	90	100	110	120	130	140	150	160	170	180	190	200	210
2 3/8" - 6.65 #	23 3/4	23 1/2	23	22	20 1/2	18 1/4	15	10 1/4														
2 7/8" - 10.40 #	19 1/2	19 1/4	19 1/4	19	18 1/2	18	17	16	14 1/2	13	10 3/4	7 3/4										
3 1/2" - 9.50 #	16	15 3/4	15 3/4	15 1/2	15	14 1/4	13 1/4	12 1/2	11	9	6 1/4											
3 1/2" - 13.30 #	16	16	15 3/4	15 3/4	15 1/2	15 1/4	15	14 1/2	14	13	12	11	10	8 1/2	6 1/4							
3 1/2" - 15.50 #	16	16	16	15 3/4	15 1/2	15 1/4	15	14 3/4	14 1/2	13 1/2	13	12 1/2	12	11	10	9	7 1/2	5 1/2				
4" - 11.85 #	14	14	14	13 3/4	13 1/2	13	12 1/2	12	11	10	8 1/2	7 1/4	5 1/4									
4" - 14.00 #	14	14	14	13 3/4	13 1/2	13 1/4	13	12 1/2	12	11 1/2	11	10	9	8	7	4 3/4						
4" - 15.70 #	14	14	14	13 3/4	13 3/4	13 1/2	13 1/2	13 1/4	12 3/4	12 1/4	11 3/4	10 3/4	10	10	9	8	7	5	2 1/4			
4 1/2" - 13.75 #	12 1/2	12 1/4	12 1/4	12	12	11 3/4	11 1/2	11	10 1/2	10	9 1/2	8 1/2	7 1/2	6 1/2	4 1/2							
4 1/2" - 16.60 #	12 1/2	12 1/4	12 1/4	12	12	12	11 1/2	11 1/4	11 1/4	10 3/4	10 1/2	10	9 1/2	8 3/4	8	7 1/4	6 1/4	5	3			
4 1/2" - 20.00 #	12 1/2	12 1/2	12 1/4	12 1/4	12	12	12	11 1/2	11 3/4	10 3/4	10 1/2	11	11	10 1/2	10	9 1/2	9	8 1/2	7 1/4	6 1/4	5 1/4	3 1/2
5" - 16.25 #	11 1/4	11 1/4	11 1/4	11	11	10 3/4	11	11 1/2	10	9 3/4	10 3/4	9	8 1/2	8	7 1/4	7 1/4						
5" - 19.50 #	11 1/4	11 1/4	11 1/4	11	11	11	10 3/4	10 3/4	10 1/4	10 1/4	10 1/2	9 3/4	9	10	8 3/4	8 1/4	7 3/4	7 1/4	6 1/4	6 1/4	5 1/2	5
5" - 25.60 #	11 1/4	11 1/4	11 1/4	11	11	11	11	10 3/4	10 1/4	10 1/2	10	10 1/4	9 1/2	8 1/2	8 3/4	9 3/4	9	8 1/2	9	8 3/4	8 1/4	8
5 1/2" - 19.20 #	10 1/4	10 1/4	10 1/4	10	10	10	9 3/4	9 3/4	9 1/2	9 1/4	10 1/2	8 3/4	8 1/2	8 1/2	8	7 1/4	6 3/4	6	7	6 1/2	5 3/4	5 1/4
5 1/2" - 21.90 #	10 1/4	10 1/4	10 1/4	10	10	10	9 3/4	9 3/4	9 1/2	9 1/2	9 1/4	9 1/4	8 3/4	8 1/2	8 3/4	7 1/4	7 3/4	7 1/2	7 3/4	6 1/2	5 3/4	7 1/4
5 1/2" - 24.70 #	10 1/4	10 1/4	10 1/4	10	10	10	10	9 3/4	9 3/4	9 1/2	9 1/2	9 1/4	8 3/4	8 1/2	8 1/4	8	8 1/4	8	7	7 1/2	7 1/2	5 1/4
6 5/8" - 25.20 #	8 1/2	8 1/2	8 1/2	8 1/4	8 1/4	8 1/4	8 1/4	8	8	8	7 3/4	7 1/2	7 1/2	7 1/4	7 1/4	7	7	6 1/2	6 1/2	6 1/4	5 3/4	5

MAXIMUM ALLOWABLE NUMBER OF TURNS WHICH CAN BE GIVEN TO 1000 m OF DRILL PIPE UNDER A GIVEN AXIAL TENSION
(Grade 105)

Nominal weight and diameter of drill pipe	NUMBER OF TURNS FOR 1000 m OF DRILL PIPE UNDER A TENSION OF (10^3 daN)																					
	0	10	20	30	40	50	60	70	80	90	100	110	120	130	140	150	160	170	180	190	200	210
2 3/8" - 6.65 #	26 1/4	26	25 1/2	25	23 1/2	21 1/2	18 1/2	15	10													
2 7/8" - 10.40 #	21 3/4	21 1/2	21	21	20 1/2	20	19 1/4	18 1/4	17 1/4	16	14	12	9 1/4									
3 1/2" - 9.50 #	17 3/4	17 1/2	17 1/4	17 1/4	16 3/4	16	15 1/4	14 1/2	13 1/4	11 3/4	9 3/4	7 1/4	2									
3 1/2" - 13.30 #	17 3/4	17 1/2	17 1/4	17 1/4	17	16 3/4	16 1/2	16 1/4	15 3/4	15	14 1/4	13 1/2	12	4 1/2								
3 1/2" - 15.50 #	17 3/4	17 3/4	17 1/2	17 1/4	17	17	16 1/2	16 1/2	16 1/2	15 3/4	15 1/4	14 3/4	14	11	9 1/2	8						
4" - 11.85 #	15 1/2	15 1/4	15	15	14 1/2	14 1/2	14	13 1/2	13	12	11	9 3/4	8	6								
4" - 14.00 #	15 1/2	15 1/2	15 1/4	15	14 1/2	14 1/2	14 3/4	14 1/2	14	13 1/4	12 1/2	12	11 1/4	10 1/2	10	9 1/4	8 1/2	7 1/2				
4" - 15.70 #	15 1/2	15 1/2	15 1/2	15	15	15	14 3/4	14 1/2	14	13 3/4	13 1/4	13	12 1/2	11 3/4	11	10 1/2	9 3/4	8	7	5 1/2		
4 1/2" - 13.75 #	13 3/4	13 1/2	13 1/2	13 1/4	13	13	12 3/4	12 1/2	12	11 1/2	11	10 1/2	9 1/2	8 1/2	7 1/2	6 3/4	4 1/4					
4 1/2" - 16.60 #	13 3/4	13 3/4	13 1/2	13 1/2	13 1/2	13 1/4	13	12 3/4	12 1/2	12 1/2	12	11 1/2	11	10 1/2	10	9 1/2	9 1/4	8 1/2	7 1/2	5	3	
4 1/2" - 20.00 #	13 3/4	13 3/4	13 3/4	13 1/2	13 1/2	13 1/4	13	13	13	12 1/2	12 3/4	12 1/4	12	11 1/2	11 1/4	11	10 3/4	10 1/4	9 3/4	9 1/4	8 3/4	8
5" - 16.25 #	12 1/2	12 1/4	12 1/4	12	12	12	11 3/4	11 1/2	11 1/2	11 1/4	11	10 1/2	10	9 1/2	9	8 1/4	7 1/4	6 1/2	5 1/2	4 1/2	2 1/4	
5" - 19.50 #	12 1/2	12 1/4	12 1/4	12 1/4	12 1/4	12	12	11 3/4	11 1/2	11 1/2	10 1/4	10 1/2	10 1/2	10 3/4	10 1/4	11	8 1/4	9	7 1/4	7 1/2	6 1/2	6 1/4
5" - 25.60 #	12 1/2	12 1/4	12 1/4	12 1/4	12 1/4	12 1/4	12	12	12	11 1/4	10 3/4	11 3/4	10 1/2	11 1/2	11 1/4	11	10 3/4	10 1/2	10 1/4	10	9 3/4	9 1/2
5 1/2" - 19.20 #	11 1/4	11 1/4	11	11	11	11	11	10 3/4	10 3/4	10 1/2	10 1/4	10	9 1/2	9 1/2	9 1/4	8 3/4	8 1/4	7 3/4	7 1/4	6 1/2	5 1/2	4 1/2
5 1/2" - 21.90 #	11 1/4	11 1/4	11 1/4	11	11	11	11	11	10 3/4	10 3/4	10 1/4	10 1/2	10 1/4	10	9 3/4	9 1/4	9 1/4	8 1/2	8 1/2	8	7 1/2	7
5 1/2" - 24.70 #	11 1/4	11 1/4	11 1/4	11	11	11	11	11	10 3/4	10 3/4	10 3/4	10 1/2	10 1/2	10 1/4	10	10	9 3/4	9	9 1/4	9	8 1/2	8 1/4
6 5/8" - 25.20 #	9 1/4	9 1/4	9 1/4	9 1/4	9	9	9	9	9	9	8 3/4	8 3/4	8 1/2	8 1/2	8 1/4	8 1/4	8	7 3/4	7 1/2	7 1/4	7	6 1/2

MAXIMUM ALLOWABLE NUMBER OF TURNS WHICH CAN BE GIVEN TO 1000 m OF DRILL PIPE UNDER A GIVEN AXIAL TENSION
(Grade 135)

NUMBER OF TURNS FOR 1000 m OF DRILL PIPE UNDER A TENSION OF (10^3 daN)

Nominal weight and diameter of drill pipe	0	10	20	30	40	50	60	70	80	90	100	110	120	130	140	150	160	170	180	190	200	210
2 3/8" - 6.65 #																						
2 7/8" - 10.40 #																						
3 1/2" - 9.50 #	22 3/4	22 3/4	22 1/2	22 1/4	22	21 1/4	20 1/4	20	19 1/2	18 1/2	17 1/4	16	14 1/2	12 1/2	9 3/4	6						
3 1/2" - 13.30 #	22 3/4	22 3/4	22 1/2	22 1/2	22 1/4	22	22	21 1/2	21 1/4	20 3/4	20 1/4	19 3/4	18 3/4	18	17 1/4	16 1/2	15 1/2	14 1/2	13 1/4	11	9	6
3 1/2" - 15.50 #	22 3/4	22 3/4	22 1/2	22 1/2	22 1/2	22 1/4	22 1/4	22	21 3/4	21 1/4	20 3/4	20 1/2	20	19 1/2	19	18 1/2	18	17 1/4	16 1/2	15 1/2	14 1/2	13 1/4
4" - 11.85 #	20	19 3/4	19 3/4	19 1/2	19 1/4	19	18 3/4	18 1/4	17 3/4	17 1/4	16 1/2	16	15	14	13	11 3/4	10	7 1/2	4 1/2			
4" - 14.00 #	20	19 3/4	19 3/4	19 1/2	19 1/2	19 1/4	19 1/4	19	18 3/4	18 1/2	18	17 1/2	17	16 1/2	15 3/4	15 1/4	14 1/4	13	12	11	9 1/2	8
4" - 15.70 #	20	19 3/4	19 3/4	19 1/2	19 1/2	19 1/2	19 1/4	19 1/4	19	18 3/4	18 1/2	18 1/4	18	17 1/2	17	16 1/2	16	15 1/4	14 1/2	13 3/4	12 3/4	11 3/4
4 1/2" - 13.75 #	17 3/4	17 3/4	17 1/2	17 1/4	17 1/4	17	17	16 3/4	16 1/2	16	15 1/2	15	14 1/2	14	13 1/2	12 3/4	12	11	10	8 1/2	6 1/2	4 1/2
4 1/2" - 16.60 #	17 3/4	17 3/4	17 3/4	17 1/2	17 1/4	17 1/4	17 1/4	17	16 3/4	16 1/2	16 1/4	16	15 1/2	15 1/4	15	14 1/2	14	13 1/2	12 3/4	12 1/4	11 1/2	10 1/2
4 1/2" - 20.00 #	17 3/4	17 3/4	17 3/4	17 1/2	17 1/2	17 1/4	17 1/4	17 1/4	17	17	16 3/4	16 3/4	16 1/2	16 1/4	16	15 3/4	15 1/2	15	14 1/2	14 1/4	14	13 3/4
5" - 16.25 #	16	15 3/4	15 3/4	15 3/4	15 1/2	15 1/2	15 1/2	15 1/4	15 1/4	15	14 3/4	14 1/2	14 1/4	14	13 1/2	13	12 1/2	12	11 1/2	11	10 1/4	9 1/4
5" - 19.50 #	16	15 3/4	15 3/4	15 3/4	15 3/4	15 1/2	15 1/2	15 1/2	15 1/2	15 1/4	15 1/4	15	14 3/4	14 3/4	14 1/2	14 1/4	14	13 1/2	13	12 1/2	12 1/4	12
5" - 25.60 #	16	15 3/4	15 3/4	15 3/4	15 3/4	15 3/4	15 3/4	15 1/2	15 1/2	15 1/2	15 1/4	15 1/4	15 1/4	15	14 3/4	14 3/4	14 3/4	14 1/2	14 1/2	14 1/4	14 1/4	14
5 1/2" - 19.20 #	14 1/2	14 1/2	14 1/2	14 1/2	14 1/4	14 1/4	14 1/4	14 1/4	14	14	14	13 3/4	13 1/2	13 1/4	13	13	12	11	11 1/4	11		
5 1/2" - 21.90 #									14	14	14	13 3/4	15 1/4	15	14 3/4	14 3/4	14 1/2	14 1/2	14 1/2	14 1/4	12 3/4	12
5 1/2" - 24.70 #													13 1/2	13 1/4	13	12 1/2	12 1/2	12	11 3/4	11 1/4	14 1/4	14
6 5/8" - 25.20 #																					10 3/4	10

BACK-OFF OF A FROZEN STRING OF DRILL PIPE

(1) Before a back-off first determine the depth of the free point (see page K 20).
(2) Make-up drill string to a maximum of 80 % of the torsional limit (see tables K 21 to K 24).
(3) Put the neutral point on a level with the joint to back-off.
The weight indicator tension is given by :

$$T = P + \frac{ph \times S}{1000}$$

T = weight indicator tension (10^3daN),
P = weight in mud of the free length of drill pipe plus travelling block, hook, etc. (10^3daN),
ph = hydrostatic pressure at the back off point (bar),
S = area of matting surface of tool joint (cm²) (see K 26).
(4) Apply rightward twist amounting to 60 to 80 % of the rightward twist used to make-up the string (see item 2 above).

Example. Back-off at 2000 m of a string of 5'' - 19,50 # 4 1/2 IF or NC50, Class S, grade 95 drill pipe, mud weight : d = 1,40 ; weight of travelling block, hook, etc. : 8.10^3daN ; free suspended weight of drill string : 90 10^3daN.

Computations :
Weight per metre of drill pipe : 31,4 kg/m (see B 23).
Buoyancy factor : 0,822 (see E 14).
Weight in mud : 31,4 × 0,822 = 25,8 kg/m.
Weight of 2000 m of DP in mud : 2000 × 25,8 # 52 10^3daN.

Hydrostatic pressure at 2000 m : ph = $\dfrac{2000 \times 1,4}{10}$ = 280 bars.

Area of matting surface of tool joint : S = 34,7 (see K 26).

First step. Make-up drill string : for a tension of 90 10^3daN (weight of drill string). Table K 22 gives 10 1/4 maximum number of turns per 1000 m of srill pipe. For 2000 m we can give 2000 × 10,25 = 20,5 turns. We will make-up the drill string with 80 % of 20,5 # 16 turns to the right.

Second step. Put the neutral point at 2000 m. The weight indicator must show :

$$T = P + \frac{ph \times S}{1000} \quad \text{where} \quad P = 52 + 8 = 60 \quad \text{and} \quad \frac{ph \times S}{1000} = \frac{280 \times 34,7}{1000} = 10$$

$$T = 60 + 10 = 70 \ 10^3 daN.$$

Third step. Apply a twist to the left : 80 % of 16 turns = 12 turns.

TOOL JOINT MATTING SURFACE AREA

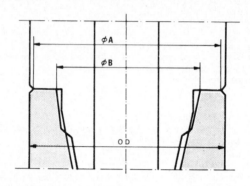

Style and size of connection		Outside diameter (in)	Diameter A (mm)	Diameter B (mm)	Area (mm²)
REG	2 3/8	3 1/8	76,60	68,26	948,23
	2 7/8	3 3/4	90,88	77,79	1735,01
	3 1/2	4 1/4	103,58	90,49	1996,28
	4 1/2	5 1/2	134,54	119,06	3082,93
	5 1/2	6 3/4	164,31	141,68	5436,57
	6 5/8	7 3/4	186,13	153,99	8587,48
FH	3 1/2	4 5/8	113,90	102,79	1891,24
	4	5 1/4	127,40	110,33	3186,34
	4 1/2	5 3/4	140,49	123,82	3460,35
	5 1/2	7	170,66	150,02	5197,72
	6 5/8	8	195,66	173,83	6334,50
IF	2 3/8	3 3/8	82,55	74,61	979,76
	2 7/8	4 1/8	100,41	87,71	1876,40
	3 1/2	4 3/4	116,28	103,58	2193,13
	4	5 3/4	140,49	124,62	3305,47
	4 1/2	6 1/8	150,42	134,94	3468,90
	5 1/2	7 3/8	181,37	163,91	4735,54

L OFFSHORE

Summary

DEFINITIONS

DEAD WEIGHT OR " DW " : total weight of shipment (tons-force) (inclusive of passengers, ballast and consumable materials) (Port en lourd).

LIGHT WEIGHT : weight of the hull with machinery (Poids lège).

LIGHT DISPLACEMENT : weight of displaced water when the vessel is empty (déplacement lège).

LOAD DISPLACEMENT : weight of displaced water when the vessel is laden (déplacement en charge).

> Light displacement + dead weight = load displacement

Note. Displacements and dead weight in metric tonnes force or in long tons-force (1016 kg).

TONNAGE : freight carrying capacity of vessel in tons-shipping US equal to 2,83 m³ or 100 cu ft (jaugeage or tonnage).

BOTTOM : part of the vessel under the waterline (carène).

DRAUGHT or DRAFT : depth of water that vessel requires to float (tirant d'eau).

STEM : metal piece to which ship's sides are joined, forming upright continuation of keel at fore-end (étrave).

STARBOARD : right side of vessel looking forward (tribord).

PORT : left side of vessel looking forward (babord).

BOW : fore-end of vessel (proue).

STERN : hind part of vessel (poupe).

FIRST ORDER WAVE (SINE WAVE)

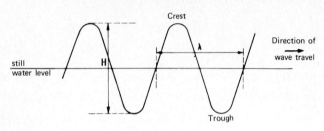

H = wave height,
λ = wave length,
T = wave period-time : in seconds to cover wave length λ,
c = water celerity : translation velocity of profile,

$$\lambda \text{ metres} = c \text{ metres per second} \times T \text{ second}$$

H_s = significant wave height,
$H 1/3$ = average height of the largest one-third of the waves,
$H 1/10$ = average height of the largest one-tenth of the waves,
$H 1/p$ = average height of the largest one-pth of the waves,
H_{max} = maximum wave height recorded.

Centennial wave : maximum wave height that might be expected to occur once in 100 years.

Decennial wave : maximum wave height that might be expected to occur once in 10 years.

   a = air gap,
   b = tide,
        + storm heightening,
        + current or wind heightening.

Zero level on charts
   France : mean level at equinoctial
low water spring (measured in metres)
USA - Great Britain : mean low water spring (MLWS)
(measured in fathoms or in feet).

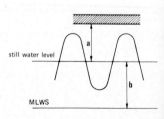

VESSEL MOTIONS

Motions due to sea conditions

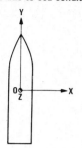

(a) *Translational motions :*
— surge : along the y-axis (cavalement),
— sway : along the x-axis (embardée),
— heave : along the z-axis (pilonnement),

(b) *Rotational motions :*
— roll : about the y-axis (roulis),
— pitch : about the x-axis (tangage),
— yaw : about the z-axis (lacet),

(c) *Horizontal motion :*
— drift (due to current, wind, etc.) (dérive).

VESSEL STABILITY

Centre of buoyancy (C) : centre of gravity of water displaced by vessel.

Centre of gravity of the vessel (G).

Metacentre (M) : point of intersection of two successive lines of action of the force of buoyancy as the vessel is inclined through a very small angle.

Distance between '' G '' centre of gravity and '' C '' centre of buoyancy : a.

$CM = \rho$ = metacentric radius,

$GM = \rho - a$ = metacentric height. Initial $(\rho - a)$ is one indication of the vessel's stability.

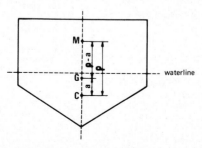

LOAD LINE MARKING
FOR OCEAN-GOING VESSELS

▬▬▬▬▬ ----- Top of deck line

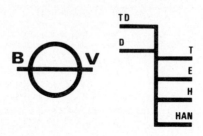

The letters BV   signify Bureau Veritas,
The letters TD   signify ligne de charge tropicale en eau douce (Tropical fresh water allowance),
The letter D   signifies ligne de charge en eau douce (Fresh water allowance),
The letter T   signifies ligne de charge tropicale (Load line in tropical zone),
The letter E   signifies ligne de charge d'été (Summer load line),
The letter H   signifies ligne de charge d'hiver (Winter load line),
The letters HAN signify ligne de charge d'hiver pour l'Atlantique Nord (Winter North Atlantic load line).

▬▬▬▬▬ ----- Top of deck line

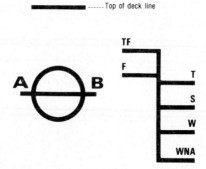

The letters AB   signify American Bureau of shipping,
The letters TF   signify Tropical Fresh water allowance,
The letter F   signifies Fresh water allowance,
The letter T   signifies load line in Tropical zone,
The letter S   signifies Summer load line,
The letter W   signifies Winter load line,
The letters WNA signify Winter North Atlantic load line.

BEAUFORT WIND VELOCITY SCALE

BEAUFORT SCALE	DESCRIPTION	WIND VELOCITY (1)			
		knots	km/h	m/s	mph
0	Calm	less than 1	less than 1,8	less than 0,5	less than 1
1	Light airs	1 - 3	1,8 - 6	0,5 - 1,7	1 - 3
2	Light breeze	4 - 6	7 - 12	1,8 - 3,3	4 - 7
3	Gentle breeze	7 - 10	13 - 18	3,4 - 5,0	8 - 12
4	Moderate breeze	11 - 16	19 - 28	5,1 - 7,8	13 - 18
5	Fresh breeze	17 - 21	29 - 38	7,9 - 10,6	19 - 24
6	Strong breeze	22 - 27	39 - 49	10,7 - 13,6	25 - 31
7	Moderate gale	23 - 33	50 - 61	13,7 - 16,9	32 - 38
8	Fresh gale	34 - 40	62 - 74	17,0 - 20,6	39 - 46
9	Strong gale	41 - 47	75 - 88	20,7 - 24,4	47 - 54
10	Whole gale	48 - 55	89 - 102	24,5 - 28,3	55 - 63
11	Storm	56 - 63	103 - 117	28,4 - 32,5	64 - 72
12	Hurricane	64 - 71	118 - 133	32,6 - 36,9	73 - 82
13	Hurricane	72 - 80	134 - 149	37,0 - 41,4	83 - 92
14	Hurricane	81 - 89	150 - 166	41,5 - 46,1	93 - 102
15	Hurricane	90 - 99	167 - 183	46,2 - 50,8	103 - 114
16	Hurricane	100 - 108	184 - 201	50,9 - 55,8	115 - 124
17	Hurricane	109 - 118	202 - 220	55,9 - 61,1	125 - 136

(1) Velocity taken 10 metres above sea level.

SEA STATE : DOUGLAS SCALE

DEGREE	SEA CONDITIONS	AVERAGE WAVE HEIGHT	
		m	ft
0	Sea like a mirror	0	0
1	Small wavelets	0 - 0,10	0 - 0,30
2	Large wavelets	0,10 - 0,50	0,3 - 2,0
3	Small waves	0,50 - 1,25	2 - 4
4	Moderate waves	1,25 - 2,50	4 - 8
5	Large waves	2,50 - 4,00	8 - 13
6	Large waves with white foam crests	4,00 - 6,00	13 - 20
7	Moderately high waves	6,00 - 9,00	20 - 30
8	High waves	9,00 - 14,00	30 - 46
9	Very high waves	higher than 14	higher than 46

VELOCITIES EXPRESSED IN km/h - m/s - mph - knots (nœuds)

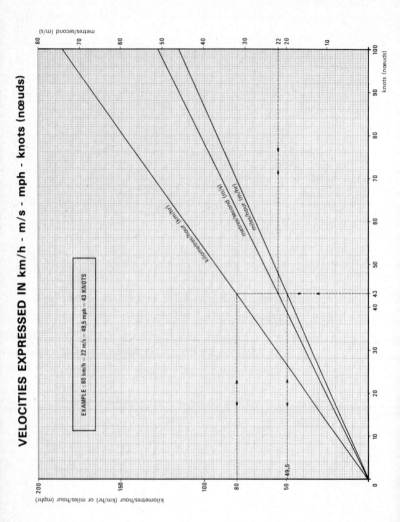

EXAMPLE : 80 km/h = 22 m/s = 49,5 mph = 43 KNOTS

WORD SPELLING CODES

Alphabet	INTERNATIONAL RADIO	France telephone
A	Alpha	Anatole
B	Bravo	Berthe
C	Charlie	Célestin
D	Delta	Désiré
E	Echo	Eugène
F	Fox-trott	François
G	Golf	Gaston
H	Hotel	Henri
I	India	Irma
J	Juliett	Joseph
K	Kilo	Kleber
L	Lima	Louis
M	Mike	Marcel
N	November	Nicolas
O	Oscar	Oscar
P	Papa	Pierre
Q	Quebec	Quintal
R	Roméo	Raoul
S	Sierra	Suzanne
T	Tango	Thérèse
U	Uniform	Ursule
V	Victor	Victor
W	Wisky	William
X	X-Ray	Xavier
Y	Yankee	Yvonne
Z	Zoulou	Zoé

INTERNATIONAL PRONUNCIATION OF NUMBERS

0	zero
1	one
2	two
3	three
4	four
5	five
6	six
7	seven
8	eight
9	nine

MOORING CHAIN PROOF AND BREAK TESTS,
LENGTH OVER FIVE LINKS, AND APPROXIMATE WEIGHT
(API Spec 2F)

Nominal diameter, d		Proof test load		Break test load		Chain length over 5 links				Approximate weight	
						in		mm			
in	mm	10³lb	kN	10³lb	kN	Mini.	Maxi.	Mini.	Maxi.	lb/ft	kg/m
1 3/8	34	158.0	668	238.0	1007	30.25	31.00	748	767	18.8	26.5
1 1/2	38	187.0	828	282.0	1248	33.00	33.80	836	857	22.3	33.0
1 5/8	42	218.0	999	329.0	1513	35.75	36.65	924	947	26.2	40.4
1 3/4	44	251.0	1097	379.0	1654	38.50	39.45	968	992	30.4	44.3
1 7/8	48	287.0	1295	432.0	1952	41.25	42.25	1056	1082	34.9	52.7
2	50	324.0	1400	489.0	2110	44.00	45.10	1100	1128	39.7	57.2
2 1/8	54	364.0	1620	548.0	2441	46.75	47.95	1188	1218	44.8	66.7
2 1/4	58	405.0	1854	611.0	2794	49.50	50.75	1276	1308	50.2	77.0
2 3/8	60	449.0	1976	676.0	2978	52.25	53.55	1320	1353	56.0	82.4
2 1/2	64	494.0	2230	744.0	3360	55.00	56.40	1408	1443	62.0	93.7
2 5/8	66	541.0	2361	815.0	3559	57.75	59.20	1452	1488	68.4	99.7
2 3/4	70	590.0	2634	889.0	3970	60.50	62.00	1540	1578	75.0	112.1
2 7/8	73	640.0	2846	965.0	4291	63.25	64.85	1606	1646	82.0	121.9
3	76	693.0	3066	1044.0	4621	66.00	67.65	1672	1714	89.3	132.2
3 1/8	79	747.0	3292	1125.0	4962	68.75	70.75	1738	1803	96.9	142.8
3 1/4	83	802.0	3603	1209.0	5430	71.50	73.25	1826	1872	104.8	157.6
3 3/8	85	859.0	3762	1295.0	5671	74.25	76.10	1870	1917	113.0	165.3
3 1/2	87	918.0	3924	1383.0	5916	77.00	78.95	1914	1962	121.5	173.2
3 5/8	92	977.0	4342	1473.0	6544	79.75	81.75	2024	2075	130.4	193.7
3 3/4	95	1039.0	4599	1566.0	6932	82.50	84.55	2090	2142	139.5	206.5
3 7/8	98	1101.0	4862	1660.0	7328	85.25	87.40	2156	2210	149.0	219.8
4	102	1165.0	5220	1756.0	7868	88.00	90.20	2244	2300	158.7	238.1
4 1/8	105	1231.0	5495	1855.0	8282	90.75	93.00	2310	2368	168.8	252.3
4 1/4	108	1297.0	5774	1955.0	8702	93.50	95.85	2376	2435	179.2	266.9
4 3/8	111	1365.0	6058	2057.0	9130	96.25	98.65	2442	2503	189.9	281.9

Mechanical properties of steel for mooring chains : Ultimate strength : 64,1 hbars (93.000 psi) - Elongation : 17 % (mini)

M WELL HEAD FLANGES BOP

Summary

PHYSICAL PROPERTIES OF STEEL FOR FLANGES
(API Spec 6A)

PROPERTY		TYPE I	TYPE 2	TYPE 3	TYPE 4
Minimum tensile strength	hbar	48,3	62,1	69,0	48,3
	psi	70 000	90 000	100 000	70 000
Minimum yield strength	hbar	24,8	41,4	51,7	31,0
	psi	36 000	60 000	75 000	45 000
Minimum elongation (% in 2'')		22	18	17	19
Minimum reduction in area %		30	35	35	32

FLANGE MATERIAL

TYPE CONNECTION	FLANGE MATERIAL Working pressure (psi)			
	2 000 3 000 5 000	10 000 (1)	15 000	20 000
Weld Neck type 6B	Type 4			
Weld Neck type 6BX		Type 2	Type 3	
Blind, threaded or integral type 6B	Type 2			
Blind or integral type 6BX	Type 2	Type 2	Type 3	
Integral type 6BX				Type 3

(1) Steel type 2 for the obsolete series 2 900 flange.

API RING-JOINT GASKET MATERIAL
AND IDENTIFICATION

MATERIAL	Brinell hardness number (max.) (1)	Rockwell " B " scale (max.) (2)	Indentification mark
Soft iron (3)	90	50	D
Low carbon steel (3)	120	68	S
Type 304 stainless steel	160	83	S304
Type 316 stainless steel	160	83	S316

(1) Brinell hardness measured with 3 000 kg load except soft iron which is measured with 500 kg load.
(2) Rockwell " B " measured with 100 kg-load and 1/16 in diameter ball.
(3) Unless otherwise specified on the purchasse order, soft iron and steel gaskets ta be cadmium plated 0,0002-0,0005 in.

MATCHING TUBULAR GOODS SIZES FOR USE WITH FLANGES
(API Spec 6A)

		SERIES 2 000 - 3 000 - 5 000 FLANGES			SERIES 10 000 - 15 000 - 20 000 FLANGES		
Nominal *	Old	Size of tubular material			Nominal	Size of tubular material	
size and bore (in)	nominal size	line-pipe nominal	fubing OD (in)	casing OD (in)	size and bore (in)	tubing OD (in)	casing OD (in)
1 13/16	1 1/2	1 1/2	1,660 and 1,900		1 11/16	1,900	
2 1/16	2	2	1,660 thr 2 3/8		11 3/16	2,063	
2 9/16	2 1/2	2 1/2	2 7/8		2 1/16	2 3/8	
3 1/8	3	3	3 1/2		2 9/16	2 7/8	
4 1/16	4	4	4 and 4 1/2	4 1/2	3 1/16	3 1/2	
7 1/16	6	6	—	4 1/2 thr 7	4 and 4 1/2	4 and 4 1/2	4 1/2
9	8	8	—	7 5/8 and 8 5/8	7 1/16		4 1/2 thr 7
11	10	10	—	9 5/8 and 10 3/4	9 (3)		7 5/8 and 8 5/8
13 5/8 (1)	12	12	—	11 3/4 and 13 3/8	11 (3)		8 5/8 and 9 5/8
13 5/8 (2)	13 5/8	—	—	11 3/4 and 13 3/8	13 5/8 (4)		10 3/4 and 11 3/4
16 3/4	16	16	—	16	16 3/4 (4)		16
16 3/4 (2)	16 3/4	—	—	16			
21 1/4 (2)	20	20	—	20			
20 3/4	20	20	—	20			

(1) This 6B flange is limited to a maximum working pressure rating of 3 000 psi (207 bars) when used over 11 3/4 13 3/8 casing.
(2) Type 6BX flanges are required for 5 000 psi (345 bars) maximum working pressure rating in these sizes.
(3) Available in 10 000 and 15 000 psi (690 and 1035 bars) rating only.
(4) Available in 10 000 psi (690 bars) rating only.
* Beginning with the Eleventh Edition of API Spec 6A, the traditional 6B flange nominal size designation is changed to a through-bore designation.

RECOMMENDED 6BX FLANGE BOLT TORQUE

BOLT SIZE	MAKE-UP TORQUE	
	m-daN	ft-lb
3/4 - 10 UNC	27	200
7/8 - 9 UNC	44	325
1 - 8 UNC	64	475
1 1/8 - 8 UN	81	600
1 3/8 - 8 UN	163	1200
1 1/2 - 8 UN	190	1400
1 5/8 - 8 UN	230	1700
1 3/4 - 8 UN	277	2040
1 7/8 - 8 UN	437	3220
2 - 8 UN	522	3850

API TYPE 6B - 960 FLANGES (1)
(API Spec 6A)

Maximum working pressure : 66,2 bars (960 psi)

Test pressure : 100 bars (1450 psi)

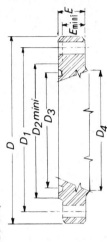

All dimensions in inches

Nominal size	Outside diameter D	Total thickness E	Basic thickness E mini	Diameter of raised face D₂ mini	Diameter of hub D	Diameter of bolt circle D₁	Number of bolts	Diameter of bolts	Length of stud bolts	Ring-joint type R or RX	Pitch diam. of groove D₃
1 1/2	6 1/8	1 1/8	7/8	3 9/16	2 3/4	4 1/2	4	3/4	4 1/2	20	2 11/16
2	6 1/2	1 5/16	1	4 1/4	3 5/16	5	8	5/8	4 3/4	23	3 1/4
2 1/2	7 1/2	1 7/16	1 1/8	5	3 15/16	5 7/8	8	3/4	5 1/4	26	4
3	8 1/4	1 9/16	1 3/4	5 3/4	4 5/8	6 5/8	8	3/4	5 1/2	31	4 7/8
4	10	1 11/16	1 3/8	6 7/8	5 3/4	7 7/8	8	7/8	6	37	5 7/8
5	11	1 13/16	1 1/2	8 1/4	7	9 1/4	8	7/8	6 1/4	41	7 1/8
6	12 1/2	1 15/16	1 5/8	9 1/2	8 1/8	10 5/8	12	7/8	6 1/2	45	8 5/16
8	15	2 3/16	1 7/8	11 7/8	10 1/4	13	12	1	7 3/8	49	10 5/8
10	17 1/2	2 7/16	2 1/8	14	12 5/8	15 1/4	16	1 1/8	8 1/8	53	12 3/4
12	20 1/2	2 9/16	2 1/4	16 1/4	14 3/4	17 3/4	16	1 1/4	9	57	15
16	25 1/2	2 13/16	2 1/2	20	19	22 1/2	20	1 3/8	9 3/4	65	18 1/2
20	30 1/2	3 1/8	2 3/4	25	23 1/8	27	24	1 1/2	10 3/4	73	23

(1) Obsolete series, presented here for information only.

API TYPE 6B - 2 000 FLANGES

Maximum working pressure : 137 bars (2 000 psi)

Test pressure : flanged 14 in and smaller 275 bars (4 000 psi)
Test pressure : flanged 16 in and higher 206 bars (3 000 psi)

All dimensions in inches

Nominal size and bore (1)	Old nominal size	Outside diameter D	Total thickness E	Basic thickness E mini	Diameter of raised face D_2 mini	Diameter of hub D_4	Diameter of bolt circle D_1	Number of bolts	Diameter of bolts	Length of stud bolts	Ring-joint Type R or RX	Pitch diam. of groove D_3
1 13/16*	1 1/2	6 1/8	1 1/8	7/8	3 9/16	2 3/4	4 1/2	4	3/4	4 1/4	20	2 11/16
2 1/16	2	6 1/2	1 5/16	1	4 1/4	3 5/16	5	8	5/8	4 1/2	23	3 1/4
2 9/16	2 1/2	7 1/2	1 7/16	1 1/8	5	3 15/16	5 7/8	8	3/4	5	26	4
3 1/8	3	8 1/4	1 9/16	1 1/4	5 3/4	4 5/8	6 5/8	8	3/4	5 1/4	31	4 7/8
4 1/16	4	10 3/4	1 13/16	1 1/2	6 7/8	6	8 1/2	8	7/8	6	37	5 7/8
5 1/8*	5	13	2 1/16	1 3/4	8 1/4	7 7/16	10 1/2	8	1	6 3/4	41	7 1/8
7 1/16	6	14	2 3/16	1 7/8	9 1/2	8 3/4	11 1/2	12	1	7	45	8 5/16
9	8	16 1/2	2 1/2	2 3/16	11 7/8	10 3/4	13 3/4	12	1 1/8	8	49	10 5/8
11	10	20	2 13/16	2 1/2	14	13 1/2	17	16	1 1/4	8 3/4	53	12 3/4
13 5/8	12	22	2 15/16	. 2 5/8	16 1/4	15 3/4	19 1/4	20	1 1/4	9	57	15
16 3/4	16	27	3 5/16	3	20	19 1/2	23 3/4	20	1 1/2	10 1/4	65	18 1/2
17 3/4*	18	29 1/4	3 9/16	3 1/4	22 5/8	21 1/2	25 3/4	20	1 5/8	11	69	21
21 1/4	20	32	3 7/8	3 1/2	25	24	28 1/2	24	1 5/8	11 3/4	73	23

* These sizes inactive : available on special order only.
(1) Beginning with the Eleven Edition of API Spec. 6A (October 1977), the traditional 6B flange nominal size designation is changed to a through-bore designation.

API TYPE 6B - 3 000 FLANGES

Maximum working pressure : 206 bars (3 000 psi)

Test pressure : flanged 14 in and smaller 412 bars (6 000 psi)
Test pressure : flanged 16 in and higher 3098 bars (4 500 psi)

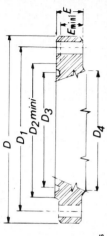

All dimensions in inches

Nominal size and bore (1)	Old nominal size	Outside diameter D	Total thickness E	Basic thickness E mini	Diameter of raised face D₂ mini	Diameter of hub D₄	Diameter of bolt circle D₁	Number of bolts	Diameter of bolts	Length of stud bolts	Ring-joint Type R or RX	Pitch diam. of groove D₃
1 13/16*	1 1/2	7	1 1/2	1 3/4	3 5/8	2 3/4	4 7/8	4	1	5 1/2	20	2 11/16
2 1/16	2	8 1/2	1 13/16	1 1/2	4 7/8	4 1/8	6 1/2	8	7/8	6	24	3 3/4
2 9/16	2 1/2	9 5/8	1 15/16	1 5/8	5 3/8	4 7/8	7 1/2	8	1	6 1/2	27	4 1/4
3 1/8	3	9 1/2	1 13/16	1 1/2	6 1/8	5	7 1/2	8	7/8	6	31	4 7/8
4 1/16	4	11 1/2	2 1/16	1 3/4	7 1/8	6 1/4	9 1/4	8	1 7/8	7	37	5 7/8
5 1/8*	5	13 3/4	2 5/16	2	8 1/2	7 1/2	11	8	1 1/4	7 3/4	41	7 1/8
7 1/16	6	15	2 1/2	2 3/16	9 1/2	9 1/4	12 1/2	12	1 1/8	8	45	8 5/16
9	8	18 1/2	2 13/16	2 1/2	12 1/8	11 3/4	15 1/2	12	1 3/8	9	49	10 5/8
11	10	21 1/2	3 1/16	2 3/4	14 1/4	14 1/2	18 1/2	16	1 3/8	9 1/2	53	12 3/4
13 5/8	12	24	3 7/16	3 1/8	16 1/2	16 1/2	21	20	1 3/8	10 1/4	57	15
16 3/4	16	27 3/4	3 15/16	3 1/2	20 5/8	20	24 1/4	20	1 5/8	11 3/4	66	18 1/2
17 3/4	18	31	4 1/2	4	23 3/8	22 1/4	27	20	1 7/8	13 3/4	70	21
21 1/4	20	33 3/4	4 3/4	4 1/4	25 1/2	24 1/2	29 1/2	20	2	14 1/2	74	23

(1) Beginning with the Eleven Edition of API Spec 6A (October 1977), the traditional 6B flange nominal size designation is changed to a through-bore designation.
These sizes inactive : available on special order only.
Note. Except for bore of welding neck flanges, dimensions for sizes 1 13/16 in to 2 9/16 in inclusive are identical with 5 000 psi flanges in table next page.

M 6

API TYPE 6B - 5 000 FLANGES

Maximum working pressure : 343 bars (5 000 psi)

Test pressure : 690 bars (10 000 psi)

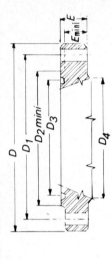

All dimensions in inches

Nominal size and bore (1)	Old nominal size	Outside diameter D	Total thickness E	Basic thickness E mini	Diameter of raised face D_2 mini	Dimater of hub D_4	Diameter of bolt circle D_3	Number of bolts	Diameter of bolts	Length of stud bolts	Ring-joint Type R or RX	Pitch diam. of groove D_1
1 13/16*	1 1/2	7	1 1/2	1 1/4	3 5/8	2 3/4	4 7/8	4	1	5 1/2	20	2 11/16
2 1/16	2	8 1/2	1 13/16	1 1/2	4 7/8	4 1/8	6 1/2	8	7/8	6	24	3 3/4
2 9/16	2 1/2	9 5/8	1 15/16	1 5/8	5 3/8	4 7/8	7 1/2	8	1	6	27	4 1/4
3 1/8	3	10 1/2	2 3/16	1 7/8	6 5/8	5 1/4	8	8	1 1/8	7 1/4	35	5 3/8
4 1/16	4	12 1/4	2 7/16	2 1/8	7 5/8	6 3/8	9 1/2	8	1 1/8	8	39	6 3/8
5 1/8	5	14 3/4	3 3/16	2 7/8	9	7 3/4	11 1/2	8	1 1/2	10	44	7 5/8
7 1/16	6	15 1/2	3 5/8	3 1/4	9 3/4	9	12 1/2	12	1 3/8	10 3/4	46	8 5/16
9	8	19	4 1/16	3 5/8	12 1/2	11 1/2	15 1/2	12	1 5/8	12	50	10 5/8
11	10	23	4 11/16	4 1/4	14 5/8	14 1/2	19	12	1 7/8	13 3/4	54	12 3/4
13 5/8 (2)	13 5/8	26 1/2	4 7/16	3 7/8	18	18 15/16	23 1/4	16	1 5/8	12 1/2	BX160	16.063
14	14	29 1/2	5 7/8	5 1/4	19 1/4	19 1/2	25	16	2 1/4	17 1/4	63	16 1/2
16 3/4 (2)	16 3/4	30 3/8	5 1/8	4 51/64	21 1/16	21 7/8	26 5/8	16	1 7/8	14 1/2	BX162	18.832

(1) Beginning with the Eleven Edition of API Spec. 6A (October 1977), the traditional 6B flange nominal size designation is changed a through-bore designation.
(2) See table page M8 for dimensions details on these sizes.
• These sizes inactive : available on special order only.

OBSOLETE SERIES 2 900 FLANGES (1)
(API Spec 6A)

Maximum working pressure : 690 bars (10 000 psi)

Test pressure : 1 030 bars (15 000 psi)

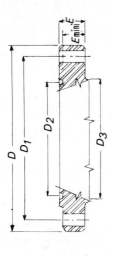

All dimensions in inches

Nominal size	Outside diameter D	Total thickness E	Basic thickness E mini	Diameter of hub D_3	Diameter of bolt circle D_1	Number of bolts	Diameter of bolts	Length of stud bolts	Ring-joint Type R or RX	Pitch diam. of groove D_2
1	6 3/4	2	1 11/16	1 7/8	4 5/8	4	1	6 3/4	82	2 1/4
1 1/2	7 1/4	2 1/16	1 3/4	2 3/8	5 1/8	4	1	6 3/4	84	2 1/2
2	7 3/4	2 3/8	2	3 1/8	5 3/4	8	7/8	7	85	3 1/8
2 1/2	8 7/8	2 3/4	2 5/16	3 3/4	6 5/8	8	1	8	86	3 9/16
3	10	3 1/16	2 5/8	4 3/8	7 1/2	8	1 1/8	9	87	3 15/16
3 1/2	11 1/2	3 3/8	2 7/8	5	8 1/2	8	1 1/4	9 3/4	89	4 1/2
4	12 1/2	3 5/8	3 1/8	5 9/16	9 1/2	8	1 3/8	10 1/2	88	4 7/8
5	14 1/4	4 3/16	3 5/8	6 13/16	11	8	1 5/8	12 1/2	90	6 1/8
10	20 3/4	5 11/16	5	11 3/4	16 3/4	12	2	16 1/2	91	10 1/4

(1) The 2 900 series flange dimensions are shown here for information only and as a convenience for former users.

API TYPE 6BX INTEGRAL FLANGES - 2 000 - 3 000 - 5 000 psi
(API Spec 6A)

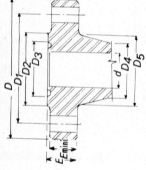

All dimensions in inches

Nominal size and bore d	Outside diameter D	Total thickness E	Basic thickness E mini	Diameter of raised face D_2	Large diam. of hub D_3	Small diam. of hub D_4	Diameter of bolt circle D_1	Number of bolts	Diameter of bolts	Length of stud bolts	Ring-joint Type BX	Groove OD D_5
Maximum working pressure 137 bars (2 000 psi)								Test pressure 206 bars (3 000 psi)				
						2 000 psi						
26 3/4	41	4 31/32	—	31 11/16	32 29/32	29 1/4	37 1/2	20	1 3/4	13 3/4	167	30,249
Maximum working pressure 206 bars (3 000 psi)								Test pressure 309 bars (4 500 psi)				
						3 000 psi						
26 3/4	43 3/8	6 11/32	—	32 3/4	34 1/4	30 9/16	39 3/8	24	2	17	168	30,481
Maximum working pressure 343 bars (5 000 psi)								Test pressure 690 bars (10 000 psi)				
						5 000 psi						
13 5/8	26 1/2	4 7/16	3 7/8	18	18 15/16	16 11/16	23 1/4	16	1 5/8	12 1/2	160	16,063
16 3/4 (1)	30 3/8	5 1/8	4 51/64	21 1/16	21 7/8	20 3/4	26 5/8	16	1 7/8	14 1/2	162	18,832
16 3/4 (2)	30 1/4	5 3/64	4 3/4	21 7/8	22 1/4	19 13/16	26 3/4	16	1 3/4	14 1/4	161	19,604
18 3/4	35 5/8	6 17/32	—	24 11/16	26 9/16	23 9/16	31 5/8	20	2	17 1/2	163	22,185
21 1/4	39	7 1/8	—	27 5/8	29 7/8	26 3/4	34 7/8	24	2	18 3/4	165	24,904

(1) This flange was adopted in June 1969 and must be marked with both working pressure (5 000 WP) and the test pressure (10 000 TP) in addition to other marking requirements.
(2) This 16 3/4 flange is obsolete (5 000 WP) and (7 500 TP), shown here for information only.

API TYPE 6BX INTEGRAL FLANGES - 10 000 psi
(API Spec 6A)

Maximum working pressure : 690 bars (10 000 psi)

Test pressure : 1 030 bars (15 000 psi)

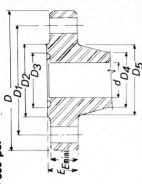

All dimensions in inches

Nominal size and bore d	Outside diameter D	Total thickness E	Basic thickness E mini	Diameter of raised face D_2	Large diam. of hub D_3	Small diam. of hub D_4	Diameter of bolt circle D_1	Number of bolts	Diameter of bolts	Length of stud bolt	Ring-joint Type BX	Groove OD D_5
1 11/16 *	7 3/16	1 21/32	1 7/16	4	3 5/16	2 13/32	5 9/16	8	3/4	5	150	2,893
1 13/16	7 3/8	1 21/32	1 7/16	4 1/8	3 1/2	2 9/16	5 3/4	8	3/4	5	151	3,062
2 1/16	7 7/8	1 47/64	1 1/2	4 3/8	3 15/16	2 15/16	6 1/4	8	3/4	5 1/4	152	3,395
2 9/16	9 1/8	2 1/64	1 3/4	4 3/4	4 3/4	3 5/8	7 1/4	8	7/8	6	153	4,046
3 1/16	10 5/8	2 19/64	2	6	5 19/32	4 11/32	8 1/2	8	1	6 3/4	154	4,685
1/16	12 7/16	2 49/64	2 7/16	7 9/32	7 3/16	5 3/4	10 3/16	8	1 1/8	8	155	5,930
7 1/16	18 7/8	4 1/16	3 5/8	11 7/8	11 7/8	10	15 7/8	12	1 1/2	11 1/4	156	9,521
9	21 3/4	4 7/8	4 3/8	14 1/8	14 3/4	12 7/8	18 3/4	16	1 1/2	13	157	11,774
11	25 3/4	5 9/16	5	16 7/8	17 3/4	15 3/4	22 1/4	16	1 3/4	15	158	14,064
13 5/8	30 1/4	6 5/8	6	20 3/8	21 3/4	19 1/2	26 1/2	20	1 7/8	17 1/4	159	17,033
16 3/4	34 5/16	6 5/8	6 19/64	22 11/16	25 13/16	23 11/16	30 9/16	24	1 7/8	17 1/2	162	18,832
18 3/4	40 15/16	8 25/32	—	27 7/16	29 5/8	26 9/16	36 7/16	24	2 1/4	22 1/2	164	22,752
21 1/4	45	9 1/2	—	30 3/4	33 3/8	30	40 1/4	24	2 1/4	24 1/2	166	25,507

API TYPE 6BX INTEGRAL FLANGES - 15 000 psi
(API Spec 6A)

Maximum working pressure : 1 030 bars (15 000 psi)

Test pressure : 1 545 bars (22 500 psi)

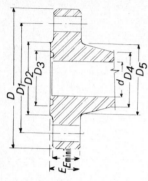

All dimensions in inches

Nominal size and bore d	Outside diameter D	Total thickness E	Basic thickness E mini	Diameter of raised face D_2	Large diam. of hub D_3	Small diam. of hub D_4	Diameter of bolt circle D_1	Number of bolts	Diameter of bolts	Length of stud bolts	Ring-joint Type BX	Groove OD D_5
1 11/16 *	7 5/8	1 3/4	1 1/2	3 13/16	3 11/16	2 11/16	6	8	3/4	5 1/4	150	2,893
1 13/16	8 3/16	1 25/32	1 9/16	4 3/16	3 27/32	2 13/16	6 5/16	8	7/8	5 1/2	151	3,062
2 1/16	8 3/4	2	1 3/4	4 1/2	4 3/8	3 1/4	6 7/8	8	7/8	6	152	3,395
2 9/16	10	2 1/4	2	5 1/4	5 1/16	3 15/16	7 7/8	8	1	6 3/4	153	4,046
3 1/16	11 5/16	2 17/32	2 1/4	6 1/16	6 1/16	4 13/16	9 1/16	8	1 1/8	7 1/2	154	4,685
4 1/16	14 8/16	3 3/32	—	7 11/16	7 11/16	6 1/4	11 7/16	8	1 3/8	9 1/4	155	5,930
7 1/16	19 7/8	4 11/16	4 1/4	12	12 13/16	10 7/8	16 7/8	16	1 1/2	12 3/4	156	9,521
9	25 1/2	5 3/4	5 1/4	15	17	13 3/4	21 3/4	16	1 7/8	15 3/4	157	11,774
11	32	7 3/8	6 11/16	17 7/8	23	16 13/16	28	20	2	19 1/4	158	14,064

API TYPE 6BX INTEGRAL FLANGES - 20 000 psi
(API Spec 6A)

Maximum working pressure : 1 380 bars (20 000 psi)

Test pressure : 2 069 bars (30 000 psi)

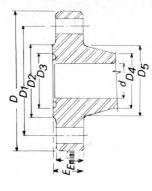

All dimensions in inches

Nominal size and bore d	Outside diameter D	Total thickness E	Basic thickness E mini	Diameter of raised face D₂	Large diam. of hub D₃	Small diam. of hub D₄	Diameter of bolt circle D₁	Number of bolts	Diameter of bolts	Length of stud bolts	Ring-joint Type BX	Groove OD D₅
1 13/16	10 1/8	2 1/2	2 9/32	4 5/8	5 1/4	4 5/16	8	8	1	7 1/2	151	3.062
2 1/16	11 5/16	2 13/16	2 37/64	5 3/16	6 1/16	5	9 1/16	8	1 1/8	8 1/4	152	3.395
2 9/16	12 13/16	3 1/8	2 55/64	5 15/16	6 13/16	5 11/16	10 5/16	8	1 1/4	9 1/4	153	4.046
3 1/16	14 1/16	3 3/8	3 5/64	6 3/4	7 9/16	6 5/16	11 5/16	8	1 3/8	10	154	4.685
4 1/16	17 9/16	4 3/16	3 57/64	8 5/8	9 9/16	8 1/8	14 1/16	8	1 3/4	12 1/4	155	5.930
7 1/16	25 13/16	6 1/2	6 1/16	13 7/8	15 3/16	13 5/16	21 13/16	16	2	17 1/2	156	9.521

API TYPE R RING-JOINT GASKETS
(API Spec 6A) (for use in 6B flanges)

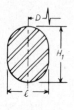

oval

All dimensions in inches

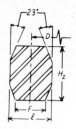

octagonal

Ring number	Pitch diam. of ring D	Width of ring ℓ	Height of ring		Width of flat octagonal ring F	Approximate distance between make-up flanges
			oval	octagonal		
R 20	2 11/16	5/16	9/16	1/2	0,206	5/32
R 23	3 1/4	7/16	11/16	5/8	0,305	3/16
R 24	3 3/4	7/16	11/16	5/8	0,305	3/16
R 26	4	7/16	11/16	5/8	0,305	3/16
R 27	4 1/4	7/16	11/16	5/8	0,305	3/16
R 31	4 7/8	7/16	11/16	5/8	0,305	3/16
R 35	5 3/8	7/16	11/16	5/8	0,305	3/16
R 37	5 7/8	7 16	11/16	5/8	0,305	3/16
R 39	6 3/8	7/16	11/16	5/8	0,305	3/16
R 41	7 1/8	7/16	11/16	5/8	0,305	3/16
R 44	7 5/8	7/16	11/16	5/8	0,305	3/16
R 45	8 5/16	7/16	11/16	5/8	0,305	3/16
R 46	8 5/16	1/2	3/4	11/16	0,341	1/8
R 47	9	3/4	1	15/16	0,485	5/32
R 49	10 5/8	7/16	11/16	5/8	0,305	3/16
R 50	10 5/8	5/8	7/8	13/16	0,413	5/32
R 53	12 3/4	7/16	11/16	5/8	0,305	3/16
R 54	12 3/4	5/8	7/8	13/16	0,413	5/32
R 57	15	7/16	11/16	5/8	0,305	3/16
R 63	16 1/2	1	1· 5/16	1 1/4	0,681	7/32
R 65	18 1/2	7/16	11/16	5/8	0,305	3/16
R 66	18 1/2	5/8	7/8	13/16	0,413	5/32
R 69	21	7/16	11/16	5/8	0,305	3/16
R 70	21	3/4	1	1	0,485	3/16
R 73	23	1/2	3/4	11/16	0,341	1/8
R 74	23	3/4	1	15/16	0,485	3/16
R 82	2 1/4	7/16	—	5/8	0,305	3/16
R 84	2 1/2	7/16	—	5/8	0,305	3/16
R 85	3 1/8	1/2	—	11/16	0,341	1/8
R 86	3 9/16	5/8	—	13/16	0,413	5/32
R 87	3 15/16	5/8	—	13/16	0,413	5/32
R 88	4 7/8	3/4	—	15/16	0,485	3/16
R 89	4 1/2	3/4	—	15/16	0,485	3/16
R 90	6 1/8	7/8	—	1 1/16	0,583	3/16
R 91	10 1/4	1 1/4	—	1 1/2	0,879	5/16
R 99	9 1/4	7/16	—	5/8	0,805	3/16

API TYPE RX PRESSURE ENERGIZED
RING-JOINT GASKETS (API Spec 6A)
(for use in 6B flanges and segmented flanges)

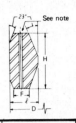

See note

All dimensions in inches

Ring number	Outside diameter of ring D	Total width ℓ	Width of flat F	Height H	Pitch diameter of groove	Approximate distance between made-up flanges
RX 20	3	11/32	0,182	3/4	2 11/16	3/8
RX 23	3 43/64	15/32	0,254	1	3 1/4	15/32
RX 24	4 11/64	15/32	0,254	1	3 3/4	15/32
RX 26	4 13/32	15/32	0,254	1	4	15/32
RX 27	4 21/32	15/32	0,254	1	4 1/4	15/32
RX 31	5 19/64	15/32	0,254	1	4 7/8	15/32
RX 35	5 51/64	15/32	0,254	1	5 3/8	15/32
RX 37	6 19/64	15/32	0,254	1	5 7/8	15/32
RX 39	6 51/64	15/32	0,254	1	6 3/8	15/32
RX 41	7 35/64	15/32	0,254	1	7 1/8	15/32
RX 44	8 3/64	15/32	0,254	1	7 5/8	15/32
RX 45	8 47/64	15/32	0,254	1	8 5/16	15/32
RX 46	8 3/4	17/32	0,263	1 1/8	8 5/16	15/32
RX 47	9 21/32	25/32	0,407	1 5/8	9	23/32
RX 49	11 3/64	15/32	0,254	1	10 5/8	15/32
RX 50	11 5/32	21/32	0,335	1 1/4	10 5/8	15/32
RX 53	13 11/64	15/32	0,254	1	12 3/4	15/32
RX 54	13 9/32	21/32	0,335	1 1/4	12 3/4	15/32
RX 57	15 27/64	15/32	0,254	1	15	15/32
RX 63	17 25/64	1 1/16	0,582	2	16 1/2	27/32
RX 65	18 59/64	15/32	0,254	1	18 1/2	15/32
RX 66	19 1/32	21/32	0,335	1 1/4	18 1/2	15/32
RX 69	21 27/64	15/32	0,254	1	21	15/32
RX 70	21 21/32	25/32	0,407	1 5/8	21	23/32
RX 73	23 15/32	17/32	0,263	1 1/4	23	19/32
RX 74	23 21/32	25/32	0,407	1 5/8	23	23/32
RX 82	2 43/64	15/32	0,254	1	2 1/4	15/32
RX 84	2 59/64	15/32	0,254	1	2 1/2	15/32
RX 85	3 35/64	17/32	0,263	1	3 1/8	3/8
RX 86	4 5/64	19/32	0,335	1 1/8	3 9/16	3/8
RX 87	4 29/64	19/32	0,335	1 1/8	3 15/16	3/8
RX 88	5 31/64	11/16	0,407	1 1/4	4 7/8	3/8
RX 89	5 7/64	23/32	0,407	1 1/4	4 1/2	3/8
RX 90	6 7/8	25/32	0,479	1 3/4	6 1/8	23/32
RX 91	11 19/64	1 3/16	0,780	1 25/32	10 1/4	3/4
RX 99	9 43/64	15/32	0,254	1	9 1/4	15/32
RX 201	2,026	0,226	0,126	0,445	—	—
RX 205	2 29/64	7/32	0,120	0,437	—	—
RX 210	3 27/32	3/8	0,213	0,750	—	—
RX 215	5 35/64	15/32	0,210	1,000	—	—

Note. The pressure passage hole illustrated in the RX ring cross-section is required in rings RX - 82 through RX - 91 only.

API TYPE BX PRESSURE ENERGIZED RING-JOINT GASKETS (API Spec 6A) (for use in 6BX flanges)

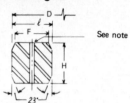

See note

All dimensions in inches

Ring number	Nominal size	Outside diameter D	Total width ℓ	Width of flat F	Height H
BX 150	1 11/16	2,842	0,366	0,314	0,366
BX 151	1 13/16	3,008	0,379	0,325	0,379
BX 152	2 1/16	3,334	0,403	0,346	0,403
BX 153	2 9/16	3,974	0,448	0,385	0,448
BX 154	3 1/16	4,600	0,488	0,419	0,488
BX 155	4 1/16	5,825	0,560	0,481	0,560
BX 156	7 1/16	9,367	0,733	0,629	0,733
BX 157	9	11,593	0,826	0,709	0,826
BX 158	11	13,860	0,911	0,782	0,911
BX 159	13 5/8	16,800	1,012	0,869	1,012
BX 160	13 5/8	15,850	0,541	0,408	0,938
BX 161 (1)	16 3/4	19,347	0,638	0,482	1,105
BX 162	16 3/4	18,720	0,560	0,481	0,560
BX 163	18 3/4	21,896	0,684	0,516	1,185
BX 164	18 3/4	22,463	0,968	0,800	1,185
BX 165	21 1/4	24,595	0,728	0,550	1,261
BX 166	21 1/4	25,198	1,029	0,851	1,261
BX 167	26 3/4	29,896	0,516	0,316	1,412
BX 168	26 3/4	30,128	0,632	0,432	1,412

(1) The BX 161 gasket is used only with the obsolete 16 3/4 5 000 psi WP 7 500 psi TP flange.
Note. All BX gaskets have a pressure passage hole.

CROSSOVER FLANGE CONNECTIONS (API Spec 6A)
CROSSOVER SPOOLS

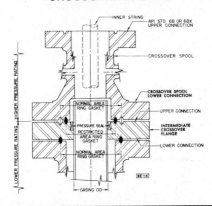

API 6B FLANGE (1)				CROSSOVER FLANGE CONNECTION		
Nominal size flange	Normal pressure rating		Normal gasket number RX or R	Restricted area gasket number RX or R	Increased pressure rating	
	bar	psi			bar	psi
8	138	2000	49	99	207	3000
8	207	3000	49	99	345	5000
8	345	5000	50	47	690	10000
10	138	2000	53	49	207	3000
10	207	3000	53	49	345	5000
10	345	5000	54	50	690	10000
12	138	2000	57	53	207	3000
12	207	3000	57	53	345	5000
13 5/8*	345	5000	160*	54	690	10000
16	138	2000	65	57	207	3000
16	207	3000	66	57	345	5000
20	138	2000	73	65	207	3000

(1) Connection which may be converted to a crossover connection by the addition of the restricted area gasket groove listed in column 5.
* API type 6BX flange and type BX ring-joint gasket.

COMPARISON OF API AND ANSI FLANGES
(ANSI was previously ASA)

(For more details see API Spec 6A and ANSI Std B16-5)

I. API versus ANSI flange dimensions

All API 6B flange dimensions agree with ANSI B16-5 flange dimensions

FLANGE TYPE	Pressure class designation		Nominal size range			
	ANSI	API	ANSI		API	
	psi	psi	in	cm	in	cm
Welding neck	600	2000	1/2 thr 24	1,3 thr 61,0	1 1/2 thr 11	3,8 thr 27,9
	900	3000	1/2 thr 24	1,3 thr 61,0	1 1/2 thr 11	3,8 thr 27,9
	1500	5000	1/2 thr 24	1,3 thr 61,0	1/2 thr 11	3,8 thr 27,9
Blind threaded	600	2000	1/2 thr 24	1,3 thr 61,0	1 1/2 thr 21 1/4	3,8 thr 54,0
and integral (1)	900	3000	1/2 thr 24	1,3 thr 61,0	1 1/2 thr 20 3/4	3,8 thr 52,7
	1500	5000	1/2 thr 24	1,3 thr 61,0	1 1/2 thr 11	3,8 thr 27,9

(1) API flanges with casing or tubing thread in certain sizes and pressure classes have hub lengths greater than required by ANSI B16-5.

II. API versus ANSI flange pressure ratings and material

FLANGE TYPE	Designation		Pressure rating				Material	
	ANSI	API	ANSI		API		ANSI	API
	psi	psi	psi	bar	psi	bar		
Welding neck	600	2000	1440	99	2000	138	ASTM	API
	900	3000	2160	149	3000	207	30 000 psi or	45 000 psi
	1500	5000	3600	248	5000	345	20,7 hbar YP	31 hbar YP
Blind, threaded	600	2000	1440	99	2000	138	ASTM	API type 2
and integral	900	3000	2160	149	3000	207	30 000 psi or	60 000 psi
	1500	5000	3600	248	5000	345	20,7 hbar YP	41,4 hbar YP

III. API versus ANSI flange physical properties

PROPERTY	ANSI				API			
	Forged ASTM A 105-1		Cast ASTM A216-WCB		Type 2		Type 4	
	psi	hbar	psi	hbar	psi	hbar	psi	hbar
Tensile strength minimal	60000	41,4	70000	48,3	90000	62,0	70000	48,3
Yeld strength minimal	30000	20,7	36000	24,8	60000	41,4	45000	31,0
Elongation in 2 in %	25		22		18		19	
Reduction in area %	38		35		35		32	

CLAMP CONNECTIONS - DIMENSIONAL DATA

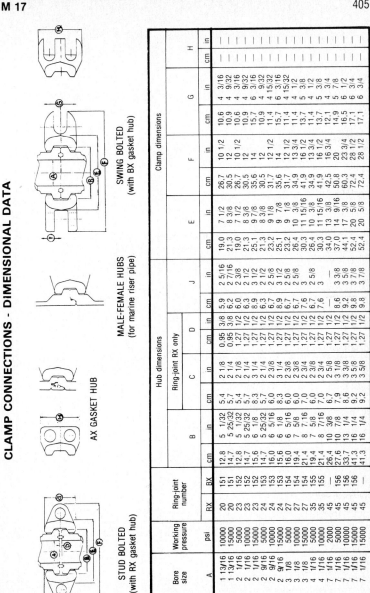

STUD BOLTED
(with RX gasket hub)

AX GASKET HUB

MALE-FEMALE HUBS
(for marine riser pipe)

SWING BOLTED
(with BX gasket hub)

Bore size	Working pressure	Ring-joint number		Hub dimensions								Clamp dimensions							
						Ring-joint RX only													
				B		C		D		J		E		F		G		H	
A	psi	RX	BX	cm	in	cm	in	cm	in	cm	in	cm	in	cm	in	cm	in	cm	in
1 13/16	10000	20	151	12.8	5 1/32	5.4	2 1/8	0.95	3/8	5.9	2 5/16	19.0	7 1/2	26.7	10 1/2	10.6	4 3/16	—	—
1 13/16	15000	20	151	14.7	5 25/32	5.7	2 1/4	0.95	3/8	6.2	2 7/16	21.3	8 3/8	30.5	12	10.9	4 9/32	—	—
2 1/16	5000	23	152	12.8	5 1/32	5.4	2 1/8	1.27	1/2	6.0	2 3/8	19.0	7 1/2	26.7	10 1/2	10.6	4 3/16	—	—
2 1/16	10000	23	152	14.7	5 25/32	5.7	2 1/4	1.27	1/2	6.3	2 1/2	21.3	8 3/8	30.5	12	10.9	4 9/32	—	—
2 1/16	15000	23	152	15.6	6 1/8	8.3	3 1/4	1.27	1/2	8.9	3 1/2	25.1	9 7/8	35.6	14	15.7	6 3/16	—	—
2 9/16	5000	24	153	15.6	6 1/8	5.7	2 1/4	1.27	1/2	6.3	2 1/2	21.3	8 3/8	30.5	12	10.9	4 9/32	—	—
2 9/16	10000	24	153	16.0	6 5/16	6.0	2 3/8	1.27	1/2	6.7	2 5/8	23.2	9 1/8	31.7	12 1/2	11.4	4 15/32	—	—
2 9/16	15000	24	153	16.0	6 5/16	8.3	3 1/4	1.27	1/2	8.9	3 1/2	25.1	9 7/8	35.6	14	15.7	6 3/16	—	—
3 1/8	5000	27	154	16.0	6 5/16	6.0	2 3/8	1.27	1/2	6.7	2 5/8	23.2	9 1/8	31.7	12 1/2	11.4	4 15/32	—	—
3 1/8	10000	27	154	19.4	7 5/8	6.0	2 3/8	1.27	1/2	6.7	2 5/8	26.4	10 3/8	34.9	13 3/4	11.4	4 1/2	—	—
3 1/8	15000	27	154	21.4	8 7/16	7.0	2 3/4	1.27	1/2	7.6	3	30.3	11 15/16	41.9	16 1/2	13.7	5 3/8	—	—
4 1/16	5000	35	155	19.4	7 5/8	6.0	2 3/8	1.27	1/2	6.7	2 5/8	26.4	10 3/8	34.9	13 3/4	11.4	4 1/2	—	—
4 1/16	10000	35	155	26.4	10 3/8	7.0	2 3/4	1.27	1/2	7.6	3	30.3	11 15/16	41.9	16 1/2	13.7	5 3/8	—	—
7 1/16	2000	45	—	27.6	10 7/8	6.7	2 5/8	1.27	1/2	—	—	34.0	13 3/8	42.5	16 3/4	12.1	4 3/4	—	—
7 1/16	5000	45	156	33.7	13 1/4	7.9	3 1/8	1.27	1/2	8.6	3 3/8	37.0	14 9/16	50.8	20	14.9	5 7/8	—	—
7 1/16	10000	45	156	41.3	16 1/4	8.6	3 3/8	1.27	1/2	9.2	3 5/8	44.1	17 3/8	60.3	23 3/4	16.5	6 1/2	—	—
7 1/16	15000	45	156	41.3	16 1/4	9.2	3 5/8	1.27	1/2	9.8	3 7/8	52.4	20 5/8	72.4	28 1/2	17.1	6 3/4	—	—
7 1/16	15000	45	—	41.3	16 1/4	9.2	3 5/8	1.27	1/2	9.8	3 7/8	52.4	20 5/8	72.4	28 1/2	17.1	6 3/4	—	—

CLAMP CONNECTIONS - DIMENSIONAL DATA (continued)

Bore size A	Working pressure psi	Ring-joint number RX	Ring-joint number BX	Hub dimensions — Ring-joint RX only B cm	B in	C cm	C in	D cm	D in	J cm	J in	Clamp dimensions E cm	E in	F cm	F in	G cm	G in	H cm	H in
9	5000	49	157	33.7	13 1/4	8.6	3 3/8	1.27	1/2	9.2	3 5/8	44.1	17 3/8	60.3	23 3/4	16.5	6 1/2		
9	10000	49	157	41.3	16 1/4	9.2	3 5/8	1.27	1/2	9.8	3 7/8	52.4	20 5/8	72.4	28 1/2	17.1	6 3/4		
9	10000	49		41.3	16 1/4	9.2	3 5/8	1.27	1/2	9.8	3 7/8	52.4	20 5/8	72.4	28 1/2	17.1	6 3/4		
11	3000	53		39.7	15 5/8	7.3	2 7/8	1.27	1/2			48.6	19 1/8	65.4	25 3/4	14.0	5 1/2		
11	5000	53	158	41.3	16 1/4	9.2	3 5/8	1.27	1/2	9.8	3 7/8	52.4	20 5/8	72.4	28 1/2	17.1	6 3/4		
1"	10000		158					1.27	1/2	13.0	5 1/8	67.6	26 5/8	84.1	33 1/8	14.0	5 1/2		
13 5/8 *	3000	57	160	46.7	18 3/8	7.6	3	1.27	1/2	11.4	4 1/2	57.0	22 7/16	75.2	29 5/8			23.5	9 1/4
13 5/8 *	5000	57	159	52.4	20 5/8	10.8	4 1/4	1.27	1/2	13.7	5 3/8	65.1	25 5/8	81.9	32 1/4				
13 5/8 *	10000	57		56.5	22 1/4	13.0	5 1/8	1.27	1/2			71.9	28 5/16	92.1	36 1/4				
16 3/4 *	2000	65		51.8	20 3/8	6.3	2 1/2	1.27	1/2			60.6	23 7/8						
16 3/4 *	3000	65		54.0	21 1/4	7.6	3	1.27	1/2	17.8	7	65.2	25 11/16	102.2	40 1/4			23.5	9 1/4
16 3/4 *	5000	65	161	65.1	25 5/8	9.5	3 3/4	1.27	1/2	20.3	8	81.3	32		46 7/8			26.0	10 1/4
18 3/4	10000		164	79.4	31 1/4							99.4	39 1/8	95.3	37 1/2		13 1/8	26.7	10 1/2
20 3/4	2000	73		63.0	24 13/16	7.3	2 7/8	1.59	5/8			74.1	29 3/16	96.2	37 7/8	14.0	5 1/2		14 1/2
20 3/4	3000	73		65.1	25 5/8	9.5	3 3/4	1.59	5/8	20.3	8	78.3	30 13/16					23.5	9 1/4
21 1/4	7500		166	86.4	34					20.3	8	109.9	43 1/4	136.8	53 3/4	40.0	15 3/4	40.0	15 3/4
21 1/4	10000		166	86.4	34					20.3	8	109.9	43 1/4	136.8	53 3/4	40.0	15 3/4	40.0	15 3/4

* Also available with AX gasket.

SPECIAL CLAMP CONNECTIONS

Bore size A	Working pressure psi	Ring joint RX	Male-female hub**	Hub dimensions — Ring-joint RX only B cm	B in	C cm	C in	D cm	D in	J cm	J in	Clamp dimensions E cm	E in	F cm	F in	G cm	G in	H cm	H in
20 3/4 +	2000	—	—	67.0	26 3/8	8.7	3 7/16	0.64	1/4	12.7	5	82.2	32 3/8	98.4	38 3/4	16.5	6 1/2	24.1	9 1/2
24 1/2	1000	—	::	72.7	28 5/8							84.3	33 3/16	91.4	36				
28	1000	—	::	88.3	34 3/4	10.2	4	0.64	1/4	10.2	4	101.1	39 13/16	109.2	43	19.4	7 5/8		
30 1/2	1000	—	::	88.3	34 3/4	10.2	4			10.2	4	101.1	39 13/16	109.2	43	19.4	7 5/8		
31	1000	—	::	96.5	38							109.2	43	129.5	51	19.4	7 5/8		
34	1000	—	::	103.5	40 3/4					13.0	5 1/8	121.4	47 13/16	140.0	55 1/8			23.5	9 1/4

** Used on underwater riser system. + Furnished with AX gasket.

VERTICAL THROUGH-BORE DIMENSIONS
OF FLANGED DRILLING-THROUGH EQUIPMENT
(API Spec 6A)

Nominal size (in)	End-flange working pressure rating (psi)	Minimum vertical through-bore of drilling-through equipment	
		in	cm
6	2000 - 3000 - 5000	7 1/16	17,94
7 1/16	10000 - 15000 - 20000	7 1/16	17,94
8	2000 - 3000 - 5000	9	22,86
9	10000 - 15000	9	22,86
10	2000 - 3000 - 5000	11	27,94
11	10000 - 15000	11	27,94
12	2000 - 3000	13 5/8	34,61
13 5/8	5000 - 10000	13 5/8	34,61
16	2000 - 3000	16 3/4	42,54
16 3/4	5000 - 10000	16 3/4	42,54
18 3/4	5000 - 10000	18 3/4	47,62
20	2000	21 1/4 (1)	53,97
20	3000	20 3/4 (2)	52,70
21 1/4	5000 - 10000	21 1/4	53,97

(1) This new bore dimension must be provided in nominal 2000 psi equipment manufactured after January 1973, instead of the former 20 1/4 in bore.
(2) Replaces 3 000 psi 20 1/4 in bore.

CAMERON TYPE "U" RAM BLOW-OUT PREVENTERS

Nominal size (in)	Working pressure bar	psi	Vertical bore cm	in	Over-all height single flanged cm	in	single clamp hub cm	in	double flanged cm	in	double clamp hub cm	in	Over-all length opened cm	in	closed cm	in	Over-all width cm	in	Fluid volume to open gal	to close gal	Closing ratio
6	207	3000	17.9	7 1/16	61.3	24 1/8	—	—	103.8	40 7/8	—	—	277.8	109 3/8	209.9	82 5/8	51.4	20 1/4	4.8	5.0	1.33	7/1
6	345	5000	17.9	7 1/16	69.5	27 3/8	63.8	25 1/8	112.1	44 1/8	106.4	41 7/8	277.8	109 3/8	209.9	82 5/8	51.4	20 1/4	4.8	5.0	1.33	7/1
7 1/16	690	10000	17.9	7 1/16	77.8	30 5/8	68.6	27	123.5	48 5/8	114.3	45	277.8	109 3/8	209.9	82 5/8	51.4	20 1/4	4.8	5.0	1.33	7/1
7 1/16	1030	15000	17.9	7 1/16	81.0	31 7/8	68.6	27	126.7	49 7/8	114.3	45	277.8	109 3/8	209.9	82 5/8	51.4	20 1/4	4.8	5.0	1.33	7/1
10	207	3000	27.9	11	74.0	29 1/8	—	—	125.1	49 1/4	—	—	372.1	146 1/2	276.5	108 7/8	63.8	25 1/8	12.1	12.7	3.36	7/1
10	345	5000	27.9	11	87.3	34 3/8	75.9	29 7/8	138.4	54 1/2	125.7	49 1/2	372.1	146 1/2	276.5	108 7/8	63.8	25 1/8	12.1	12.7	3.36	7/1
11	690	10000	27.9	11	90.8	35 3/4	81.9	32 1/4	141.9	55 7/8	133.0	52 3/8	372.1	146 1/2	276.5	108 7/8	63.8	25 1/8	12.1	12.7	3.36	7/1
11	1030	15000	27.9	11	109.9	43 1/4	—	—	173.4	68 1/4	—	—	407.7	160 1/2	281.3	110 3/4	73.7	29 *	12.1	12.7	3.4	.7/1
12	207	3000	34.6	13 5/8	78.1	30 3/4	—	—	134.0	52 3/4	—	—	435.6	171 1/2	325.4	128 1/8	74.3	29 1/4	20.6	22.0	5.8	7/1
13 5/8 *	345	5000	34.6	13 5/8	85.7	33 3/4	81.3	32	141.6	55 3/4	137.2	54	435.6	171 1/2	325.4	128 1/8	74.3	29 1/4	20.6	22.0	5.8	7/1
13 5/8 *	690	10000	34.6	13 5/8	106.0	41 3/4	83.5	32 7/8	169.2	66 5/8	146.7	57 3/4	440.7	173 1/2	330.5	130 1/8	74.3	29 1/4*	20.6	22.0	5.8	7/1
13 5/8 *	1030	15000	34.6	13 5/8	112.4	44 1/4	—	—	183.5	72 1/4	—	—	495.3	195	366.4	144 1/4	100.3	39 1/2	42.8	44.2	11.7	6 6/1
16 3/4 (1)	207	3000	42.5	16 3/4	101.9	40 1/8	80.6	31 3/4	167.3	65 7/8	146.1	57 1/2	502.3	197 3/4	369.6	145 1/2	90.8	35 3/4	37.0	40.1	10.6	7/1
16 3/4 (1)	345	5000	42.5	16 3/4	109.5	43 1/8	88.9	35	174.9	68 7/8	154.3	60 3/4	509.9	200 3/4	374.7	147 1/2	90.8	35 3/4	37.0	40.1	10.6	7/1
18 3/4	690	10000	47.6	18 3/4	—	—	127.8	50 3/8	—	—	209.2	82 3/8	615.0	242 1/8	462.3	182	106.0	41 3/4	87.1	94.5	24.90	7 4/1
20 (3)	138	2000	52.7	20 3/4	87.0	34 1/4	76.8	30 1/4	144.1	56 3/4	134.0	52 3/4*	575.3	226 1/2	422.0	166 1/8	97.8	38 1/2	29.7	31.8	8.4	7/1
20	207	3000	52.7	20 3/4	95.3	37 1/2	81.3	32	152.4	60	138.4	54 1/2	575.3	226 1/2	422.0	166 1/8	97.8	38 1/2	29.7	31.8	8.4	7/1
21 1/4 (2)	517	7500	54.0	21 1/4	142.6	56 1/8	115.6	45 1/2	252.7	99 1/2	205.7	81	626.1	246 1/2	469.9	185	118.1	46 1/2	66.6	76.8	20.3	5/1
21 1/4	690	10000	54.0	21 1/4	—	—	134.6	53	—	—	221.0	87	635.3	250 3/8	479.7	188 7/8	118.2	46 1/2	91.0	100.3	26.50	7 2/1
26 3/4	138	2000	67.9	26 3/4	101.6	40	—	—	167.0	65 3/4	—	—	689.3	271 3/8	495.6	195 1/8	117.5	46 1/4	37.3	39.4	10.4	7/1
26 3/4	207	3000	67.9	26 3/4	122.2	48 1/8	—	—	199.1	78 3/8	—	—	699.5	275 3/8	505.8	199 1/8	117.5	46 1/4	37.3	39.4	10.4	7/1

• Flange diameter: 32" (81.3 cm) for 11" - 15,000 BOP and 30 1/4" (76.8 cm) for 13 5/8" - 10,000 BOP exceeds body width.

(1) Dimensions applicable to 16 3/4" preventers sold after January 1968.

(2) Also available a 21 1/4" - 2,000 psi flanged or clamp hub model with RX-73 ring gasket.

(3) Since January 1973 the vertical bore of the 20" - 2,000 psi must be 21 1/4" (see page M19).

HYDRIL RAM BLOW-OUT PREVENTERS

I. Range and size

Nominal size (in)	Type	Pressure rating (bar)	(psi)	Vertical bore (cm)	(in)	Over-all Height Single flanged (cm)	(in)	Single hub (cm)	(in)	Double flanged (cm)	(in)	Double hub (cm)	(in)	Over-all length (cm)	(in)	Over-all width (cm)	(in)	Weight single (kg)	(lb)	Weight double (kg)	(lb)	Maximum ram size (in)
6	V	207	3000	17.9	7 1/16	64.5	25 3/8	60.0	23 5/8	112.4	44 1/4	108.0	42 1/2	184.2	72 1/2*	67.9	26 3/4	1360	3000	2721	6000	5 1/2
6	V	345	5000	17.9	7 1/16	70.2	27 5/8	58.4	23	118.1	46 1/2	112.4	44 1/4	184.2	72 1/2*	67.9	26 3/4	1451	3200	2812	6200	5 1/2
10	V	207	3000	27.9	11	73.3	28 7/8	58.4	23	127.3	50 1/8	112.4	44 1/4	241.3	95*	90.3	35 9/16	2721	6000	5442	12000	8 5/8
10	V	345	5000	27.9	11	87.3	34 3/8	61.6	24 1/4	141.2	55 5/8	116.2	45 3/4	241.3	95*	90.3	35 9/16	2721	6000	5442	12000	8 5/8
11	X	690	10000	27.9	11	97.2	38 1/4	75.9	29 7/8	158.4	62 3/8	137.2	54	279.4	110	88.9	35	3175	7000	5442	12000	8 5/8
12	V	207	3000	34.6	13 5/8	81.6	32 1/8	76.2	30	142.2	56	137.2	54	285.8	112 1/2	97.3	38.3	3628	8000	6803	15000	10 3/4
13 5/8	V	345	5000	34.6	13 5/8	92.1	36 1/4	76.5	30 1/8	152.8	60 1/8	137.2	54	285.8	112 1/2	97.3	38.3	3628	8000	6803	15000	10 3/4
13 5/8	X	690	10000	34.6	13 5/8	106.0	41 3/4	79.4	31 1/4	169.6	66 3/4	142.9	56 1/4	289.6	114	105.4	41 1/2	4989	11000	8617	19000	10 3/4
16 3/4	X	690	10000	42.5	16 3/4	114.0	44 7/8	96.2	37 7/8	185.4	73	167.0	65 3/4	336.6	132 1/2	119.4	47	9524	21000	19501	43000	13 3/8
18 3/4	X	690	10000	47.6	18 3/4	143.5	56 1/2	115.6	45 1/2	231.1	91	203.2	80	360.7	142	129.5	51	19501	43000	24036	53000	16
20	V	138	2000	54.0	21 1/4	—	—	—	—	—	—	—	—	—	—	—	—	6349	14000	12245	27000	16

* Manual lock, other : automatic lock.

II. Operating characteristics for one set of automatic locking rams

Nominal size and type	Volume to close pipe ram (l)	(gal)	shear ram (l)	(gal)	Volume to open pipe ram (l)	(gal)	shear ram (l)	(gal)	Closing ratio pipe ram	shear ram	Operating pressure pipe ram (bar)	(psi)	shear ram (bar)	(psi)
6 V - 3 000	4.5	1.2 (1)	—	—	4.9	1.3 (1)	—	—	5.32 (1)	—	44.8	650 (1)	—	—
6 V - 5 000	4.5	1.2 (1)	—	—	4.9	1.3 (1)	—	—	5.32 (1)	—	72.4	1050 (1)	—	—
10 V - 3 000	11.7	3.1 (1)	—	—	11.7	3.1 (1)	—	—	6.0 (1)	—	41.4	500 (1)	—	—
10 V - 5 000	11.7	3.1 (1)	—	—	11.7	3.1 (1)	—	—	6.0 (1)	—	65.5	950 (1)	—	—
11 X - 10 000	48.8	12.9	48.8	12.9	44.7	11.8	44.7	11.8	10.56	10.56	72.4	1050	72.4	1050
12 V - 3 000	22.3	5.9	—	—	18.5	4.9	—	—	5.2	—	46.6	675	—	—
13 5/8 V - 5 000	22.3	5.9	—	—	18.5	4.9	—	—	5.2	—	72.4	1050	—	—
13 5/8 X - 10 000	48.8	12.9	45.4	12.0	44.7	11.8	42.4	11.2	10.56	10.56	72.4	1050	24.0	350
16 3/4 X - 10 000	48.8	12.9	45.4	12.0	44.7	11.8	42.4	11.2	10.56	10.56	72.4	1050	39.6	575
18 3/4 X - 10 000	59.0	15.6	59.0	15.6	53.4	14.1	53.4	14.1	10.56	10.56	72.4	1050	72.4	1050
20 V - 2 000	64.7	17.1	64.7	17.1	59.0	15.6	59.0	15.6	10.56	10.56	72.4	1050	72.4	1050

(1) For manual locking.

RUCKER-SHAFFER LWS BLOW-OUT PREVENTERS

Nominal size (in)	Locking system	Pressure bar	Pressure psi	Vertical bore cm	Vertical bore in	Height single flanged cm	Height single flanged in	Height single hub cm	Height single hub in	Height double flanged cm	Height double flanged in	Height double hub cm	Height double hub in	Length cm	Length in	Width cm	Width in	To close l	To close gal	To open l	To open gal	closing ratio	opening ratio	Maximum ram size (in)
4 1/6	manuel	690	10 000	10.3	4 1/16	52.7	20 3/4	—	—	—	—	—	—	107.3	42 1/4	39.9	15 11/16	2.2	0.59	2.0	0.52	8.45	4.74	2 7/8
6	manuel	345	5 000	17.9	7 1/16	70.2	27 5/8	—	—	100.3	39 1/2	80.3	31 5/8	147.4	58	54.5	21 7/16	4.5	1.19	3.7	0.99	4.45	1.82	5 9/16
7 1/16(1)	manuel	690	10 000	17.9	7 1/16	60.3	23 3/4	—	—	110.5	43 1/2	—	—	189.2	74 1/2	80.0	31 1/2	24.0	6.35	22.3	5.89	10.63	19.40	5 9/16
7 1/16(1)	manuel	1 035	15 000	17.9	7 1/16	60.3	23 3/4	—	—	110.5	43 1/2	—	—	189.2	74 1/2	80.0	31 1/2	24.0	6.35	22.3	*5.89	10.63	19.40	7
8	manuel	345	5 000	22.9	9	—	—	53.3	21	114.6	45 1/8	94.0	37	201.0	79 1/8	65.4	25 3/4	9.8	2.58	8.6	2.27	5.57	3.00	8 5/8
10	manuel	207	3 000	27.9	11	—	—	—	—	106.7	42	93.7	36 7/8	183.2	72 1/8	66.4	26 1/8	6.6	1.74	5.5	1.45	4.45	1.16	8 5/8
10	manuel	345	5 000	27.9	11	87.3	34 3/8	—	—	128.3	50 1/2	—	—	226.0	89	74.0	29 1/8	11.3	2.98	9.9	2.62	5.57	2.09	8 5/8
11 (2-3)	Poslock	690	5 000	27.9	11	100.3	39 1/2	—	—	161.6	63 5/8	—	—	242.2	95 3/8	79.4	31 1/4	31.2	8.23	28.4	7.50	10.85	5.25	8 5/8
11 (1)	Poslock	1 035	15 000	27.9	11	—	—	—	—	—	—	—	—	294.6	116	133.7	52 5/8	35.6	9.4	30.7	8.1	7.11	2.8	8 5/8
13 5/8(2-3)	Poslock	345	5 000	34.6	13 5/8	85.4	33 5/8	73.7	29	126.1	49 5/8	121.9	48	263.2	105 5/8	85.1	33 1/2	40.0	10.56	36.4	9.61	10.85	6.28	10 3/4
13 5/8 (2)	Poslock	690	5 000	34.6	13 5/8	125.7	49 1/2	97.8	38 1/2	164.4	64 3/4	143.6	56 1/2	277.5	109 1/4	109.2	43	43.8	11.56	39.8	10.52	10.85	3.48	10 3/4
16 3/4(2-3)	Poslock	345	5 000	42.5	16 3/4	108.0	42 1/2	95.9	37 3/4	171.4	67 1/2	159.4	62 3/4	301.1	118 1/2	103.7	40 13/16	52.9	13.97	48.1	12.71	10.85	3.61	13 3/8
16 3/4 (2)	Poslock	690	10 000	42.5	16 3/4	—	—	—	—	—	—	172.0	67 3/4	325.1	128	140.6	55 3/8	54.8	14.47	47.3	12.50	7.11	2.06	13 3/8
18 3/4 (2)	Poslock	690	10 000	47.6	18 3/4	—	—	130.8	51 1/2	209.0	82 1/4	177.2	69 3/4	328.9	129 1/2	128.3	50 1/2	57.9	15.30	50.0	13.21	7.11	1.83	16
20 (2-3)	Poslock	138	2 000	54.0	21 1/4	98.8	38 7/8	88.0	34 5/8	—	—	154.4	60 3/4	335.9	132 1/4	104.8	41 1/4	63.9	16.88	41.1	15.35	10.85	2.52	16
20 (2-3)	Poslock	207	3 000	54.0	21 1/4	105.7	41 5/8	—	—	172.0	67 3/4	—	—	335.9	132 1/4	104.8	41 1/4	63.9	16.88	41.1	15.35	10.85	2.52	16
21 1/4 (2)	Poslock	690	10 000	54.0	21 1/4	—	—	—	—	—	—	186.0	73 1/4	346.1	136 1/4	158.2	62 5/16	60.8	16.05	52.5	13.86	7.11	1.63	16

(1) With studded flanges.
(2) Available with manual locking.
(3) Specifications given for 14" cylinder ID, for 8 1/2 or 10" cylinder ID see Rucker - Shaffer catalog.

CAMERON TYPE D BLOW-OUT PREVENTERS

Size	Working pressure		Vertical bore		Over-all height		Over-all diameter		Volumes of fluid to operate			
									opening		closing	
(in)	bar	psi	cm	in	cm	in	cm	in	l	gal	l	gal
6	345	5 000	17,9	7 1/16	64,8	25,5	73,0	28,75	5,26	1,39	6,40	1,69
7 1/16	690	10 000	17,9	7 1/16	87,0	34,25	94,9	37,375	8,97	2,37	10,41	2,75
10	345	5 000	27,9	11	88,1	34,687	109,9	43,25	17,75	4,69	21,39	5,65
11	690	10 000	27,9	11	110,1	43,343	134,6	53,00	34,30	9,06	38,42	10,15
13 5/8	345	5 000	34,6	13 5/8	99,7	39,25	134,0	52,75	39,14	10,34	45,88	12,12
13 5/8	690	10 000	34,6	13 5/8	132,1	52,00	159,1	62,625	61,13	16,15	68,51	18,10

HYDRIL ANNULAR BLOW-OUT PREVENTERS

Type	Size	Working pressure		Vertical bore		Over-all height flanged		Diameter		Volume of chambers	
	(in)	bar	psi	cm	in	cm	in	cm	in	l	gal
GL	13 5/8	345	5 000	34,6	13 5/8	132,7	52 1/4	132,7	52 1/4	74,8	19,76*
	16 3/4	345	5 000	42,5	16 3/4	144,8	57	167,6	66	127,9	33,8*
	18 3/4	345	5 000	47,6	18 3/4	160,0	63	175,3	69	166,6	44,0*
	21 1/4	345	5 000	54,0	21 1/4	184,8	72 3/4	182,2	71 3/4	219,5	58,0*
GK	6	207	3000	17,9	7 1/16	81,3	32	81,3	32	10,8	2,85
	6	345	5000	17,9	71/16	93,7	36 7/8	90,8	35 3/4	14,6	3,86
	7 1/16	690	10000	17,9	7 1/16	121,0	47 5/8	125,7	49 1/2	35,7	9,42
	8	207	3000	22,7	8 15/16	96,2	37 7/8	87,6	34 1/2	16,4	4,33
	8	345	5000	22,7	8 15/16	106,0	41 3/4	104,1	41	25,9	6,84
	9	690	10000	22,9	9	141,6	55 3/4	144,1	56 3/4	60,2	15,90
	10	207	3000	27,9	11	101,0	39 3/4	101,6	40	28,1	7,43
	10	345	5000	27,9	11	121,3	47 3/4	112,4	44 1/4	37,1	9,81
	11	690	10000	27,9	11	161,3	63 1/2	158,1	62 1/4	95,0	25,10
	12	207	3000	34,6	13 5/8	114,9	45 1/4	120,7	47 1/2	43,0	11,36
	13 5/8	345	5000	34,6	13 5/8	137,5	54 1/8	132,7	52 1/4	68,1	17,98
	13 5/8	690	10000	34,6	13 5/8	184,2	72 1/2	172,7	68	130,7	34,53
	16	138	2000	42,5	16 3/4	125,7	49 1/2	135,3	53 1/4	65,9	17,42
	16	207	3000	42,5	16 3/4	136,8	53 7/8	141,0	55 1/2	79,6	21,02
	16 3/4	345	5000	42,5	16 3/4	155,6	61 1/4	151,1	59 1/2	108,6	28,70
	18 (2)	138	2000	45,4	17 7/8	135,9	53 1/2	141,0	55 1/2	79,8	21,09
MSP	6	138	2000	17,9	7 1/16	65,4	25 3/4	75,9	29 7/8	10,8	2,85
	8	138	2000	22,7	8 15/16	76,8	30 1/4	81,3	32	17,3	4,57
	10	138	2000	27,9	11	80,6	31 3/4	94,6	37 1/4	28,1	7,43
	20 (3)	138	2000	52,7	20 3/4	133,4	52 1/2	149,2	58 3/4	117,5	31,05
	29 1/2 (1)	34,5	500	74,09	29 1/2	173,4	68 1/4	210,8	83	227,1	60,00

*   GL Type BOPS have secondary chambers the volumes of which are : 8,24 gal (31,2 l) for 13 5/8'', 17,3 gal (65,5 l) for 16 3/4'', 20 gal (75,7 l) for 18 3/4'', 29,5 gal (111,7 l) for 21 1/4''.
(1) Regularly furnished with 30'' - 300 ANSI series flange.
(2) Replaced by MSP 20'' - 2 000 psi.
(3) Since January 1973 the open bore is 21 1/4'' instead of 20 3/4''.

412

M 24

HYDRIL TYPE GK BLOW-OUT PREVENTERS
Average closing pressure required to establish pack-off (surface installation)

Type BOP	Completely shut-off bar	psi	6 5/8 bar	psi	5 1/2 bar	psi	5 bar	psi	4 1/2 bar	psi	3 1/2 bar	psi	2 7/8 bar	psi	2 3/8 bar	psi	1.900 bar	psi	1.660 bar	psi
6" - 3 000	69	1 000	—	—	—	—	—	—	24	350	28	400	31	450	35	500	41	600	48	700
6" - 5 000	69	1 000	—	—	—	—	—	—	24	350	28	400	31	450	35	500	41	600	48	700
7 1/16-10 000	79	1 150	—	—	—	—	—	—	24	350	38	550	52	750	59	850	62	900	69	1 000
8 - 3 000	72	1 050	—	—	—	—	24	350	28	400	31	450	38	550	45	650	52	750	59	850
8 - 5 000	79	1 150	—	—	—	—	28	400	31	450	38	550	45	650	52	750	59	850	66	950
9 -10 000	79	1 150	—	—	24	350	24	350	26	380	39	570	52	760	59	860	63	920	69	1 000
10 - 3 000	79	1 150	—	—	24	350	28	400	31	450	38	550	45	650	52	750	59	850	66	950
10 - 5 000	79	1 150	—	—	—	—	28	400	31	450	38	550	45	650	52	750	59	850	66	950
11 -10 000	79	1 150	—	—	35	500	26	380	29	420	41	600	54	780	60	870	67	960	69	1 000
12 - 3 000	79	1 150	31	450	41	600	36	525	38	550	41	600	48	700	55	800	62	900	69	1 000
13 5/8 - 5 000	79	1 150	38	550	—	—	43	625	45	650	48	700	52	750	55	800	62	900	69	1 000
13 5/8 -10 000	79	1 150	—	—	28	400	32	460	36	525	44	640	56	815	61	885	68	990	72	1 050
16 - 2 000	79	1 150	24	350	35	500	31	450	35	500	41	600	48	700	55	800	62	900	69	1 000
16 - 3 000	79	1 150	31	450	—	—	36	525	38	550	41	600	48	700	55	800	62	900	69	1 000
16 3/4 - 5 000	79	1 150	—	—	38	550	38	550	41	600	45	650	52	750	59	850	66	950	72	1 050
18 - 2 000	76	1 100	35	500	—	—	40	575	41	600	45	650	48	700	52	750	59	850	66	950

Note. Maximum packing unit life will be realized by use of lowest closing pressure that will maintain a seal.

RUCKER-SHAFFER ANNULAR BLOW-OUT PREVENTERS

Nominal size	Working pressure		Vertical bore		Over-all height				Diameter		Operating volumes			
					flanged		hubbed				to close		to open	
in	bar	psi	cm	in	cm	in	cm	in	cm	in	l	gal	l	gal
6	207	3 000	17,9	7 1/16	74,0	29 1/8	—	—	73,7	29	17,7	4,57	12,4	3,21
6	345	5 000	17,9	7 1/16	78,4	30 7/8	70,5	27 3/4	73,7	29	17,7	4,57	12,4	3,21
7 1/16	690	10 000	17,9	7 1/16	107,3	42 1/4	99,1	39	109,2	43	66,3	17,11	54,1	13,95
8	207	3 000	22,9	9	82,6	32 1/2	—	—	90,2	35 1/2	28,0	7,23	19,5	5,03
8	345	5 000	22,9	9	92,7	36 1/2	85,7	33 3/4	101,6	40	42,8	11,05	33,8	8,72
10	207	3 000	27,9	11	83,5	32 7/8	76,4	30 1/16	101,3	39 7/8	42,6	11,00	26,3	6,78
10	345	5 000	27,9	11	105,4	41 1/2	95,4	37 9/16	113,7	44 3/4	72,4	18,67	56,5	14,59
11 (1)	690	10 000	27,9	11	134,6	53	125,1	49 1/4	144,8	57	118,5	30,58	95,6	24,67
12	207	3 000	34,6	13 5/8	103,4	40 11/16	95,9	37 3/4	117,8	46 3/8	91,1	23,50	56,9	14,67
13 5/8	345	5 000	34,6	13 5/8	114,1	44 15/16	108,6	42 3/4	127,0	50	91,4	23,58	67,5	17,41
13 5/8 (1)	690	10 000	34,6	13 5/8	168,3	66 1/4	157,5	62	152,4	60	198,6	51,24	165,4	42,68
16 3/4 (1)	345	5 000	42,5	16 3/4	131,9	51 15/16	125,3	49 5/16	152,4	60	128,9	33,26	99,2	25,61
18 3/4 (1)	345	5 000	47,6	18 3/8	152,4	60	144,8	57	168,3	66 1/4	186,6	48,16	145,8	37,61
20	138	2 000	54,0	21 1/4	117,2	46 1/8	112,4	44 1/4	124,5	49	126,3	32,59	65,6	16,92
21 1/4 (1)	345	5 000	54,0	21 1/4	167,6	66	161,3	63 1/2	180,3	71	237,8	61,37	185,1	47,76

(1) Wedge cover model.

ACHEVÉ D'IMPRIMER
EN FÉVRIER 1984
PAR L'IMPRIMERIE CHIRAT
42540 SAINT-JUST-LA-PENDUE
Dépôt légal: 1er trimestre 1984
N° d'impression 6692
N° d'éditeur 667 (Réimpression)

IMPRIMÉ EN FRANCE